Cellular Biology of the Lung

ETTORE MAJORANA INTERNATIONAL SCIENCE SERIES

Series Editor:
Antonino Zichichi
European Physical Society
Geneva, Switzerland

(LIFE SCIENCES)

Cellular Biology of the Lung

Edited by

G. Cumming

The Midhurst Medical Research Institute
Midhurst, West Sussex, United Kingdom

and

G. Bonsignore

Institute of Respiratory Pathophysiology
University of Palermo
Palermo, Italy

Plenum Press · New York and London

Library of Congress Cataloging in Publication Data

International School of Thoracic Medicine (5th: 1981: Erice, Italy)
  Cellular biology of the lung.

  (Ettore Majorana international science series. Life sciences; 10)
  Bibliography: p.
  Includes index.
  1. Lungs—Diseases—Congresses. 2. Lungs—Congresses. 3. Cytology—Congresses. I. Bonsignore, G. II. Cumming, Gordon. III. Title. IV. Series. [DNLM: 1. Lung—Cytology—Congresses. W1 ET712M v. 10/ WF 600 C393 1981]
  RC756.I585   1981   616.2'407   81-23407
  ISBN-13: 978-1-4613-3405-7   e-ISBN-13:978-1-4613-3403-3
  DOI: 10.1007/978-1-4613-3403-3
                                                                                   AACR2

Proceedings of the Fifth Course of the International School
of Thoracic Medicine, held March 1 — 6, 1981, in Erice, Sicily

© 1982 Plenum Press, New York
Softcover reprint of the hardcover 1st edition 1982
A Division of Plenum Publishing Corporation
233 Spring Street, New York, N.Y. 10013

PREFACE

This volume records the proceedings at the Fifth School of
Thoracic Medicine held at the Ettore Majorana School of
International Scientific Culture in March 1981.

Foregathered there were a heterogeneous group comprising
clinicians, pathologists, ultra-microscopists, biochemists,
immunologists, cellular biologists and physiologists and they
presented the twenty seven papers seen in the contents list.
The discussion which followed each paper was faithfully
recorded (and where necessary translated) and may be found
after each author's presentation. This free discussion is
perhaps the most valuable part of the School of Thoracic
Medicine, and most clearly defines the present boundaries of
knowledge, and the directions in which enquiry is being
pursued.

The collaboration of many people made the production of
this book possible - for the translation and the discussion
typescript Miss Guiliana de Ferio, for the final typing and
layout Miss Corinne Wade ably assisted by Miss Karen Wadey
and Miss Kim Lekstrom. The illustrations have been dealt
with where necessary by Mr. John Griffiths and the production
of the book was done at The Midhurst Medical Research
Institute prior to its delivery to Plenum Press.

CONTENTS

THE INNERVATION OF BRONCHIAL MUCOSA

Peter K. Jeffery

Department of Lung Pathology
Cardiothoracic Institute, Brompton Hospital
London,   SW3 6HP

The airways which supply the respiratory portion of the lung receive both motor (efferent) and sensory (afferent) fibres. The efferent innervation has in general, excitatory and inhibitory components supplying smooth muscle, blood vessels and submucosal glands. Preganglionic parasympathetic fibres from vagal nuclei descend in the vagus nerve (cranial X) to ganglia located in the

Table 1.   Sensory Innervation

Site	Type-Receptor Reflex	Consequence
Larynx	I — expiration $\longrightarrow$	expiratory effort
Extrapulmonary airway	I — cough (ep) $\longrightarrow$	inspiration followed by expiration blast
Intrapulmonary airway	I — irritant (ep) $\longrightarrow$	bronchoconstriction hyperventilation unpleasant sensation
	II— stretch (subep) $\longrightarrow$ "Hering-Breuer"	increasing inhibition of inspiratory center
Lung	III—J-receptor $\longrightarrow$	rapid shallow breathing bradycardia pulmonary hypotension

airway walls: arising from these ganglia are postganglionic fibres which innervate specific effector structures. Postganglionic fibres originating from paraverterbral sympathetic ganglia (stellate) innervate similar effectors. Afferent endings are of broadly three types based on their position, firing pattern, adaptation to maintained stimulus, and nerve supply (whether myelinated or not) (Widdicombe 1954 a & b; Widdicombe & Sterling 1970; Widdicombe 1974). The results of their stimulation are summarised in table 1.

The distribution of the gross nerve supply to the tracheobronchial tree is similar in many species and has been largely determined by silver staining methods (Berkley 1893; Spencer & Leof 1964; Nagaishi 1972). In the trachea and main bronchus nerve bundles and ganglia are found largely in the posterior membranous portion of the airway (Fisher 1964). On entering the lung the nerve bundles divide to form distinct peribronchial and perivascular plexi (species differences exist in the extent and immediacy with which this happens). The peribronchial plexus then further divides to form extra- and endochrondrial (Subepithelial) plexi: this division depending on the quantity and extension of supporting cartilage in succeeding generations of intrapulmonary airway (e.g. little in rat: more in man). In general, the more distal single bronchiolar plexus has fewer fibres and ganglia than plexi of airway generations of a higher order: nerve bundles are, however, found in the bronchiolar region and early light microscopic reports of alveolar innervation (Hirsch et al 1968) have now been confirmed by electron microscopy in both animals and man and show the depth to which nerve fibres penetrate the lung (Meyrich & Reid 1971; Fox, Bull & Guz 1980). The brief of the present paper is, however, to review the innervation of airway mucosa and the respiratory region will not be further considered. Similarly the innervation of bronchial and pulmonary vessels will be excluded from this review except to say that they receive a dual innervation also: i.e. rich adrenergic with species differences in the degree of cholinergic.

We are left therefore to consider the motor and sensory supply to the (i) epithelium lining the airways of the lower respiratory tract (with its numerous cell types), (ii) submucosal glands (with serous and mucous acini) and (iii) muscle responsible for airways calibre (fig. 1). Lastly a consideration of the distribution, structure and innervation of ganglia is of importance in the light of new findings and thoughts on the control of bronchial effectors. For clarity each section is broken into two areas: 1) an introductory light microscopic section dealing with the results of early silver strains, histochemical (acetylcholinesterase), fluorescence (for catecholamines) studies and the very recent results of immunocytchemistry (antibodies to specific transmitters): and 2) electron microscopic studies which are complementary to the above show the precise localization of nerve terminals (this will form the bulk of illustration due only to the author's own bias).

EPITHELIUM

Light microscopy

The early use of silver stains showed a rich epithelial innervation at all levels of airway in mouse (Honjin 1956), dog (Elftman 1943) rabbit, monkey and man (Fisher 1964), in each case this being derived from a subepithelial plexus. All but Fisher, who suggested a <u>motor</u> supply to epithelial goblet cells, believed the supply to be exclusively sensory. The results of the vital stains methylene blue (injected or perfused) give support to the presence of intraepithelial nerve fibres occurring as far peripherally as the respiratory bronchiole in a number of species (Larsell 1921; Fillenz and Woods 1970). Using histochemical techniques Fillenez and Woods (1970) failed to localize either acetylcholinesterase or catecholamine fluoresence to airway epithelia of guinea pigs, rabbits or dogs implying a lack of motor

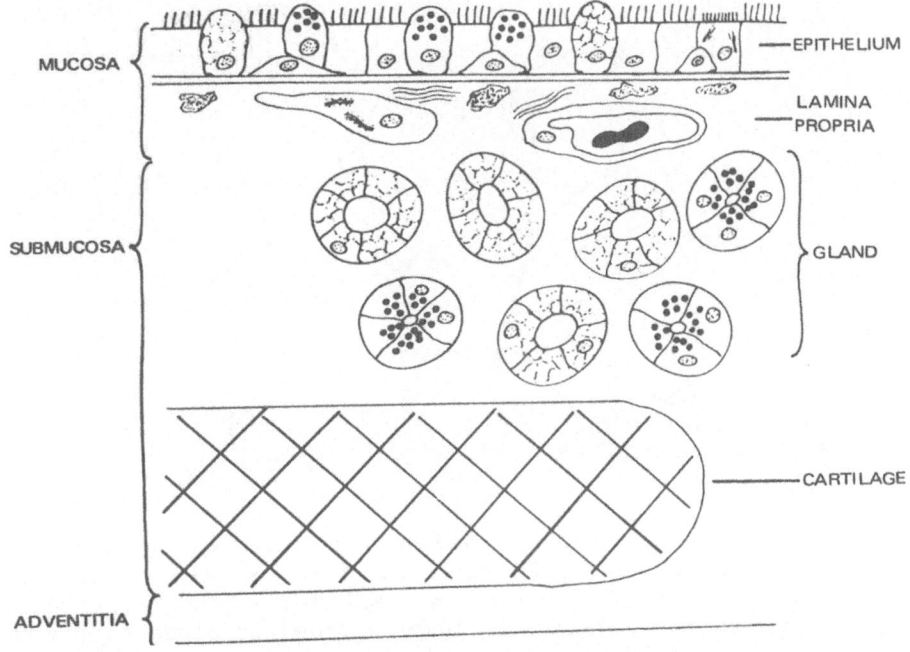

Fig. 1.   Diagrammatic representation of airway wall.

innervation at this site. However, Ericson et al (1969) did
describe occasional such fluorescent fibres within the epithelium
of the upper trachea of mice.

The afferent nature of many of the intra-epithelial nerve
fibres of rat has recently been shown using the techniques of
cobalt chloride diffusion and precipitation following its 24 hr in
vitro application to the distal cut ends of pulmonary vagi:
swellings and hook-shaped terminations were found in bronchial and
bronchiolar epithelium (Lacy 1980). As no evidence of transynaptic
diffusion of cobalt was found the author considered all extra-
ganglionic cobalt-filled nerve fibres as sensory.

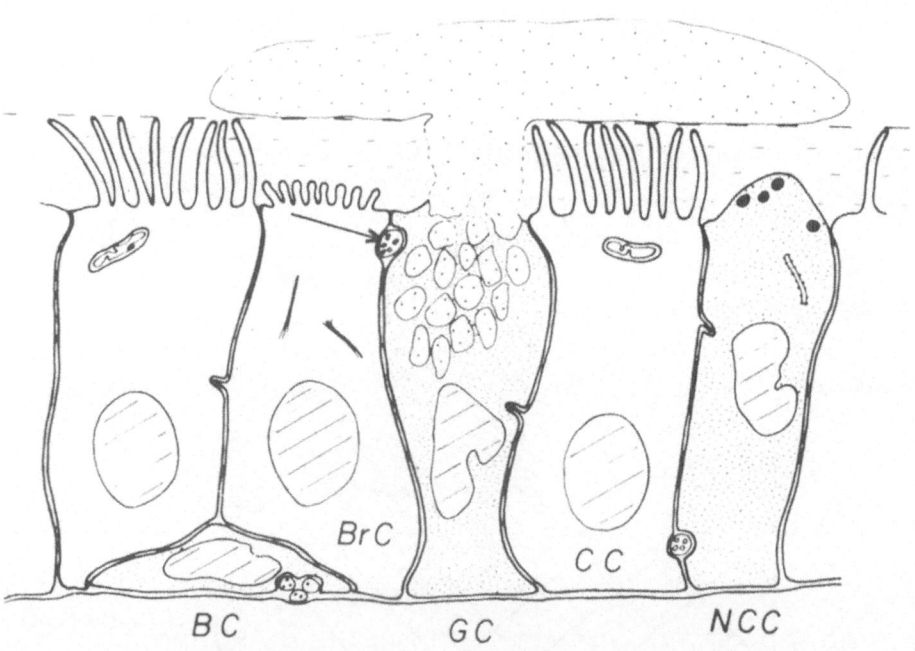

Fig. 2.   Diagramatic representation of airway lining epithelium
          showing ciliated (CC), goblet (GC), serous (NCC), brush
          (BrC) and basal (BC) cells. Epithelial nerve fibres are
          seen either near to the lumen (arrow) or close to basement
          membrane (87%).

Electron microscopy

A number of electron-microscopic studies have now shown unequivocally that nerve fibres pierce the epithelial basement membrane and come to lie in apposition to a number of distinct epithelial cell types (see Jeffery & Reid 1973, Breeze & Wheeldon 1977; Richardson 1979). There appears, however, to be great species variation in their morphology, distribution and number at each airway level studied. For this reason a selection of well-studied species will now be considered:

Rat Few studies of epithelial innervation have been quantitative but in one such study of the rat lung by Jeffery and Reid (1973) intra-epithelial nerve fibres were found in extra- but not intra-pulmonary airway epithelium. There was a significantly higher concentration of intraepithelial nerve fibres in the upper (117 mm^{-1} epithelium) than in the lower (70 mm^{-1}) trachea or main bronchus (23 mm^{-1}). In each case more nerves were found in the anterior than in the posterior membranous wall. Most (87%) of the epithelial nerve fibres were close to the basement membrane and associated particularly with basal cells: 13% were superficial, many within 1 um of the airway lumen (fig. 2). All were without myelin, Schwann cell sheath or basement membrane and were surrounded by the plasma membranes of epithelial cells and separated from them by a gap of about 15 nm: no specialized synaptic complex was ever seen. The mean diameter of fibres was 0.4 μm and each was found either alone or in a group of up to seven fibres (fig. 3). Thirty three percent had dense-cored neurosecretory vesicles (fig. 4), most usually accompanied by a large number of agranular vesicles, 17% had agranular vesicles only, with the remaining 50% without vesicles but containing neurotubules and mitochondria. A few fibres larger than normal (1-2 μm) contained accumulations of β-glycogen and vesicles of both types (fig. 5). Nerve fibres were frequently associated with ciliated (fig. 4) mucous (fig. 6) and basal cells (fig. 3) but few were found with brush cells. Kultschitsky (syn-interalia Feyrter or granulated) cells were infrequently found: one showed a multiple innervation of nerve fibres some fibres with and others without vesicles (Jeffery & Reid 1973). The innervation of groups of such cells (called neuroepithelial bodies) has also been shown in a number of other species including man (Lauweryns et al 1970). These Kultschitsky cells are believed to be part of the APUD (amine precursor uptake and decarboxylation) system (Pearse 1969) and when grouped together as neuroepithelial bodies have (among many suggestions) thought to have a chemoreceptor function releasing their contents in response to hypoxia (interalia Cook and King 1969; Lauweryns et al 1973; Lauweryns & Cokelaere 1973; Moosavi et al 1973; see also Taylor of this volume).

The functional significance of neurosecretory vesicles in

nerve fibres (fig. 4) is still controversial but if they do
indicate motor function then the possibility that ciliated, mucous
and Kultschitsky cells of the rat epithelium receive a motor
innervation must be considered as likely. With regard to species
variation in epithelial innervation, it is perhjaps somewhat
surprising to see that mouse tracheal epithelium lacks an intra-
epithelial nerve supply; with only the occasional nerve found in
the epithelium of the larynx (Pack et al. 1980).

   Goose Of relevance to motor innervation of airway epithelium
is the finding of neurosecretory vesicles in intra-epithelial
endings of goose tracheal epithelium, a species lacking submucosal
glands but with abundant stimulation of the peripheral cut ends of
the descending eosophageal nerves results in an outpouring of mucus
suggesting a motor supply to an neural control of epithelial mucous
cells at least in this species: 75% of the response was blocked by
atropine with the mediator for the remaining atropine-resistant
response not known.

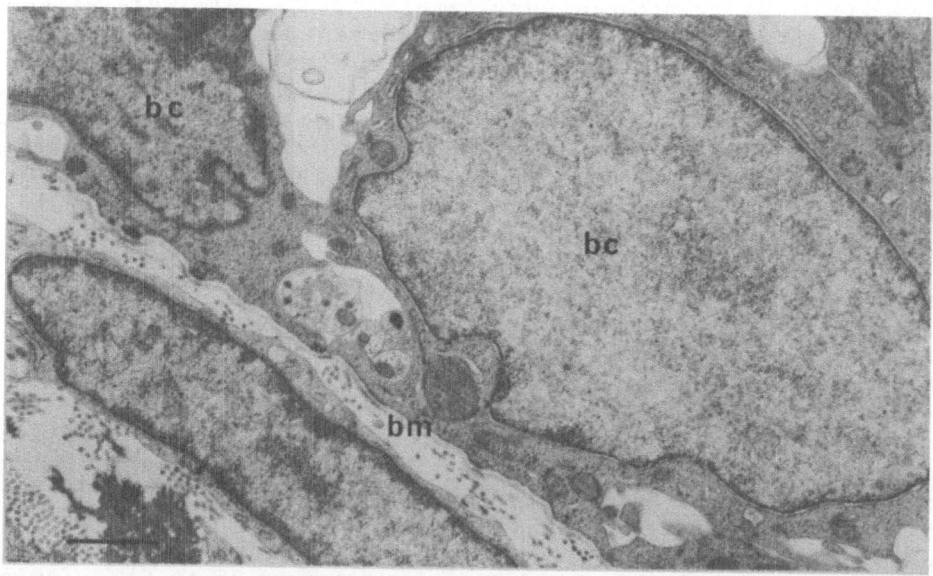

Fig. 3.   Electronmicrograph (EM): a bundle of six intra-epithelial
          fibres in rat trachea. The largest fibre contains a
          single mitochondrion, four dense-cored vesicles and a
          large number of small agranular vesicles. Epithelial
          basement membrane (bm) and basal cells (bc). All electron
          micrographs are of tissue fixed by glutaraldehyde and
          osmium tetroxide: stained with uranyl acetate and lead
          citrate.   1 µm marker at bottom left corner.

Cat   The results of another quantitative study in the cat showed similarities with rat in the epithelial distribution of nerve fibres (Das, Jeffery & Widdicombe 1978).   Whilst intra-epithelial nerves were found at the hilum they were absent from a more peripheral intrapulmonary airway.   Additionally the main carina was examined for intra-epithelial nerves and was found to have the highest concentration of nerves of all the airways examined.   The nerve to cell ratios was 1:5, 1:3 and 1:12 for the lower trachea, carina and hilum respectively.   In constrast to the rat most of the intra-epithelial fibres were without neurosecretory vesicles suggesting the majority had a sensory function.

In a further study these authors set out to assess the extent to which these intra-epithelial fibres were sensory by selective nerve section followed by an electron microscopic examination for evidence of nerve degeneration (Das, Jeffery & Widdicombe 1979). As physiological studies had shown that vagal block abolishes "irritant receptor" reflexes and that irritant receptor impulse

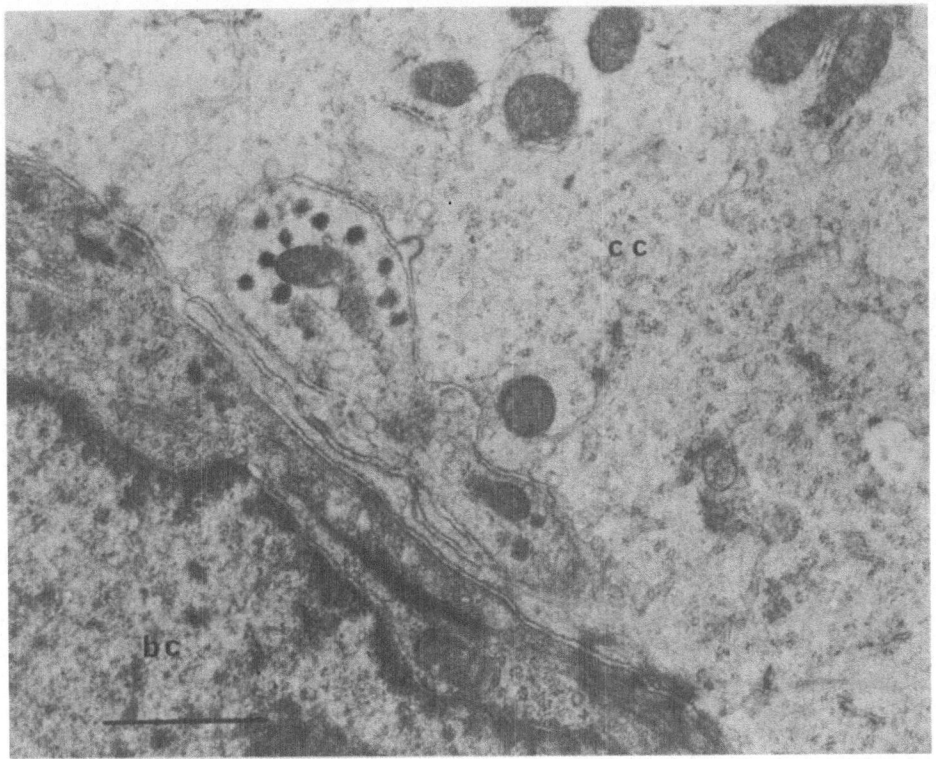

Fig. 4.   EM of a single intraepithelial nerve fibre lying between a basal (bc) and ciliated (cc) cell.   The fibre contains mitochondria and only vesicles of the dense-cored type. Rat trachea.

traffic was present in vagal fibres, infra-nodose unilateral cervical vagal section was carried out in a number of cats. This surgical intervention assumes that cervical vagotomy will cause degeneration of afferent fibres (whose cell bodies lie in the nodose ganglion) leaving intact the post-ganglionic parasympathetic motor fibres (whose cell bodies are in the airway wall).

Between 70 and 90% of the intraepithelial nerve fibres had degenerated five days following vagotomy, the degeneration being more complete in the hilum than in the trachea: this supports the conclusion of predominantly sensory epithelial innervation in the cat. As some degenerating nerve profiles were found on the contralateral (i.e. intact and innervated) side it is likely there is some "cross over" of nerve fibres, a suggestion which has recently received some support from studies in the dog (Dixon et al 1980).

<u>Human</u> Since Rhodin's (1966) study of the epithelium of human trachea in which he demonstrated a single intra-epithelial fibre (lacking neuro-secretory vesicles), little attention has been paid to this tissue. We have recently begun an examination of the innervation of human airways (4 levels) the epithelium being obtained by biopsy and from lung resection in a number of individuals. We have found that the lamina propria has a rich innervation with bundles of both myelinated and unmyelinated fibres. To date intra-epithelial nerves have been found in main stem extrapulmonary bronchus and trachea but not in any segmental airway examined. They are infrequently found, often close to basement membrane (fig. 7) but sometimes extremely close to the airway lumen (fig. 8). No fibre has been found to contain neurosecretory vesicles and they are thus likely to represent sensory fibres perhaps giving rise to "irritant" receptor discharge in response to inraluminal mechanical or chemical stimulation.

SUBMUCOSAL GLANDS

Light microscopy

The early studies of Larsell (1922) and Elftman (1943) showed, by silver staining techniques, that the submucosal glands of rabbit, dog and man were associated with nerve fibres derived from nearby autonomic ganglia, but the nature of such fibres was uncertain. Using histochemical methods (i.e. acetylcholinesterase) Wardell et al (1970) showed that these fibres were cholinergic in the dog: Falk-Hillarp fluorescent techniques indicative of catecholaminergic innervation were negative in this species. In contrast histochemical and fluorescence techniques applied to the tracheal airways of cat and monkey demonstrated the presence of both cholinergic and catecholaminergic innervation (Silva & Ross 1974, El-Bermani 1978). In cats fluorescent nerve fibres were seen passing close to and between glands and, by electron microscopy, their terminals were found to enter gland acini. Studies of human resection or post-mortem material have shown the presence of cholinergic fibres close to glands but there is a notable absence of catecholamine fluorescence in all but the most cranial aspect of the trachea. More recently Uddman et al (1978) have shown the presence of nerve cell bodies and fibres with vasoactive intestinal peptide (VIP)-like immuno-fluorescence close to submucosal glands.

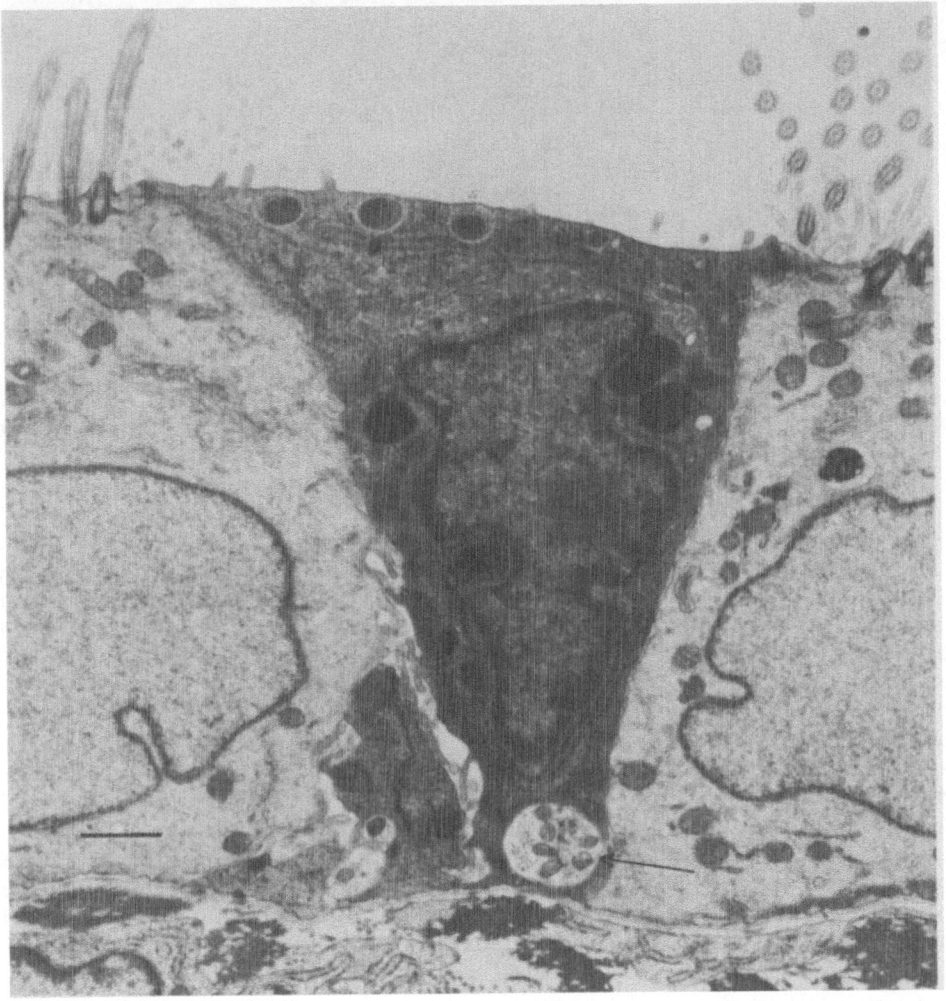

Fig. 5. EM of a single nerve ending close to epithelial basement membrane. The larger than usual profile has an accumulation of mitochondria and β-glycogen. Rat trachea.

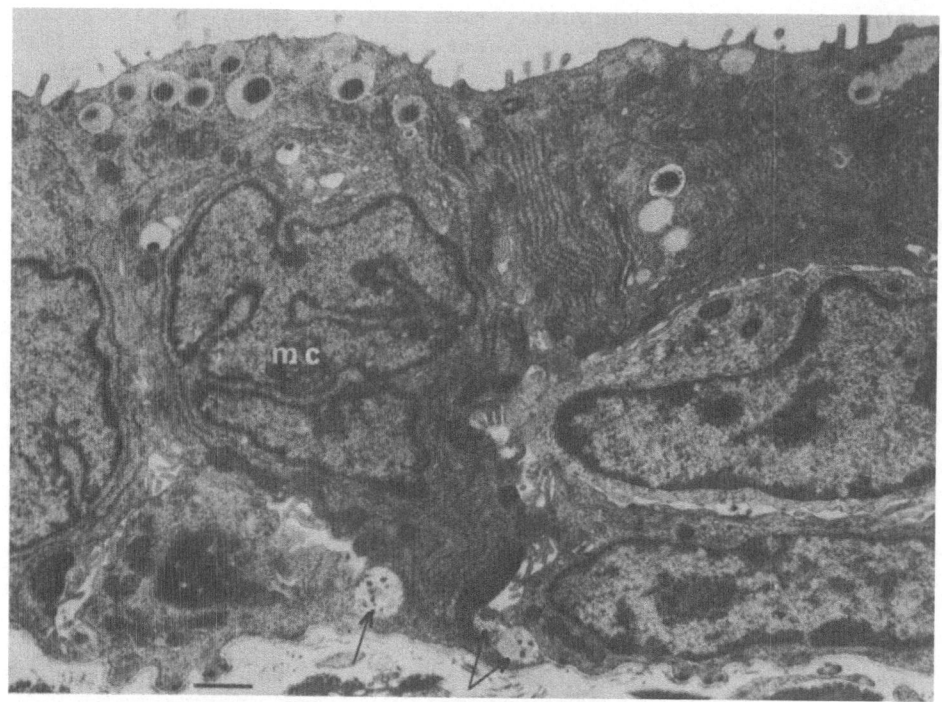

Fig. 6.   EM of epithelial mucous cells (mc) with three nerve
           fibres at their bases (arrows). Rat trachea.

Electron microscopy

        The innervation of submucosal glands has been examined also by
electron microscopy.   Their close proximity to gland acini is not
disputed but the extent to which individual serous, mucous, or duct
cells are innervated (i.e. intra-acinar/duct fibres) is not known.
In man, Bensch et al (1965) and Meyrick & Reid (1970) showed intra-
acinar varicose nerve endings with either agranular or granular
vesicles, each structure indicative of motor function.   Our studies
have been concerned with the innervation of cat submucosal glands
as this animal is much used to investigate the neural control of
airway's secretion.   In contrast to the human airway, cat submucosal glands occupy a larger portion of the airway wall: the Reid
index for human is 0.33 and for cat is 0.75 (Jeffery 1978).   Like
the human they are comprised of both serous and mucous acini but
the largest proportion (75%) of acini show a mixture of cell types
with the serous cell most predominant.

        In a recent study (Le Blond and Jeffery - in preparation)
nerve fibre bundles were found surrounding cat gland acini (fig. 9)
or as single fibres piercing the acinar basement membrane when they
became closely invested by individual secretory cells (fig. 10).
Whilst intra-acinar nerve fibres were found with all three types of

acini, when counted, there were fewer intra-acinar nerves than
expected (by random distribution) with serous and more with mixed
acini.   Conversely those nerve fibres outside but adjacent to acini
were particularly associated with serous rather than with mixed
acini.   Seventy one percent of intra-acinar nerves contained
neurosecretory vesicles of the agranular type suggestive of motor
function.   Nerve fibres with only dense-cored vesicles (5%) were
found exclusively outside the acini: the transmitter associated
with this latter type of vesicle is not yet known but it may well
be a catecholamine or peptide (e.g. VIP).   In summary the acini of
cat sub-mucosal glands have been shown to be innervated by single
fibres and surrounded by bundles or nerve fibres many of which
would appear to have a motor function.   The unequal distribution of
nerves to distinct types of acini may well be important in
influencing the volume and type of secretion which is discharged
from airway submucosal glands in response to neural stimulation.

    Our preliminary studies with human airways show that an intra-
acinar innervation is lacking but that many nerve bundles lie
outside and adjacent to acini (fig. 11).   There is thus once again
species variation in the innervation of secretory glands and care
must be taken in extrapolating from animal studies to man.

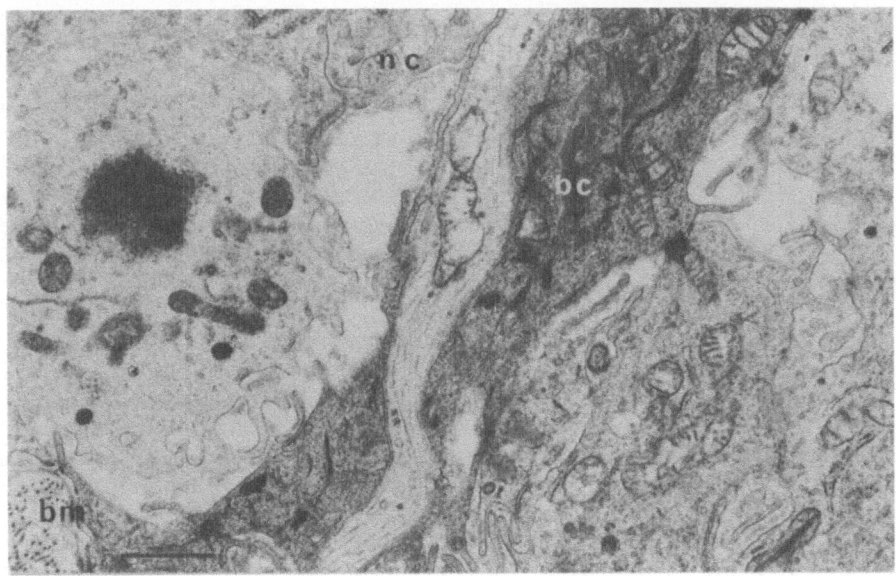

Fig. 7.   EM of human main stem bronchus showing an intraepithelial
          nerve fibre cut in longitudinal section.   The neurotubules
          of the fibre are clearly seen and the fibre, orientated at
          right angles to the basement membrane (bm), is lying
          between a basal (bc) and non-ciliated (nc) cell.

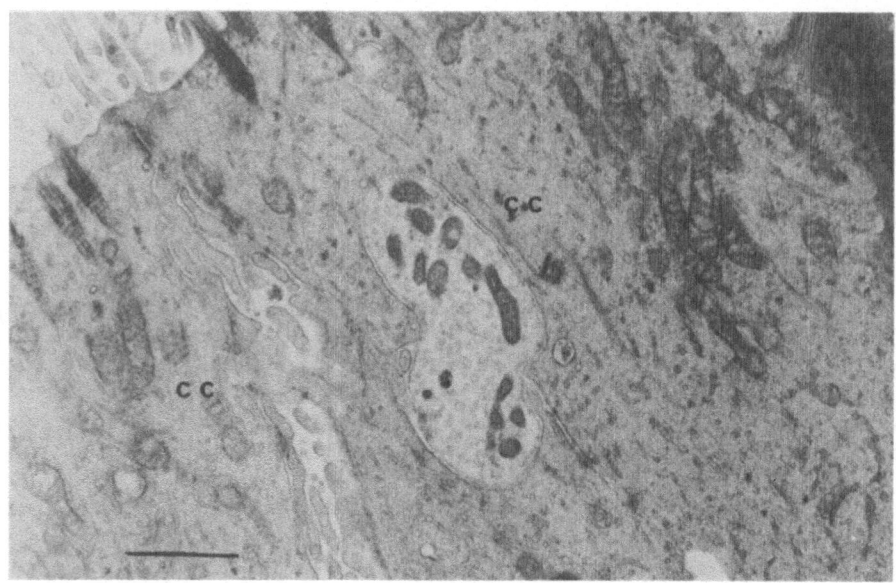

Fig. 8.   EM human airway showing a nerve fibre close to the lumen
and between two ciliated cells (cc).

AIRWAY MUSCLE

Light microscopy

The innervation of tracheo-bronchial muscle has been
recognised since the early days of silver staining (interalia
Larsell 1922).   Most studies indicate that fibres from the
peribronchial plexus penetrate the muscle and terminate in a simple
end knob (? motor) or, in Larsells study, a "smooth muscle spindle"
(? sensory receptor).

Acetylcholinesterase positive finres have been found in most
species studied including man (El-Bermani et al 1970; El-Bermani
1973; Mann 1971; Richardson & Ferguson 1979).   In contrast the
presence of catecholamine fluorescence in bronchial smooth muscle
is species dependent and is abundant in the cat, lacking or minimal
in many other species or restricted solely to the upper trachea as
in man. (Richardson & Beland 1976; El-Bermani 1978; O'Donnell et
al 1978; Richardson & Ferguson 1979).   VIP and, in the human
substance P also, have been localised in and around the bronchial
smooth muscle of a number of species (Said et al 1980).   With
regard afferent endings the recent cobalt and horse-radish
peroxidase diffusion studies of Lacy (1980) convincingly show
sensory innervation of smooth muscle the afferent fibres of which
run in the vagus.

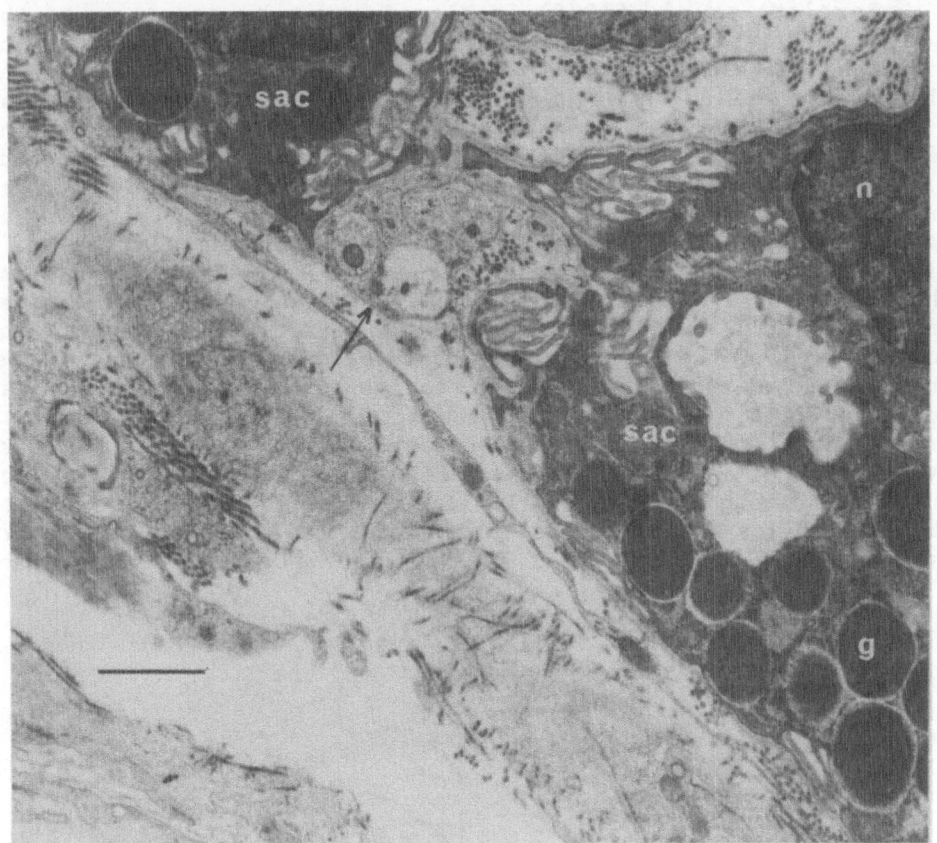

Fig. 9.   EM of cat submucosal glands illustrating a nerve bundle of
          four fibres lying close to and between two serous acini
          (sac).   Nuclei (n) and secretory granules (g) of acinar
          cells.

Electron microscopy

    Sensory endings – In 1970 Fillenz and Woods described in
rabbits nerve profiles with numerous mitochondria and glycogen
lying close to bronchial muscle.   In 1974 Von During et al examined
1 μm serial sections of rat intrapulmonary bronchi tracing the
course of nerve bundles with myelinated fibres and their branches.
The branches were found to penetrate the muscle as "leaf-like"
expansions each devoid of myelin or Schwann cell.   By electron
microscopy they showed these endings to be orientated parallel to
muscle fibres, embedded in collagen and to contain mitochondria and
"receptor matrix".

Bartlett et al. (1976) localized stretch receptors to the mebranous portion of the trachea and main bronchi of dogs by recording the electrical discharge of the receptors whilst probing the airway wall. On each occasion the localization was to an airway lining surface area of about 3 mm square. Dissection and removal of lining epithelium and lamina did not interrupt receptor discharge whilst the removal of deep submucosal tissue did. Ten such pieces were examined by electron microscopy and all showed nerve fibre bundles and bronchial muscle. Serial sections of 5 blocks did not, however, reveal a specific receptor structure and the search for a specific stretch receptor with known physiological function still continues. The study confirmed, however, the localization of the stretch receptor to a region containing muscle. This is of interested as not all studies agree with a submucosal localization and suggest such stretch recpetor activity in rabbits arises from receptors in the epithelial lining of airways (Bitensky et al 1975). A more recent electron microscopic study of trachealis muscle of mouse following cervical vagotomy indicates that as many as 80% of fibres within this muscle are of the sensory type: many of these are filled with mitochondria (Pack & Widdicombe - personal communication).

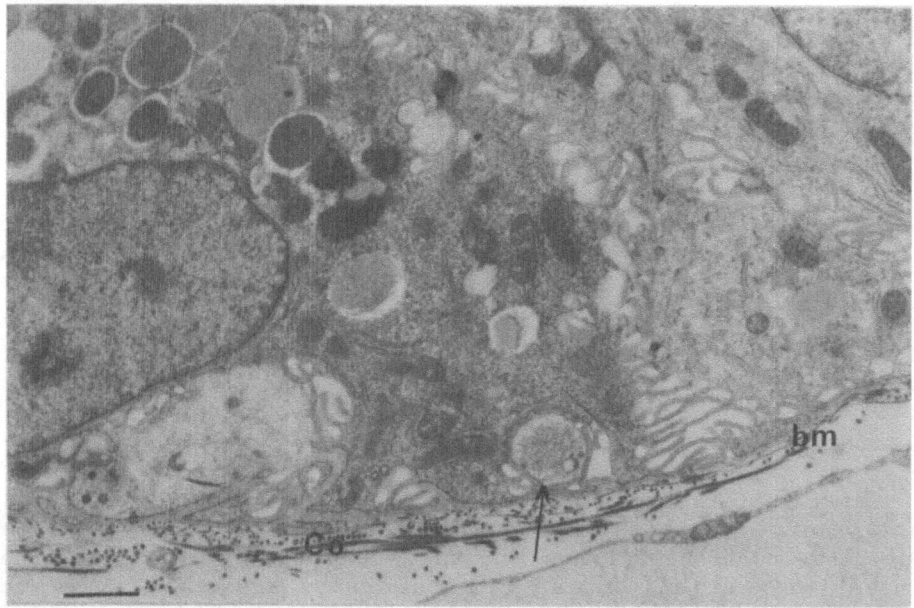

Fig. 10. EM of a secretory acinus of the mixed type (i.e. with both serous and mucous granules) having a single intra-acinar nerve fibre at its base (arrow). The fibre is filled with neurosecretory vesicles most of which are of the agranular type. Basement membrane (bm) and external collagen (Co).

Thus, while still controversial, most studies indicate a submucosal position for the "stretch" receptors of the mammalian lung but the exact structure, orientation and association of such receptors with airways muscle is not known.

Motor endings – With regard the motor innervation of muscle (reviewed by Richardson & Ferguson 1979) electron microscopic studies fail to show synaptic contact of nerve with muscle. While some nerve varicosities lie further than a micron away from the muscle others may share a common investment of basement membrane with muscle yet show no further synaptic specialization. There are also variations in muscle to muscle contacts: muscle fibres are separated from each other in guinea pig and dog airways while frequent connections of the nexus (gap injunction) type are seen in those of the human. The latter type of junction implies electrical coupling and is similar to that often found in the gatrointestinal tract. Nerve profiles with 50 nm agranular vesicles (? cholinergic) are seen in the airways of many species, and those of

Fig. 11. EM of human submucosal gland showing part of a mucous
          acinus (mac) and its myoepithelial cell (m). External to
          the acinus lie nerve fibres each with an abundance of
          neurosecretory vesicles (arrows). The nerves are
          surrounded by electron–dense collagen.

similar size but with eccentric dense cores (? adrenergic) in
others (e.g. cat). Those endings with larger (180-200 nm) dense-
cored vesicles have been found in cat, guinea pig and human
trachealis muscle and this is supportive evidence for the presence
of a transmitter other than acetyl-choline or nor-adrenaline.
Complementary electrophysio/pharmacological studies now indicate
that other transmitters (e.g. purines such as ATP and peptides such
as VIP) are indeed present and when released in response to vagal
stimulation result in bronchial relaxation: i.e. the so-called nonadrenergic component of bronchial relaxation (Coburn & Tomita 1973;
Richardson & Bouchard 1973; Richardson & Beland 1976; Chesrown et
al 1980; Matsuzaki et al 1980).

     Much more needs to be done to characterise these endings
morphologically and in a quantitative way at each of several airway
levels of the tracheobronchial tree. It is also becoming clear
that the type of innervation and control of airway muscle not only
varies with species but also with airway generation in any one
individual.

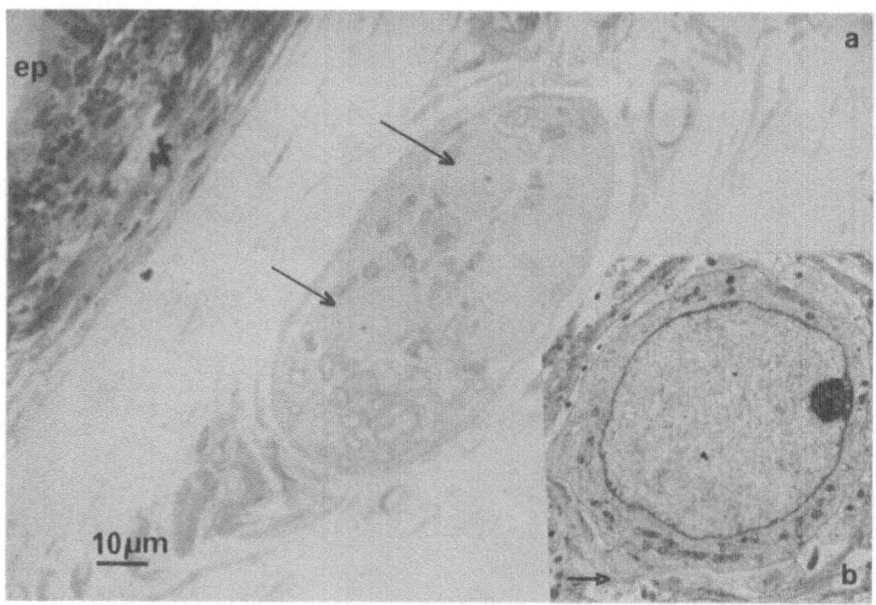

Fig. 12a Light micrograph showing rat bronchial epithelium and
          (ep) ganglion. The soma of five neurons is shown: two
          have nuclei with nucleoli (arrows). The remainder
          contains myelinated fibres and accessory cells.

Fig. 12b EM of neuronal soma to show dense nucleolus, Golgi zone
          (G) and endoplasmic reticulum (ER). The soma is
          surrounded by nerve fibres many showing neurosecretory
          vesicles (arrow).

GANGLIA

Light microscopy

Silver impregnation methods demonstrate well the ganglia present in the airway walls of several species: in general their number decreases with airway generation but they are present in most species at least as far peripherally as small bronchi (Spencer & Leof 1964; Fisher 1964). In monkey and rat they appear to be particularly abundant in the connective tissue elements of the hilum (Zussman 1966; El-Bermani & McCarthy 1980). Whilst there is a commonly held view that ganglia located within the airway wall are parasympathetic (i.e. contain neurons giving rise to postganglionic cholinergic motor fibres) (Nagaishi 1972) cell clusters exhibiting catecholamine 1971; Jacobowitz et al 1973; Knight 1980), and immunocytochemical reactivity for peptides also (Uddman et al 1978). It has recently been suggested that while accumulation of sensory neurons are located mainly in the nodose ganglion (i.e. at a distance from pulmonary tissue) some may also lie locally within the airway wall (Richardson 1979).

Thus it would appear that there is heterogeneity in both the structure and function of pulmonary ganglia.

Electron microscopy

The early light microscopic studies showed each ganglion with about 20-30 cells, the larger ganglia enclosed by an epineurium and connective tissue capsule (Honjin 1956). By electron microscopy neuronal cell bodies are seen within the capsule, each cell body up to 60 μm in diameter. Accessory (or satellite) cells, bundles of unmyelinated and myelinated nerves, blood vessels and mast cells may be additionally found (fig. 12a) in most of the ganglia examined in the airway wall. The neural soma contains a rounded nucleus with distinct nucleolus (fig. 12b), a Golgi zone, neuro-secretory vesicles, endoplasmic reticulum and ribosomes. El-Bermani & McCarthy (1980) demonstrate two types of synaptic complex in monkey airway ganglia: a) somal invagination (which is the most common) where the axon terminal is received in a deep depression or invagination of the neural soma and the axon terminal contains a large number of clear vesicles, mitochondria and asymmetric presynaptic thickenings. b) somal protrusions ("buttes") - where axon terminals show symmetrical (i.e. on both pre- and postsynaptic membranes) thickenings with the lateral surfaces of the somal protrusions. They suggest that the arrangement of the first type of ending might be responsible for isolating a large number of preganglionic terminals from the effect of ganglionic blocking drugs.

In recent electron microscopic studies of rat, cat and human airway ganglia (Richardson & Ferguson 1976; Knight 1980; Jeffrey, unpublished) granulated cells resembling the Kultschitsky (syn.-Feyrter) cells of airway epithelium have been identified (fig. 13). Their ultrastructural appearance is similar to cells described in myenteric and mesenteric ganglia (Oosaki 1970; Elfvin 1971) superior cervical ganglia (Matthews & Raisman 1969; Williams & Palay 1969) and abdominal and aortic paraganglia (Morgan et al 1976).

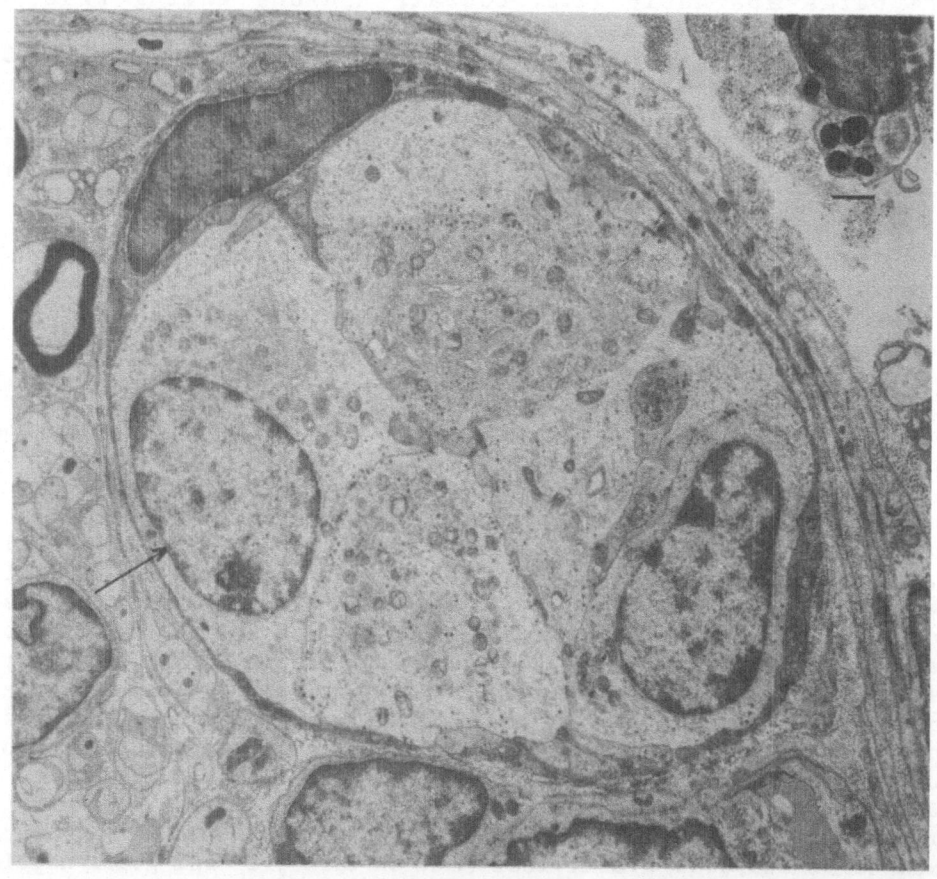

Fig. 13. EM of a group of granulated cells (Kultschitsky-like) at the edge of the ganglion but close to unmyelinated fibres. The granulated cells (arrow) have electron-lucent cytoplasms containing an abundance of large dense-cored neurosecretory-like vesicles more often located close to their plasma membranes. Rat trachea.

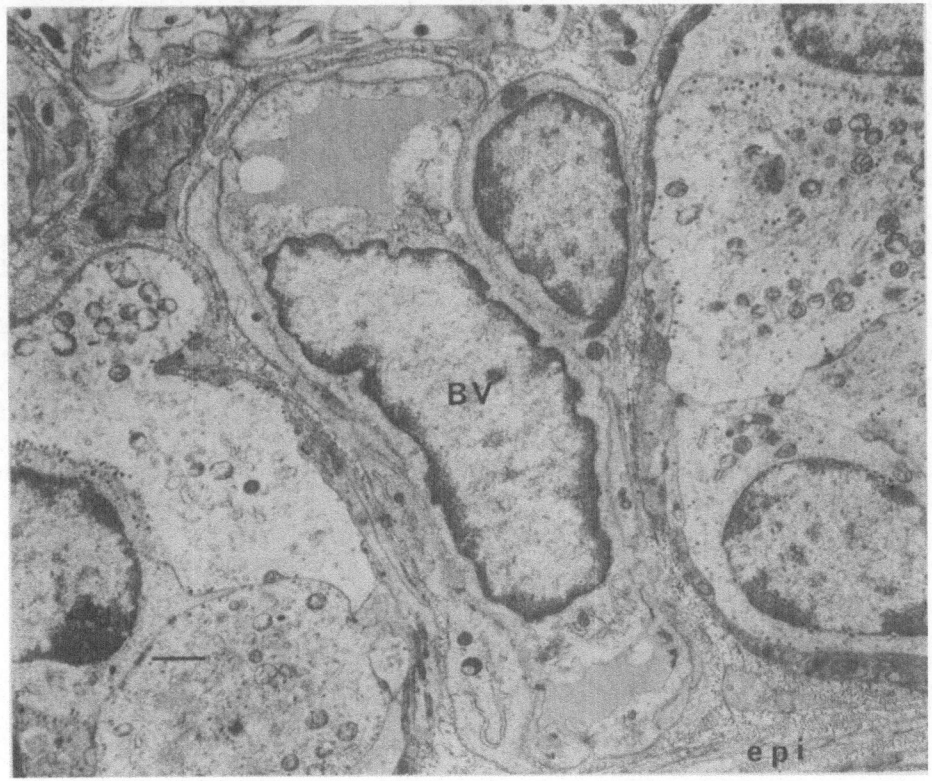

Fig. 14. EM of granulated cells lying on each side of an intra-
          ganglionic blood vessel (BV). Epineurium (epi.)

They are smaller than adjacent neurones and are usually present as
groups of cells, each cell with electron-lucent cytoplasm
containing neurosecretory-like vesicles of the large dense-cored
type. The peripheral localization of granules within the cell is
characteristic. They may be adjacent to a neuronal soma but no
specialized zones of cell to cell contact have yet been found:
analysis of serial sections are needed to determine the spatial
relationships of neuron and granulated cells. The latter may be
close to intraganglionic blood vessels (fig. 14) and thus the
release of their secretion may well diffuse to parts of the nerve
bundle at a distance from their primary location. The intraganglionic K or granulated cells may be responsible for the
catecholamine amine or immuno-fluorescence found in ganglia (see
above) and it has been suggested they may function as inhibitory
catecholamine-secreting interneurons (Jacobowitz et al 1973) or be

the cell bodies of the peptidergic nerves found innervating
bronchial effectors (Uddman et al 1978).

    Thus the neural control of bronchial epithelium, submucosal
glands and muscle appears to be more complex than originally
proposed (fig. 15).   Cholinergic (parasympathetic) innervation
seems to be the most important drive to effectors although species
differences do exist in the extent to which this is so.   The role
of the adrenergic (sympathetic) system is now less clear.   There
may be: 1) direct inhibition of bronchial effectors acting via
 β adreno-receptors, 2) indirect inhibition acting via a-adreno
receptors located on postganglionic parasympathetic cholinergic
neruons of mucosal ganglia, or 3) the main inhibitory system
(especially in guinea pig, and man) may be purinergic or
peptidergic with only a minor role played by the sparse adrenergic
innervation of airways.   Clearly much more needs to be done to
clarify the physiological significance of these histological and
ultrastructural findings.

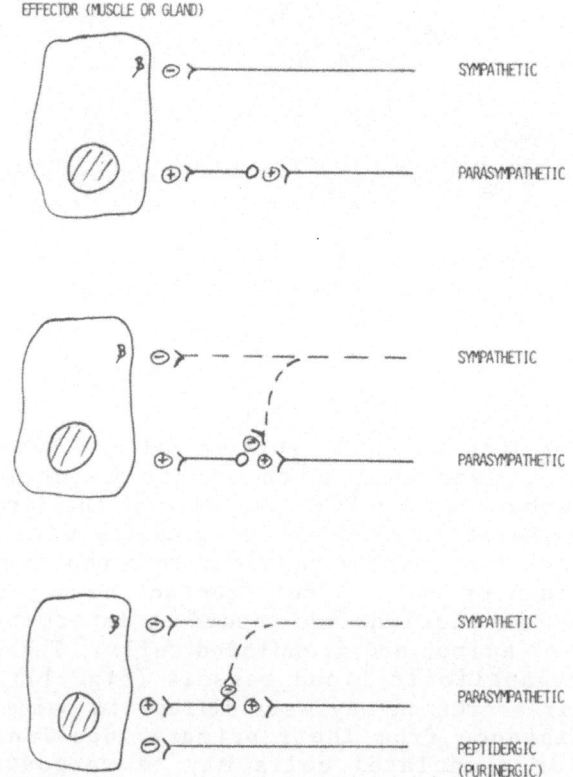

Fig. 15.   Three possible pathways for neural control of bronchial
         effectors.

SUMMARY AND CONCLUSIONS

1)   The nerves which enter the lung form extra- and endochondrial
      plexi and are comprised of both sensory and motor fibres
      derived from the vagus, and cervical sympathetic ganglia.

2)   Sensory receptors include:-

      a) irritant and cough receptors located in the epithelium
      b) stretch receptors associated with airway muscle
      c) juxta-capillary (J) receptors in the respiratory zone.

3)   Motor terminals are found in:

      a) airway epithelia (depending on species)
      b) submucosal gland acini (unequivocal in cat)
      c) airway muscle (but not as neuromuscular junctions)

4)   The transmitters vary with species and may include:

      a) acetylcholine
      b) catecholamines
      c) peptides (e.g. vasoactive intestinal peptide-VIP)
      d) purines (e.g. ATP - still controversial)

5)   Airway ganglia are more complex than previously thought and
      may contain:

      a) neuronal cell bodies
      b) myelinated and unmyelinated fibres
      c) granulated (K-like) cells which may modify efferent traffic
      d) blood vessels
      e) mast cells

      These neural pathways influence the beating of cilia;
secretion of mucus, epithelial permeability as well as their
classical actions on vasculature and tracheobronchial smooth
muscle.

ACKNOWLEDGEMENTS

      I am grateful to Christine Le Blond of the E.M. unit of
Chelsea College for her collaboration in the submucosal gland
section of this review, and to Sonia Neil for typing the
manuscript.

REFERENCES

Bartlett, D., Jeffery, P., Sant'Ambrogio, G., and Wise, J.C.M., 1976. Location of stretch receptors in the trachea and bronchi of the dog. J. Physiol. 258, 409.

Bensch, K.G., Gordon, G.B., Miller, L.R., 1965. Studies on the bronchial counterpart of the Kultschitzky (argentaffin) cell and innervation of bronchial glands. J. Ultrastruct. Res. 12, 668.

Berkley, H.J., 1893. The intrinsic pulmonary nerves by the silver method. J. Comp. Neurol. 3, 107.

Bitensky, L., Chambers, D.J., Chayen, J., Cross, B.A., Guz, A., Jain, F.K., and Johnston, J.J., 1975. Evidence concerning the site of receptors mediating the Hering-Breuer reflex. J. Physiol. 249, 30-31P.

Blumke, S., 1968. Experimental and morphological studies on the efferent bronchial innervation I. The peribronchial plexus. Beit. Pathol. Anat. & Path. 137, 239.

Breeze, R.G., and Wheeldon, E.B., 1977. State of the art: The cells of the pulmonary airways. Am. Rev. Resp. Dis. 116, 705.

Burnstock, G., 1972. Purinergic nerves. Pharmacological Rev. 24, 509.

Cabezas, G.A., Grof, P.D., and Nadel, J.A. 1971. Sympathetic versus parasympathetic nervous regulation of airways in dogs. J. Appl. Phys. 31, 651.

Chesrown, S.E., Venugopalan, C.S., Gold, W.M., and Drazen, J.M., 1980. In vivo demonstration of noradrenergic inhibitory innervation of the guinea pig trachea. J. Clin. Invest. 65, 314.

Coburn, R.F., and Tomita, T., 1973. Evidence for nonadrenergic inhibitory nerves in the guinea pig trachealis muscle. Am. J. Physiol. 224, 1072.

Cole, G.A., Polak, J.M., Wharton, J., Marangos, P., and Pearse, A.G.E., 1980. Neuron specific enolase as a useful histochemical marker for the neuroendocrine system of the lung. J. Path. Soc. 132, 351.

Cook, R.D., and King, A.S., 1969. Nerves of the Avian lung: electron microscopy. J. Anat. 105, 202.

Das, R.M., Jeffery, P.K., and Widdicombe, J.G., 1978. The epithelial innervation of the lower respiratory tract of the cat. J. Anat. 126, 123.

Das, R.M., Jeffery, P.K., and Widdicombe, J.G., 1979. Experimental degeneration of intra-epithelial nerve fibres in cat airways. J. Anat. 128, 259.

Dixon, M., Jackson, D.M., and Richards, I.M., 1980. A study of afferent and efferent nerve distribution to the lungs of dogs. Respiration 39, 144.

v.During, M., Andres, K.H., and Iravani, J. 1974. The fine structure of the pulmonary stretch receptor in the rat. Z. Anat. Entwickl-Gesh, 143, 215.

El-Bermani, A.I., 1973. Innervation of the rat lung. Acetylcholmesterase-containing nerves of the bronchial tree. Am. J. Anat. 137, 19.

El-Bermani, A-W., I, 1978. Pulmonary noradrenergic innervation of rat and monkey: a comparative study. Thorax, 33, 167.

El-Bermani, A-W., and McCarthy, L.F., 1980. Synaptic specialization of pulmonary parasympathetic ganglia: a three dimensional study. Acta Anat., 107, 361.

El-Bermani, A-W., I., McNary, W.F., and Bradley, d.E., 1970. The distribution of acetylcholinesterase and catecholamine containing nerves in the rat lung. Anat. Rec. 167, 205.

Elftman, A.G., 1943. The afferent and parasympathetic innervation of the lungs and trachea of the dog. Amer. J. Anat. 72, 1.

Elfvin, L-G., 1971. Ultrastructural studies on the synaptology of the inferior mesenteric ganglion of the cat. J. Ultrastruct. Res. 37, 432.

Ericson, L.E., Hakanson, R., Larson, B., Owman, C., Lundler, F., 1972. Fluorescence and electron microscopy of amine-storing enterochromaffin-like cells in tracheal epithelium of mouse. S. Sellforsch, 124, 532.

Fillenz, M., and Woods, M.J., 1970. Sensory innervation of airways. In "Breathing: Hering-Breuer Centenary Symposium". (Ciba Foundation Symp. ed. R. Porter), pp. 101-109.

Fisher, A.W.F., 1964. The intrinsic innervation of the trachea. J. Anat. 98, 117.

Fox, B., Bull, T.B., and Guz, A., 1980. Innervation of alveolar walls in the human lung: an electron microscopic study. J. Anat. 131, 683.

Hirsch, E.F., Kaiser, A.C., Borner, H.B., Cooper, T., and Rains, J.J., 1968. Innervation of the mammalian lung I - The afferent receptors, Arch. Path., 85, 51.

Honjin, R., 1956. On the nerve supply of the lung of the mouse with special reference to the structure of the peripheral vegetative nervous system. J. Comp. Neurol., 105, 587.

Jacobowitz, D., Kent, K.M., Fleisch, J.H., and Cooper, T., 1973. Histofluorescent study of catecholamine-containing elements in cholinergic ganglia from the calf of dog lung. Proc. Soc. Exp. Biol. Med. 144, 464.

Jeffery, P.K. 1978. Structure and function of mucus-secreting cells of cat and goose airway epithelium. In "Respiratory Tract Mucus", Ciba Foundation Symposium 54 (new series) ed. R. Porter pp. 5-24 Elsevier.

Jeffery, P.K., and Reid, L. 1973. Intra-epithelial nerves in normal rat airways: a quantitative electron microscopic study. J. Anat. 114, 35.

Knight, D.S., 1980. A light and electron microscopic study of feline intrapulmonary ganglia. J. Anat. 131, 413.

Lacy, M.G., 1980. Pulmonary afferent innervation in rat demonstrated by intra-axonal diffusion of cobalt. Proc. Phys. Soc. 63P.

Larsell, O., 1921. Nerve terminations in the lung of the rabbit. J. Comp. Neurol. 33, 105.

Larsell, O., 1922. The ganglia, plexuses, and nerve-teminations of the mammalian lung and pleura pulmonalis. J. Comp. Neurol. 35, 97.

Lauweryns, J.M., and Cokelaere, M. 1973. Intrapulmonary neuro-epithelial bodies: hypoxia-sensitive neuro (chemo-) receptors. Experientia 29, 1384.

Lauweryns, J.M., Cokelaere, M., and Theunynck, P., 1973. Serotinin producing neuroepithelial bodies in rabbit respiratory mucosa. Science 180, 410.

Lauweryns, J.M., Peuskens, J.C., and Cokelaere, M. 1970. Argyrophil, fluorescent and granulated (paptide and amine producing?) AGF cells in human infact bronchial mucosa. Light and electron microscopic studies. Life Sciences 9, 1417.

Mann, S.P., 1971. The innervation of mammalian bronchial smooth muscle: the localization of catecholamines and cholinesteroses. Histochem. J., 3, 319.

Matsuzaki, Y., Hamosaki, Y., and Said, S.I., 1980. Vasoactive intestinal peptide: a possible transmitter of nonadrenergic relaxation of guinea pig airways. Science 210, 1252.

Matthews, M.R., and Raisman, G., 1969. The ultrastructure and somatic efferent synopses of small granule-containing cells in the superior cervical ganglion. J. Anat. 105, 255.

Meyrick, B., and Ried, L., 1970. Ultrastructure of cells in the human bronchial submucosal glands. J. Anat. 107, 281.

Meyrick, B., and Reid, L., 1971. Nerves in rat intra-acinar alveoli: an electron microscopic study. Resp. Phys., 11, 367.

Moosavi, H., Smith, P., and Heath, D., 1973. The Feyrter cell in hypoxia. Thorax, 28, 729.

Morgan, M., Pack, R.J., and Howes, A., 1976. Structure of cells and nerve endings in abdominal vagal paraganglia of the rat. Cell & Tissue Res., 169, 467.

Nagaishi, C., 1972. Functional anatomy and histology of the lung. University Park Press, Balt.

O'Donnell, S.R., Saar, N., and Wood, L.J. 1978. The density of adrenergic nerves at various levels in the guinea pig lung. Clin. Exp. Pharmacol. Physiol. 5, 325.

Oosaki, T., 1970. A granular vesicle-containing ganglion cell in the Auerbach's plexus of the rat small intestine. Fukushima J. Med. Sci. 17, 41.

Pack, R.J., Al-Ugaily, L.H., Morris, G., and Widdicombe, J.G., 1980. The distribution and structure of cells in the tracheal epithelium of the mouse. Cell and Tissue Research, 208, 65.

Pearse, A.G.E., 1969. The cytochemistry and ultrastructure of polypeptide hormone-producing cells of the APUD series and the embryologic, physiologic and pathologic immplications of the concept. J. Histochem. Cytochem. 17, 303.

Phipps, R.J., Richardson, P.S., Corfield, A., Gallagher, J.T., Jeffery, P.K., Kent, P.W., and Passatore, M., 1977. A physiological, biochemical, and histological study of goose tracheal mucus and its secretion. Phil. Trans. Roy. Soc. (B) 279, 513.

Rhodin, J., 1966. Ultrastructure and function of the human tracheal mucosa. Am. Rev. Resp. Dis., 93, 1.

Richardson, J.B., 1979. Nerve supply to the lungs. American review of respiratory disease, 119, 785.

Richardson, J.B., and Beland, J., 1976. Nonadrenergic inhibitory nerves in human airways. J. Appl. Physiol. 41, 764.

Richardson, J.B., and Bouchard, T., 1973. Demonstration of a nonadrenergic inhibitory nervous system in the trachea of the guineapig. J. Allergy Clin. Immunol. 56, 473.

Richardson, J.B., and Ferguson, C.C., 1976. The fine structure of the ganglia in the human lung. J. Cell. Biol. 70, 48a.

Richardson, J.B., and Ferguson, C.C., 1979. Neuromuscular structure and function in the airways. Fed. Proc. 38, 202.

Said, S.I., Mutt, V., and Erdos, E.G., 1980. The lung in relation to vasoactive polypeptides, In: "Metabolic Activities of the Lung" Ciba Foundation Symp. 78 (new series) (ed. R. Porter) Excerpta Med.

Silva, D.G., and Ross, G., 1974. Ultrastructural and fluorescence histochemical studies on the innervation of the tracheo-bronchial muscle of normal cats and cats treated with 6-hydroxydopamine. J. Ultrastruct. Res. 47, 310.

Spencer, H., and Leof, D., 1964. The innervation of the human lung. J. Anat. (Lond) 98, 599.

Uddman, R., Alumets, J., Densert, O., Hakanson, R., and Sundler, F. 1978. Occurrence and distribution of VIP nerves in the nasal mucosa and tracheobronchial wall. Acta. Otolaryngol, 86, 443.

Wardell, J.R., Chabrin, L.W., and Payne, B.J., 1970. The canine tracheal pouch: a model for use in respiratory mucus research. Am. Rev. Resp. Dis., 101, 741.

Widdicombe, J.G., 1954. Receptors in the trachea and bronchi of the cat. J. Physiol. (Lond), 123, 71.

Widdicombe, J.G., 1954. Respiratory reflexes from the trachea and bronchi of the cat. J. Phys. 123, 55.

Widdicombe, J.G., 1974. "Enteroceptors", In the Peripheral Nervous System (ed. J.I. Hubbord) Plenum.

Widdicombe, J.G., and Sterling, G.M., 1970. The autonomic nervous system and breathing. Arch. Internal. Med. 126, 311.

Williams, T.H., and Palay, S.L., 1969. Ultrastructure of the small neurons in the superior cervical ganglion. Brain Res. 15, 17.

Zussman, W.V., 1966. Fluorescent localization of catecholamine stores in the rat lung. Anat. Rec., 156, 19.

DISCUSSION

LECTURER: Jeffery   CHAIRMAN: Cumming

CUMMING: What is unique about nerves, which produces
               reduction of silver salts to silver metal?

JEFFERY: The rationale behind the silver staining is not
               well-known, but I presume that the reduction in
               silver is working through the same mechanism as the
               reduction of osmium tetroxide and in that case it
               is the unsaturated fatty lipids with their double
               bonds which reduces the silver.

CUMMING: So this is really lipid identification rather than
               nerve identification.

JEFFERY: It is more complex than that.   That is the
               rationale behind part of the staining, it would not
               be the rationale behind the staining for reticulum.
               For that I would have to bounce the comment to
               others in the audience, and by all means anyone who
               is working on collagen, or has indeed worked with
               silver stains, please comment.   I think one further
               caution though is that all workers using silver
               stains realise the capricious nature of these
               stains and the results must be interpreted always
               with some caution.

CUMMING: Thank you.   Herbert Spencer.

SPENCER: I was most interested to hear Jeffery's description
               because many years ago I was interested in the
               nerve supply of the lung, we knew nothing then of
               these newer systems, like the purinergic and the
               others he mentioned, we only had the silver stains,
               though I also used vital methylene blue.   I'd like
               to ask him: has he used that himself?   Because we
               found all these stains extremely capricious.   The
               distribution in the nerve is as he just described.
               Of course we had no means of determining the nature
               of these nerves, whether they were parasympathetic
               or sympathetic.   We made some intelligent guesses,
               I think, and I was intrigued to see the
               continuation of the earlier work which we had done
               on purely morphological grounds.   The silver stains
               are capricious in the extreme, they very seldom
               worked in my experience, and they work, of course,

only with myelinated fibers. Vital methylene blue, I found quite useful, for those who are going to use light microscopy, but again extremely capricious, working well with fetal tissue, for no very clear reason. Nothing has been mentioned about the nerve supply of the vessels and this was something which particularly interests me.

JEFFERY: Thank you very much. In answer to your specific question whether we used methylene blue, in fact we have not, but it has been much used by other workers who concluded from their studies with methylene blue that there was only a sensory innervation of the epithelium. Can I also thank Spencer for the comment with regard to the silver staining and our continuation of the work. What I find so fascinating is that the early workers used these stains, and by observation only suggested functions for the nerves, which we later confirmed using electron microscopy.

BIENENSTOCK: I have a question on VIP. In the clinical syndromes in which diarrhoea characterizes VIP secretion, is there any change, either clinically or practically on biopsy of the lungs at the ultrastructural level.

JEFFFERY: I know of no studies where there has been an examination in those patients of the possibility of exudate in the lungs. There has been a suggestion that there may be abnormalities in the lung rather similar to those found in the gut for Hirschsprung's disease, the suggestion being that the chronic disease affects the ganglia and the submucosa and leads to constriction in peristalsis and it has been suggested (by J. B. Richardson working in Canada) that a similar mechanism may act in the lung, leading to bronchial asthma.

BIENENSTOCK: According to Said VIP is present in mast cells in the lung, and more recently others have shown it also in polymorphonuclear leukocytes, and the question arises if you find VIP surrounding submucosal glands can you be certain that it is in neurones or beside them.

JEFFERY: Again, I can only mirror the comments of the workers in the field. I think quite rightly Said's work refers to the VIP in the lung as VIP lag material, it has VIP immunoreactivity. Whereas

Julia Polak seems to come on a much firmer statement and says it is actually VIP. I think the localization of VIP as an immunofluorescence in mast cells is worrying from the point of view of its localisation in the lung, and more has to be done to make sure that it is actually VIP. The distribution around the glands, however, from the morphological point of view, would suggest that it was in the fibres rather than in isolated cells, such as polymorphs or mast cells. Furthermore Julia Polak has now been looking at the distribution of a neuronespecific enzyme, which is specific for nerve fibers and neurons and correlating that picture with the distribution of VIP. So perhaps we will have some more data on that too. I think a word of caution is very proper, we should stick to talking about VIP-like material rather than saying that it is confirmed in the lung.

CORRIN: I have two points regarding the Kultschitzsky type, first of all the one in the surface epithelium, I think you showed a single such cell innervated and I think this is a new finding, is it not? I believe Lauwryns has emphasized the difference between the clusters of cells and the single ones, and he calls the clusters neuroepithelial bodies because he believed the single ones were not innervated. I was interested to see them in the nerve ganglia and you suggest they may modify efferent traffic. The question would be: in response to what? – what would trigger these cells to modify efferent traffic in the ganglia.

JEFFERY: I don't think that Lauwryns think categorically states that single cells are not innervated. I think he says in general they are not. You are quite right that he does talk about a number of cells clustered together to form a neuroepithelial body. Of course we can't be sure in the section that we took through that one cell that there weren't more cells further on deep in the section. With regard what may trigger the release of substances from the granulated cells within the ganglia, perhaps it's sympathetic fibres, perhaps not. We have no data on that.

TAYLOR: I too was very impressed with that early slide with the Kultschitzsky cell which you showed with well demonstrated innervation, I didn't quite catch what animal it was in. Was it a rabbit or a rat?

JEFFERY: Rat.

TAYLOR: I'm not quite sure whether there is such a clear distinction in the rat between individual cells and groups of cells or neuro-epithelial bodies as Lauwryns has called them. Certainly if that had been in a rabbit it would have been of very considerable interest I think, to demonstrate innervation of individual cells. As you say Lauwryns makes a clear distinction between the two and regards them as being entirely separate structures. As to what you said earlier I would entirely agree about the difficulty of the silver stains, although I think the stain is a considerable advance. But the ultimate basis of the silver stains I think is just not understood, is it?

COHEN: Can I ask you the consequences of the different types of innervation in terms of drug development for treatment of asthmatics?

JEFFERY: I'd rather not answer that. I have been careful to keep away from the pharmacological aspects; it becomes a little more complex if one considers the pharmacology. I think that there is no question that for example beta receptors are on the muscle. On the other hand we can't draw conclusions on the effects of say propanolol, a beta-blocker, with regard to the innervation of the muscle, and I think we have to make a distinction between studies which indicate clearly the particular type of innervation to muscle and other studies which indicate that they have beta receptors, which might not say that they are sympathetically innervated. It could well be of course that the tone of the muscle is being affected by circulating catecholamines and on top of this there is neural modification or modulation of that response.

CUMMING: I wonder if I may make a comment on the clinical aspects of asthma. It's a general belief I think that the nature of asthma is that of air flow limitation due to smooth muscle contraction and that characterizes the disease. There is good evidence that that is an inadequate definition. If we attempt to explain the mechanis of action on a rather poorly based hypothesis I think we might very well run into difficulties. I wonder if I might ask a morphological point. I was fascinated to find that the nerve fibers get right out into the alveolar walls and therefore I find myself at somewhat of a loss when you showed us the axon count progressively falling as you went peripherally. Now, since the alveoli are increasing by a geometric progression as you go outwards and the nerve fibers are diminishing by what looked like an arithmetic progression, how do you explain this.

JEFFERY: The very easy answer to that is that if you look at the nerve fibers in the bronchial mucosa in the lamina propria and the submucosa, of course they extend right to the periphery to the terminal respiratory bronchiole, which suggests that they will go on. The falling count that I referred to from level to level as one proceeds or from generation to generation as one proceeds peripherally of course only related to the epithelial innervation and not to the innervation within the subepithelial tissue. The other thing to bear in mind is that the innervation of deep alveolar structures can occur via the pleura, nerves which diverge at the hilum, at the root of the lung and go out to innervate the pleura and perhaps the subpleural deep alveolar tissue.

SOUROUR: In our practice we sometimes get the problem of bronchorhoea, excessive secretions associated with some pulmonary diseases. Which part of the pulmonary innervation is involved in the pathogenesis of this bronchorrhoea? And can you suggest any modifications for the treatment of bronchorhrea?

JEFFERY: In terms of bronchorrhoea, by which we mean a secretion in excess of 100 ml per day, and if we are taking about mucoid secretion rather than an increase in epithelia permeability we are discussing the larger airways in the lung, in other words the airways with cartilage in their wall, since they have submucosal glands which contribute about 40 times more to the volume of mucus than do goblet cells. So we are talking about large, central airways. With regard to treatment I must reserve any comment and bounce the comments to my clinical colleagues who have much more experience on that side.

KAY: Is there any evidence that intraepithelial submucosal mast cells are innervated?

JEFFERY: A very interesting question. Because of the suggested relationship of nerves with mast cells, notably by Salvato, I have never found nerves against mast cells, intraepithelial mast cells, in our studies, that would of course be of great interest. They wouldn't have to be next or adjacent to the cells, but at least close to the cells. In that context it's interesting that in the rat studies we found intraepithelial granulated cells resembling mast cells, the so-called globular leukocytes, in the largest numbers in the upper trachea, 14% of the cells were globular leukocytes, granulated cells. And it is in that part of the airway where one finds most of the fibers which I talked about this morning. That's about as near as I can get to it.

CUMMING: Thank you, Peter. The last question.

ROUSSOUW: Mr. Chairman, first a comment. The possibility of the nature of the nerve-like structures in the alveolar walls, is it not possible that they can be button-like structures of nerve fibers on their way to supplying capillary bed sites. And the second question: what does Jeffery consider the minimum size of the gap between a nerve terminal and a muscle cell or an epithelial cell to imply functional relationship?

JEFFERY: To take the last part of your question first, with regard the epithelium, the gap between nerve and cell, was of the order of 15nm. That's about the right sort of size for a synaptic gap, but of course I'll remind you that there were no synaptic complexes or specializations. So it would suggest I think in this sort of innervation that release ofneurotransmitter would affect a number of cells by its diffusion rather than a specific cell. In terms of muscle the distance between nerve and muscle is much much greater, very variable, and it is difficult to see how there could be a one nerve to one cell relationship.

ROUSSOUW: We need to consider the possibility of fixation shrinkage in this specific relationship with muscle fibers but not with epithelial cells.

JEFFERY: Yes, the comment on shrinkage is of course an
                important one, one which particularly applies to
                light microscopy, where we are talking in terms of
                about a 10% shrinkage with formalin fixation. In
                terms of the new epoxy resins that are being used
                for electron microscopy there is extremely little
                shrinkage and I think we are preserving the
                anatomical relationships of nerve with cell as near
                as possible, considering the sorts of studies that
                are being done.

CUMMING: Thank you Peter, I think we should now move on to
                our second presentation, continuing on the same
                theme, by Margaret Ayres.

CELL DIVISION AND DIFFERENTIATION IN BRONCHIAL EPITHELIUM

Margaret Ayers and Peter K. Jeffery

Department of Lung Pathology, Cardiothoracic Institute
Brompton Hospital, London, England

The division and differentiation of the cells of the bronchial epithelium can be studied with pulse labelling and autoradiography which gives information on mitosis and the cell cycle, and these will be described in this paper.

I Cells in the Bronchial Epithelium

During the past few years the cellular constituents of bronchial epithelium in different mammalian species, including human, have been studied and characterised (Jeffery and Reid, 1975; McDowell et al, 1978; Jeffery and Reid, 1977; Breeze and Wheeldon, 1978; Becci et al, 1978). Whilst most of the information discussed here refers to rodent tracheo-bronchial epithelium, wherever possible reference to human studies has been made.

Bronchial epithelial cells can be divided into two compartments (Fig. 1), basal and superficial. The basal cell, the most common cell in the basal compartment, was regarded by earlier workers as the most important progenitor cell for the epithelium (Rhodin and Dalham, 1956; Rhodin, 1959; Blenkinsopp, 1967). The superficial compartment contains ciliated cells, intermediate cells without seretory granules, and of particular interest, the secretory cells (Fig. 2). Of the secretory cells there are three types, distinguished both by the nature of their secretory granules and the airway level at which they occur: serous cells (Fig. 2), containing dense granules and mucous (goblet) cells (Fig. 3) containing lucent granules occur in the proximal airways; Clara cells (dense granules) are found distally in small airways. Cells in the superficial compartment were thought to have only limited capacity for division and to arise from basal cells by division and

33

subsequent differentiation (Condon, 1942; Rhodin, 1963; Blenkinsopp, 1967; Bolduc and Reid, 1976): few superficial cells were observed in mitosis relative to the number of basal cells.

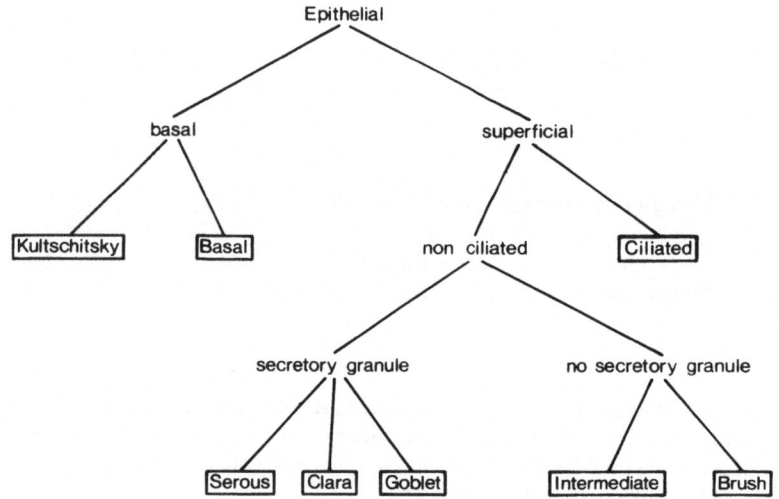

Fig. 1.   Cell types present in basal and superficial compartments of rat airway epithelium.

II Cell Division

a) Mitosis

Classically, cells in division indicate growth, maintenance and/or repair. The proliferative capacity of a tissue was initially judged by the proportion of cells in mitosis in the population at any one time. This ratio was called the mitotic index:

$$\text{Mitotic Index (M.I.)} = \frac{\text{no. cells in mitosis}}{\text{total cells counted}}$$

Stathmokinetic agents e.g. colchincine or more recently the Vinca alkaloids Vinblastine and Vincristine, arrest dividing cells in metaphase, allowing mitotic figures to accumulate with time after administration of the agent. Thereby the rate of entry into mitosis can be calculated under ideal conditions and on average is:

$$\text{Rate of Entry into mitosis} = \frac{\text{number of cells found in metaphase}}{\text{total cells counted x duration of arrest}}$$

TABLE I

STUDIES OF MITOSIS IN RESPIRATORY TRACT

Author	Animal	Sex	Age	Tissue cell type	Colch-icine	Mitotic Index*	Mitotic Index***
Bertalanffy 1953	Rat	M	Adult	Alveolar wall cell	+	3.6% in 24 hr**	0.002
Simmnett & Hepplestone 1966	Mice (A/Gr$_b$)	M	90 days	Alveolar wall cell	+	.00397 nuclei (4h**)	0.004
Harris 1973	Hamster	M	63 days	Trachea	+	2.6×10^{-5}u length epithelium	
Bolduc 1976	Rat	M	33 days	Trachea-basal cells	+	.0078 (4h**) nuclei	0.008
		F	33 days	Trachea-basal cells	+	.0033 (4h**) nuclei	0.003
Boren & Paradise 1978	Hamster	M	63-91 days	Trachea-basal cells	+	0.001 (4h**)	0.0003
	Hamster	M	63-91 days	Trachea-basal cells	-	0.00043	0.0004

*   Mitotic Index expressed in a variety of ways by different workers.

**   Time after Colchicine administration, during which mitotic figures accumulate.

*** M.I. from each study expressed as defined in text for comparison.

Furthermore the Turnover Time which equals the time taken to replace all cells of the population, can be derived.

Tissues of young and growing animals have been classified by Le Blond (1964), according to both the fraction of their cells entering mitosis and the increase in DNA content in 24 hours, as "static" (e.g. neurones), "expanding" (e.g. liver) or "renewing" (e.g. rapidly as in epidermis and slowly as in the respiratory tract). In adult bronchial epithelium, cells in division are rarely found and their occurrence was disputed by early workers (Waller and Bjorkman, 1882). In the nineteen twenties and thirties, however, mitotic figures in organ cultures of bronchus and in alveolar "dust" cells were observed and confirmed (Carleton, 1924; Strelin, 1930; Clara, 1936).

The first quantitative studies of cell division in the respiratory tract were reported by Bertalanffy (1951) and Bertalanffy and Le Bond (1953). These workers observed rat alveolar cells and, using colchicine to cause accumulation of metaphases found that 3.5% of alveolar cells entered mitosis in 24 hours. Other studies of cell division in respiratory epithelium have been done in a variety of species (rat, mouse, hamster) at airway levels from the trachea to the alveolus: comparison between these results is made difficult by differences in the way the rate of cell division is expressed (Table I). A recent study by Boren and Paradise (1978) of hamster trachea illustrates the very small number of cells in division at any one time: the mitotic index for basal cells is calculated to be 0.001 after metaphase accumulation for 4 hours due to colchicine. Without colchicine the mitotic index is 0.00043 (i.e. 4 cells in every 10,000 are dividing).

The mitotic index in the respiratory tract is subject to many sources of variation, namely animal species, sex and age, airway level at which the observation is made, position in the epithelium (i.e. either basal or superficial), diurnal variation, and whether the observation is made in vitro or in vivo.

While the methods used to determine mitotic index are adequate, the information gained and the conclusions which may be drawn are limited because the turnover time of a tissue depends not only on the number of dividing cells but also on the time taken for each division. Thus a tissue in which mitosis is of long duration may have a high mitotic index but be proliferating slowly. Also variations in both these parameters occur in different sites of the same tissue and so sampling must be adequate.

b) Cell Cycle

Further information regarding cell proliferation was made accessible when in 1953 Howard and Pelc observed that DNA synthesis

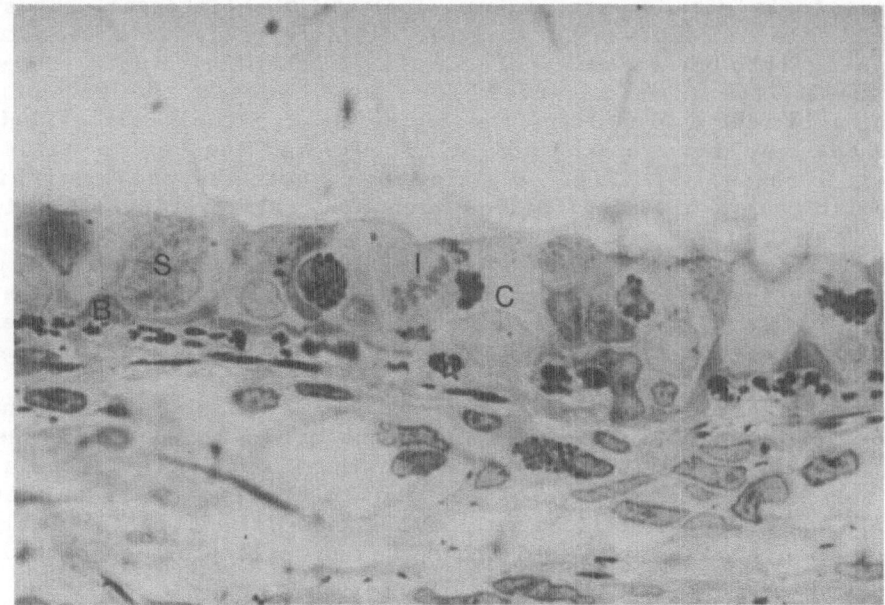

Fig. 2. Main bronchial epithelium of normal rat showing ciliated
         (c), serous (s) and basal (b) cells; the intermediate cell
         (I) is in mitosis. 1μ section, stained with toluidine
         blue X 1000.

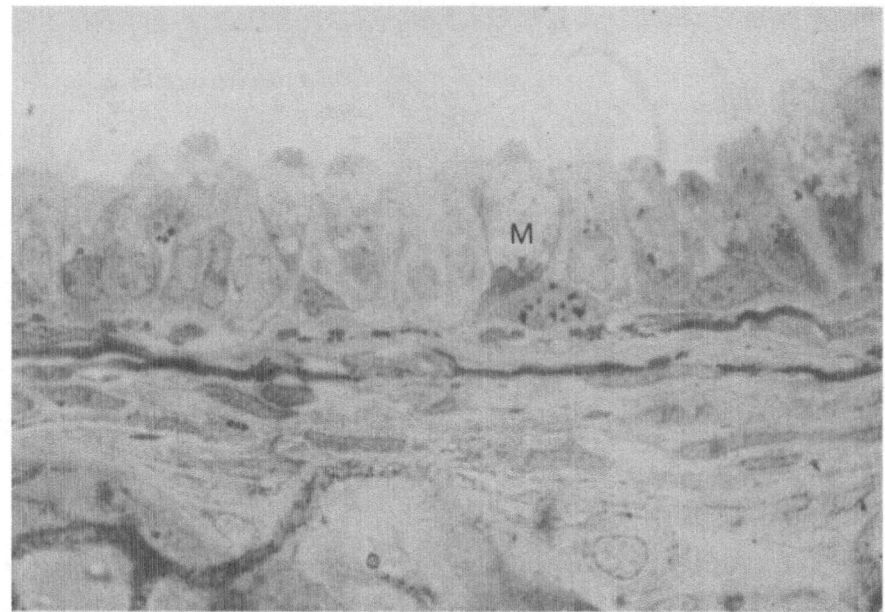

Fig. 3. Main bronchial epithelium of rat exposed to tobacco smoke
         for 14 days, showing mucous cells (m), containing lucent
         granules. 1μ section stained with toluidine blue, X 1000.

and mitosis were separated by a discrete time interval.   Thus the
Cell Cycle concept was formulated: dividing cells were seen to be
passing through a series of complex biochemical events preceding
mitosis; the time interval, or 'gap' between the end of DNA
synthesis (the S phase) and the beginning of mitosis was called G_2,
and the 'gap' between the end of mitosis and the beginning of the
next S phase, G_1 (Fig. 4).   In vivo, not all cells within a
population are dividing; cells which are proliferating are said to
be "cycling" : it is these cells which constitute the
"growth fraction".   There are several alternatives for the non-
proliferating fraction (Fig. 4) : they can a) permanently leave the
cycle, perform their appropriate function and subsequently die; b)
leave the cycle temporarily and enter a resting non-functional
state (Go) yet be available to re-enter the cycle upon the
appropriate stimulus (e.g. liver parenchymal cells which respond to
partial hepatectomy), or c) leave the cycle and die because of
abnormality or malfunction.   The growth rate of a cell population
is the result of the size of the growth fraction (i.e. number of
cells proliferating), the cell cycle time (the time for cells to
traverse the cell cycle) and the number of cells lost through death
or migration from the site of interest.

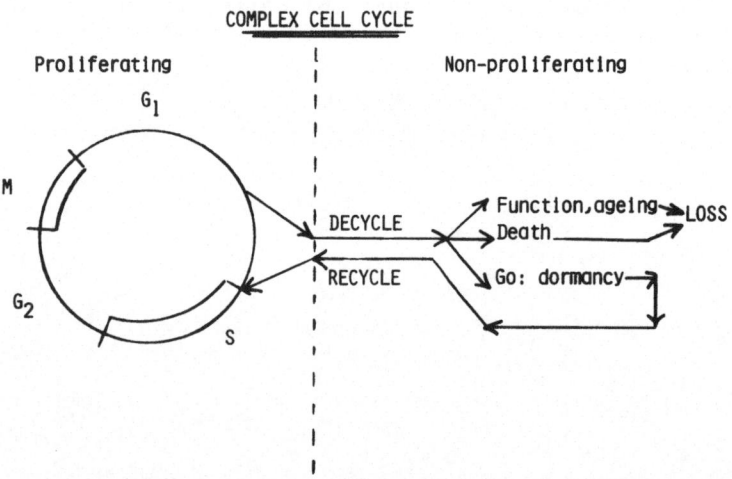

Fig. 4. Diagrammatic model of cell population showing division
        into:- the proliferating compartment - cells in cell cycle;
        and the non-proliferating compartment - cells in different
        functional states.

Thus there are a number of kinetic parameters which can be
measured and, together with histological characteristics of the
component cells, enable populations to be divided into compartments
(sub-populations).   To summarize these kinetic parameters are:-

growth fraction

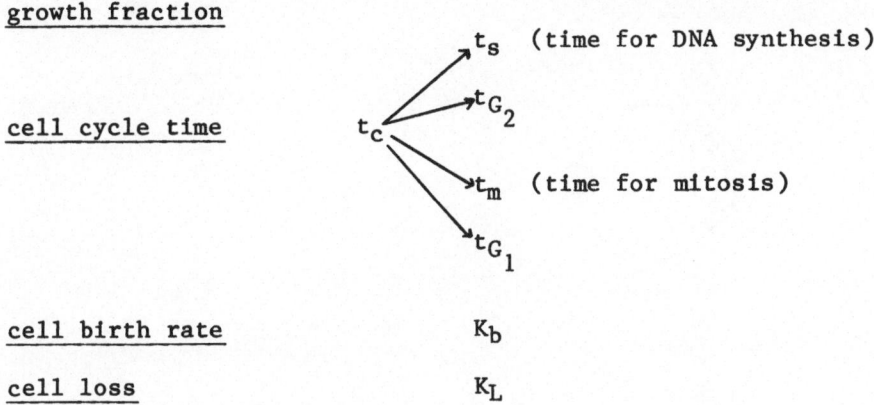

cell cycle time t_c

cell birth rate K_b

cell loss K_L

The above can be used to calculate the population growth rate (i.e. K_G - K_L). For more detailed information with references, the reader is referred to Aherne et al. (1977).

c) Pulse labelling and autoradiography

There are several techniques available to measure the above kinetic parameters, all based on the commercial availability of radioactively-labelled DNA precursors. The molecule most often used is thymidine labelled with the β particle-emitter tritium. When this molecule is presented to a population of cells, those in the S phase (i.e. synthesizing DNA) will incorporate the ^{3}H-thymidine (^{3}H-T) which binds irreversibly within the DNA molecule. Thereafter that cell is "labelled" and can be visualized using autoradiographic techniques (Fig. 5). This involves covering thin sections of appropriately processed tissue with silver halide emulsion which is sensitive to β particle emission. Silver halide crystals which have interacted with β particles can be reduced to silver grains after incubation in the dark for a suitable time, and subsequent development. These grains, localized over nuclei are easily visible and can be quantified microscopically (Fig. 6). Details of the applications and associated problems of this technique are available in Rogers (1979). This procedure enables the fraction of cells synthesizing DNA (i.e. the Labelling Index) to be calculated for a tissue (L.I.$_{epi}$) or for a particular cell type (L.I.$_{cell}$) within a tissue.

Thus: The Labelling Index (L.I.) = $\dfrac{\text{number of labelled cells}}{\text{total cells counted}}$

There are also several other techniques available for measuring and studying the kinetic parameters of dividing cell population. Further information and discussion is provided in Aherne et al. (1977).

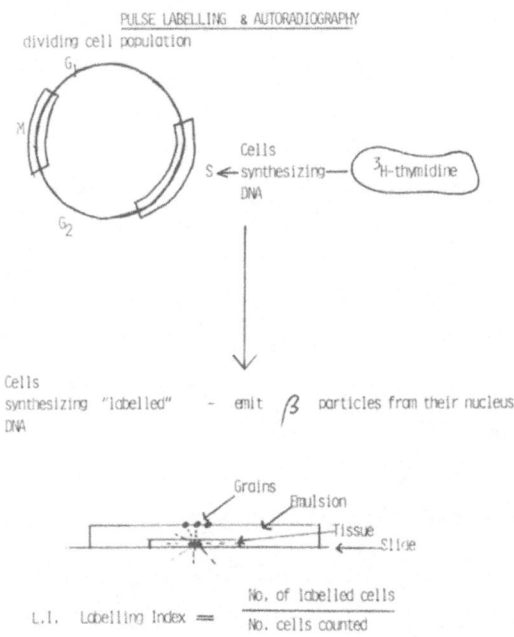

CELL DIVISION - TECHNIQUES

PULSE LABELLING & AUTORADIOGRAPHY

dividing cell population

Cells
S ← synthesizing — ₃H-thymidine
DNA

Cells
synthesizing "labelled" - emit β particles from their nucleus
DNA

Grains
Emulsion
Tissue
Slide

L.I.   Labelling Index = No. of labelled cells / No. cells counted

Fig. 5. Diagramatic representation of pulse labelling and
autoradiography.

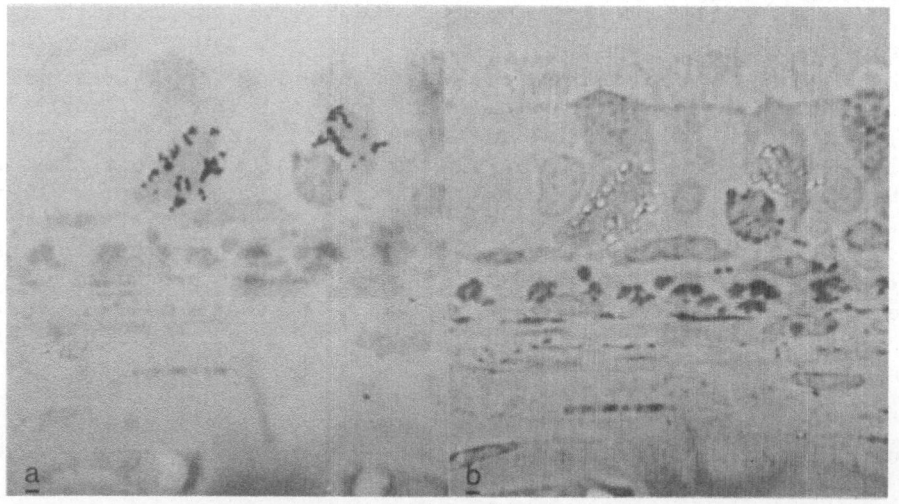

Fig. 6. Autoradiographs of epithelium from main bronchus of normal
rat showing grains localized over nuclei of 2 serous cells
a) bright field microscopy b) phase contrast microscopy.
1μ, toluidine blue, 1000 X.

The techniques of ^{3}H-T labelling have been usefully applied to study cell proliferation in the respiratory tract. Since 1960 labelling indices in the airway epithelium and lung parenchyma have been studied in mice, rats, hamsters and human tissue in vitro. Most authors have observed that after a single pulse label of ^{3}H-T the greatest proportion of labelled cells lie in the basal layer of the epithelium although a small number of superficial cells also take up the label (Messier and Le Blond, 1960; Spencer and Shorter, 1962; Shorter et al. 1964; Blenkinsopp, 1967; Bindreitter et al. 1968; Divertie et al. 1968; Ayers and Jeffery, 1981). Estimates of the percentage of cells labelled show that 1 hour after exposure to ^{3}H-T between 0.5% and 4% of tracheobronchial epithelial cells, and between 0.2% and 1% of bronchiolar epithelial cells are labelled in mice and rats (Shorter et al. 1964; Shorter et al. 1966; Divertie et al. 1968). Kury et al. 1969 studying human lung in vitro with ^{3}H-T found 0.2% - 0.7% bronchiolar epithelial cells were labelled. Blenkinsopp (1967) concluded from his data that basal cells divide to give rise to one basal and one superficial cell while subsequent division of a superficial cell results in 2 superficial progeny.

More specific and detailed kinetic information has been published by Blenkinsopp (1969), Harris et al (1975), Gordon and Lane (1977; 1980) and Boren and Paradise (1978) including information on cell cycle times (t_c), and the time taken for DNA synthesis (t_s) for both basal and superficial compartments of mouse and rat trachea and for basal, mucous and ciliated cell types in hamster tracheal epithelium.

It can be seen from the literature that the amount of cell proliferation, as determined by labelling indices, shows variation between animal species, strain, (Meyer zum Gottesberge and Koburg 1963; Simnett and Hepplestone, 1966) age, sex, disease status of the animals, (Wells, 1970), and from proximal to distal levels of the lung. The last mentioned has been illustrated by the work of Bolduc and Reid (1976) who looked at mitotic and labelling indices at 5 airway levels in male and female rats of 3 different ages. It was found that labelling indices decreased from proximal to distal levels, and were greater at each airway level in males than in females (but only in the youngest age group). The L.I. in young animals was greater than in older ones as might be expected from their growing status.

In our laboratory we have used the techniques of pulse labelling and autoradiography to observe cell division in normal bronchial epithelium of male specific-pathogen-free rats, and the effect of irritation by tobacco smoke (Fig. 7). To label dividing cells all animals were given a pulse label of ^{3}H-T at a dose of 1μCi/gm body weight (i.p.) 1 hour before death. From each animal samples of main bronchus were selected and at least 20 lμm resin-embedded sections were cut and processed for autoradiography.

These sections, when stained with toluidine blue and viewed by light microscopy allow accurate identification of at least 6 cell types (ciliated, serous, transitional, mucous, intermediate and basal) and the localization of autoradiographic grains to the nuclei of the dividing cell (Fig. 6). The total numbers of each cell type present and the numbers of each cell labelled were counted and their labelling indices calculated.

Table II shows the percentage, the L.I. and percentage labelled cells (i.e. ratio: number of each cell type labelled/total cells labelled) of each epithelial cell type from main bronchi of 9 healthy unexposed rats. These results show that:-

(i)   the percentage of mucous cells in unexposed epithelium is low and none of the mucous cells present (during the 1 hr. in which ^{3}H-T is available) are dividing.

(ii) while 67% of the dividing cell population is basal, 27% are serous (secretory) cells.

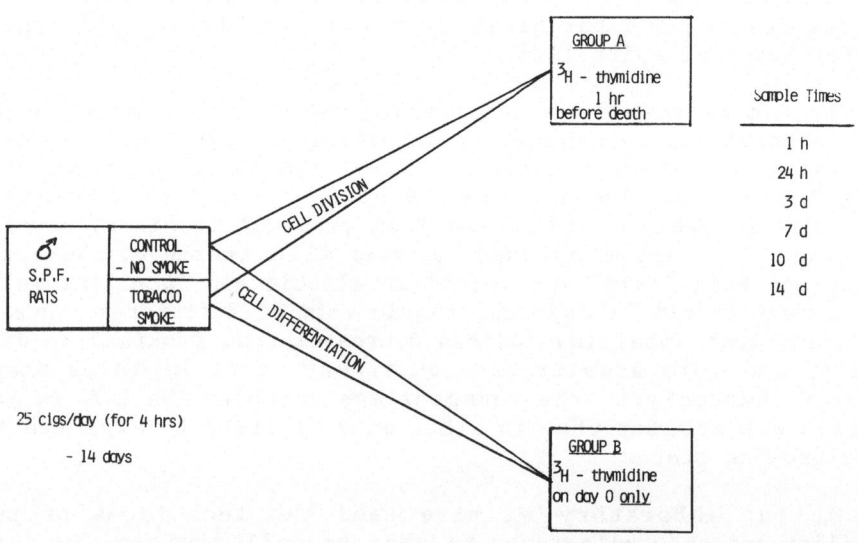

Fig. 7.   Experimental plan for study of cell division and cell differentiation.

TABLE II

CONTROL: Cell Division 1 hour after ^{3}H-T

Cell type	% cell type in epithelium	L.I. cell (X 10^3)	% labelled cells
ciliated	41.2	-	-
serous	29.8	3.0($\pm$0.6)	27
transitional	-	-	-
mucous goblet	2.9	-	-
intermediate	6.8	4.4($\pm$1.9)	7
basal	22.1	10.1($\pm$0.3)	66

N = 9 animals and 153 sections counted

TABLE III

SMOKED % labelled cells

Cell type	CONTROL Group A (day 0)	SMOKED Group A (day 14)
ciliated	0	2.6
serous	27	10.3
transitional	0	0
mucous goblet	0	15.4
intermediate	7	0
basal	66.6	64.1

Another group of animals (Group A) were exposed to tobacco smoke for 14 days (25 cigarettes in 4 hours daily). Animals were then killed 1 hour after a ^{3}H-T pulse (administered as above), at times of 24 h, 3, 7, 10 and 14 days of exposure. Compared with the data from normal animals the results of exposure showed: (i) there was a highly significant rise in L.I. which is observed at 24 h., maintained until 3 days and falls to control levels by 7 days (Fig. 8); (ii) During the 14 days of exposure, tobacco smoke caused a proliferative response in the transitional and intermediate cells (maxima at 3 days) and a marked proliferative response in the mucous (goblet) cells (maximum at 7 days). Interestingly there was no significant proliferative response seen in ciliated cells, and a small decrease only, in basal cells (Figs. 9a and b).

When the results are expressed as percent distribution of label for each cell type at 14 days of exposure (Table III), it can be seen that the percentage of dividing cells which are basal cells remains the same, but that there is a shift from dividing serous to dividing mucous cells.

In summary, in normal SPF rat secretory as well as basal cells form a significant progenitor cell population. With irritation by tobacco smoke there is a change in the dividing cell population, the most prominent response being seen in the mucous secretory cell.

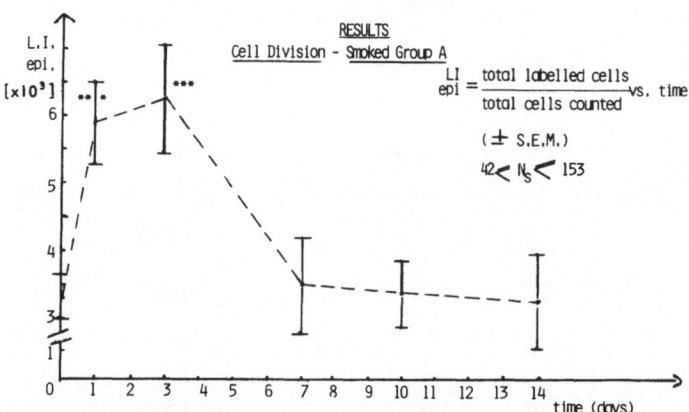

* peaks significantly different from control levels

Ns=no. sections counted/time point

Fig. 8.   Graph showing the effect of tobacco smoke on labelling index (cell division) of whole epithelium from rat main bronchus – L.I.$_{epi}$ vs. time (days).

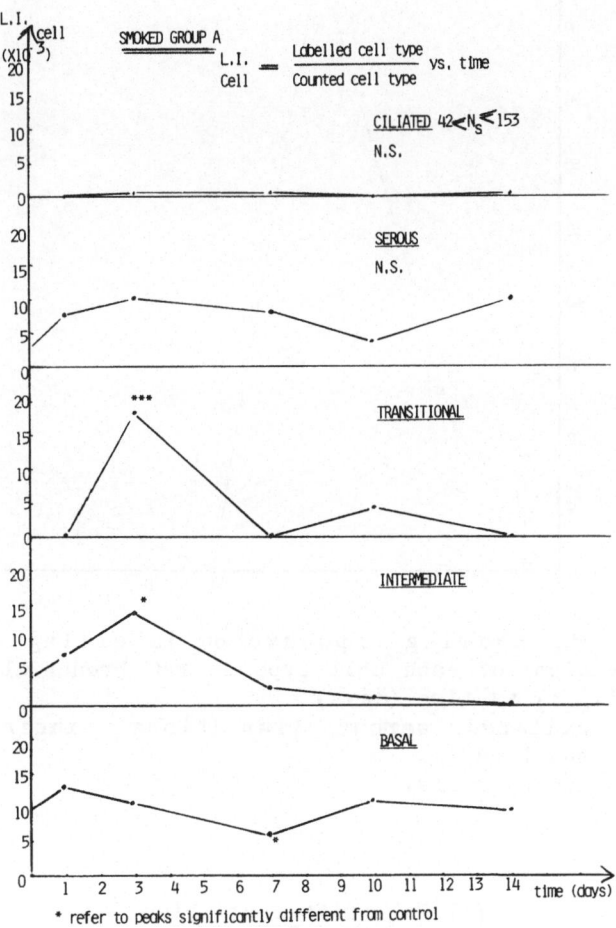

Fig. 9a

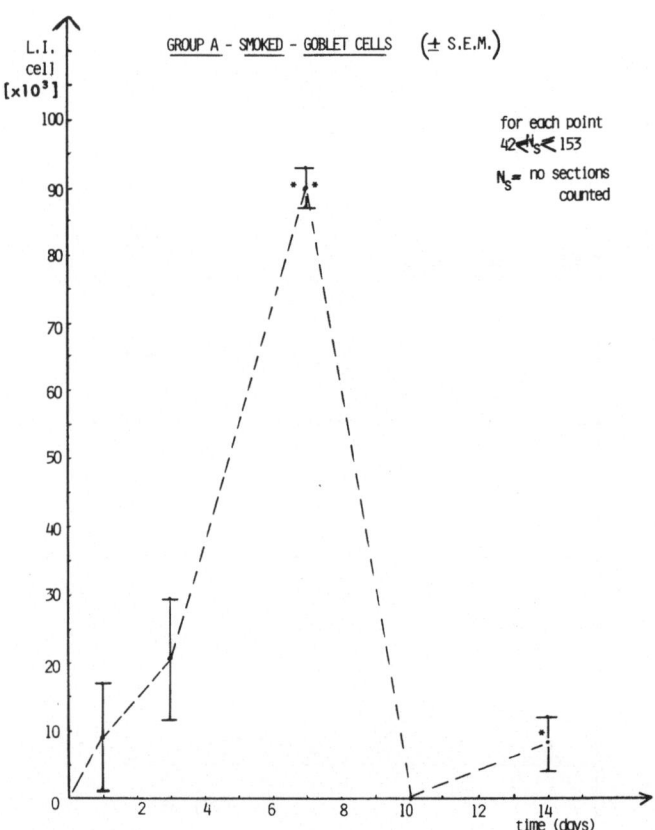

Fig. 9b.   Graphs showing exposure on labelling index (cell
           division) of each cell type in rat bronchial epithelium:
           L.I.$_{cell}$ vs.time (days)
           a)   ciliated, serous, transitional, intermediate and
                basal cells.
           b)   mucous cells.

III Cell Differentiation

Besides division, 'new' cells may appear via differentiation
(i.e.   transformation of one cell type into another without
division).   Drasch (1881) observed dividing cells in the
respiratory tract and suggested that the daughter cells
differentiated first into mucous then into ciliated cells without
further division.   An alternative suggestion arose from
observations of human epithelia from asthmatics and the results of
experimental work in normal and treated rabbits and rats (Condon,
1942; Hilding, 1943): it was thought that dividing basal cells

differentiated into ciliated columnar cells which further differentiated to produce mucous cells.

The availability of a single pulse of ^{3}H-T given only at the beginning of the experiment made it possible to label cells irreversibly and thereafter observe how the labelling pattern changes from one cell type to another with time. Spencer et al. (1962), Shorter et al. (1964; 1966), and Divertie et al. (1968) observed that following such administration of ^{3}H-T cells "migrated" from the basal to the superficial layer and thereafter disappeared. The time taken for this to occur varied from study to study and was termed the "life-span" for that cell type. It was suggested that 2 sub-populations of cells existed in both bronchial and alveolar epithelium each with different life spans (Shorter et al. 1964; Divertie et al. 1968). Calculated turnover times for epithelia of trachea, large bronchi and small bronchi varied from 5 - 8 days in mice and rats (Shorter et al, 1964; 1966; Divertie et al. 1968) to 24-41 days for rats with minimal disease (Wells, 1970). Bindretter et al (1968) after following change in the distribution in each cell type with time suggested that differentiation progressed from basal cells to mucous cells and then on to ciliated cells, which would support the early suggestions of Drasch (1881).

Gordon and Lane (1974; 1977) used a gentle mechanical stimulus to initiate a wave of mitosis in the cells which were left intact after the procedure. Mitotic division occurred 24-48 hours after the stimulus and this was followed at 60-90 hours by appearance of mucous and ciliated cells without further division. In further work where ^{3}H-T was not incorporated McDowell et al. (1978; 1979) and Becci et al. (1978) in elegant morphological studies of respiratory epithelium and its response to injury identified and described in human and hamsters basal, ciliated, mucous, "small granule mucous" and "indifferent" cells. They proposed that the "indifferent" cell which appeared as an immature and undifferentiated cell acts as a progenitor cell and by differentiation gives rise to the other epithelial cell types.

After experiments in which rats were exposed to tobacco smoke for 14 days, Jeffery and Reid (1980) reported a 2% to 18% increase in the numbers of mucous cells in bronchial epithelium with a concurrent decrease in serous cells (i.e. secretory cells containing dense granules). They also observed "transitional" secretory cells showing features of both granular types (serous and mucous) and suggested that a transformation of serous cells occurred in the response to the irritant. We therefore decided to study further the normal pathway of differentiation in rat bronchial epithelium and its modification, induced by tobacco smoke, using pulse labelling combined with autoradiographic techniques (Fig. 7).

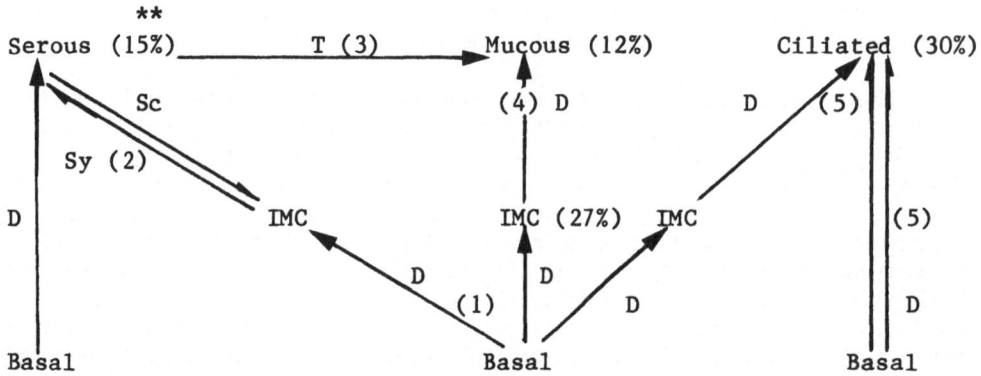

IMC = "Intermediate" cell
D   = differentiation
T   = transformation via transitional cell type
Sc = discharge of secretory granules
Sy = synthesis
** = these figures represent % labelled cell type at 14 days after
      single pulse of ^{3}H-T.

Fig. 10 Preliminary interpretation of cell differentiation -
         results for normal bronchial epithelium.

     Male SPF rats (Group B) were divided into control and smoked
groups and given a single pulse of ^{3}H-T at a dose of 1μ Ci/gm body
weight on day 0 of the experiment, thereby labelling a cohort of
cells in the S phase. By calculation of L.I. (epi) and L.I. (cell)
with time and tobacco smoke exposure, the fate of this labelled
group was followed. Animals from control and smoked groups were
killed sequentially and tissue samples processed, as described
above. In both the control and smoked groups there was a
significant decrease in the proportion of cells containing label,
suggesting a gradual loss of cells from the epithelium. The
results can be expressed as a change in % labelled cells with time
(tables IV and V) - i.e. the shift of label from one cell type to
another indicates the direction of cellular differentiation (Fig.
10).

     The schema for normal epithelium in Fig. 10 shows that there
are several pathways of differentiation which may operate, the
intermediate cell having a central role. The percentage increase
of labelled cells which are intermediate (corresponding with a
decrease in basal) is partly due to basal cell differentiation
pathway (1). We suggest also that a contribution is made by the
discharge of secretory granules from serous cells (2) which then
are counted as "intermediate". Labelled serous cells are then
replaced by subsequent synthesis of granules. Serous cells may

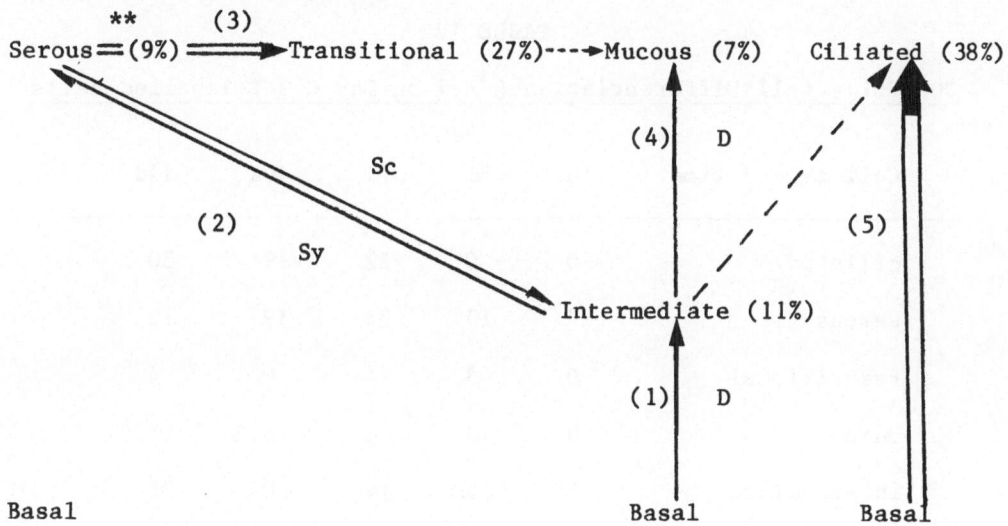

D = differentiation
T = transformation via transitional cell type
Sc = discharge of secretory granules
Sy = synthesis
** = these figures represent % labelled cell type at 14 days after single pulse of ^{3}H-T.

Fig. 11   Preliminary interpretation of cell differentiation – results for tobacco smoke – exposure bronchus.

also contribute to the increase in percent of labelled cells which are mucous by synthesis of altered secretory granules (i.e. transformation – (3)).   There also appears to be a contribution to the mucous cell by differentiation via the intermediate (pathway (4)).   Differentiation of basal mainly favours the production of ciliated cell either directly or via the intermediate cell ((5)).

The schema in Fig. 11 shows the pathways of differentiation which may operate after irritation by tobacco smoke.   Whereas pathways (1), (2), and (5) appear to be operating normally (see Table V), irritation emphasizes pathway (3) in favour of the transitional cell which is after 14 days 27% of labelled cells. The implications are that the increase in mucous cell number observed after tobacco smoke exposure is due largely to their self-replication accompanied by a contribution from transformation of existing serous cells.

TABLE IV

CONTROL: Cell Differentiation (^{3}H-T on Day 0.) % labelled cells

Cell type / time	1h	3d	7d	10d	14d
ciliated	0	0	22	19	30
serous	27	10	33	19	15
transitional	0	3	4	0	4
mucus	0	0	0	6.2	11.5
intermediate	6	15	14	0	27
basal	67	71	30	38	19

TABLE V

SMOKED: Cell Differentiation (^{3}H-T on Day 0.) % labelled cells

cell type / time	1h	3d	7d	10d	14d
ciliated	0	3	24	25	38
serous	27	27	10	21	9
transitional	0	0	0	7	27
mucus	0	3	5	7	7
intermediate	6	22	29	4	11
basal	67	46	38	32	14

SUMMARY

In this paper, studies of cell division and differentiation in the respiratory tract using 2 basic techniques have been reviewed, and new experimental findings have been presented.

Firstly, information can be gained by counting the number of cells entering mitosis with time. These measurements enable the Mitotic Index, and rate of entry into mitosis to be determined, and thus the Turnover Time of a cell type (life span) or tissue to be estimated.

Secondly, ^{3}H-thymidine may be used to establish the dividing cell population (i.e. progenitor cells) of bronchial epithelium in the normal and the kinetic parameters of the cell cycle, for each cell type. Furthermore, given as a single pulse it may be used to elucidate the pathways of differentiation.

Early work on cell division using both methods produced numerical results at variance but there was agreement that bronchial epithelium contained few dividing cells and most of these were basally situated.

Later studies expanded these findings and suggested that a small fraction of both basal and superficial-secretory cell groups are continuously dividing (i.e. there is a slow turnover time). Experimental studies show that these cells also respond to injury by dividing; either to restore the integrity of the epithelium or, in the case of more chronic stimuli, to change the tissue so that it can cope more adequately with irritation, e.g. the increase in numbers of secretory-mucous cells by division formed the most significant part of the bronchial epithelial response to tobacco smoke exposure.

Studies designed to observe change in the proportions of originally labelled cells with the time suggest that dividing cells, some of which are superficial and functional, can also give rise to other epithelial cells by differentiation and transormation. This normally occurring process can be altered by an irritative stimulus.

We acknowledge the support given to M.A. by the Cystic Fibrosis Research Trust.

REFERENCES

Aherne, W.A., Camplejohn, R.S., and Wright, N.A. 1977.   An introduction to cell population kinetics. Arnold.

Becci, J., McDowell, E.M., and Trump, B.F., 1978. The respiratory epithelium II Hamster trachea, bronchus and bronchioles. J. Natl. Cancer. Inst. 61, 551.

Bertalanffy, F.D. 1951. Mitotic activity and renewal of the lung. M.Sc. Thesis, McGill University, Montreal, Canada.

Bertalanffy, F.D., and LeBlond, C.P., 1953. The continuous renewal of the two types of alveolar cells in the lung of the rat. Anat. Res. 115, 515.

Bindretter, M., Schuppler, T., Stockinger, L., 1968. Zell proliferation und Differenzierung im Trachealepithel der Ratte. Exp. Cell Res., 50, 377.

Blenkinsopp, W.K., 1967. Proliferation of respiratory tract epithelium in the rat. Exp. Cell Res., 46, 144.

Blenkinsopp, W.K., 1969. Cell proliferation in the epithelium of the oesophagus trachea and ureter in mice. J. Cell. Sci., 5, 393.

Bolduc, P., and Reid, L., 1976. Mitotic index of the bronchial and alveolar lining of the normal rat lung. Am. Rev. Resp. Dis., 144, 1121.

Boren, H.G., and Paradise, L.J. 1978. Cytokinetics of lung. In:- Pathoenesis & Therapy of Lung Cancer, Ed. Harris C.C. Ch. 7. N.Y. Dekker, p.369.

Breeze, R.G., and Wheeldon, E.B., 1978. The cells of the pulmonary airways. Am. Rev. Resp. Dis. 116, 705.

Carleton, H.M., 1923 – 1924. The pulmonary lesions produced by the inhalation of dust in guinea pigs. J. Hyg. Lond. 22, 438.

Clara M., 1936. Vergleichende Histobiologie des Nierenglomerulus und der Lungenalveole. Nach untersuchangen beim Menschen and Beim Kanichen. Z. Mikroskop, Anat. Forsch., 40, 147.

Condon, W.B., 1942. Regeneration of tracheal and bronchial epithelium. J. Thorac. Surg. 11, 333.

Drasch, O., 1881. Zur Frage der regeneration des tracheal – epithels mit Rucksicht auf die Karyokinese und die beodentung der Bechersellen. Zitzber-ksl, Akad. Wiss., Math-naturwiss, Kl, Abt., III, 83, 341.

Divertie, M.B., Shorter, R.G., and Titus, J.L., 1968. Cell kinetics and tumour formation – Cell turnover in the lungs of mice with hereditory lung tumours. Thorax, 23, 83.

Gordon, R.E., and Lane, B.P., 1977. Cytokinetics of rat tracheal epithelium stimulated by mechanical trauma. Cell Tissue Kinet., 10, 171.

Gordon, R.E., and Lane, B.P., 1980. Wound repair in rat tracheal epithelium. Division of G_1 and G_2-arrested cells following injury. Lab. Invest. 42, 616.

Harris, C.C., Frank, A., Barrett, L., McDowell, E., Trump, B., Paradise, L., and Boren, H. 1975. Cytokinetics in the respiratory epithelium of the hamster, cow, and man. J. Cell. Biol., 67, 158a.

Hilding, A.C., 1943. The relation of ciliary insufficiency to death from asthma and other respiratory diseases. Ann. Otol. Rhin. Laryngol, 52, 5.

Howard, A., and Pelc, S.R., 1953. Synthesis of deoxyribonucleic acid in normal and irradiated cells and its relation to chromosome breakage. Heredity, (suppl) 6, 261.

Jeffery, P.K., and Reid, L., 1975. New observations of rat airway epithelium: a quantitative and electron microscopic study. J. Anat., 120, 295.

Jeffery, P.K., and Reid, L., 1977. The respiratory mucous membrane. In: Respiratory Defence Mechanisms, ed. Brain, J. et al, Dekker, New York, 193.

Jeffery, P.K., and Reid, L., 1981. The effect of tobacco smoke, with or without phenylmethyloxadiazole (PMO), on rat bronchial epithelium: a light and electron microsopic study. J. Path. 133, 341.

Kury, G., Rev-Kury, L.H., and Carter, H.W. 1969. Radioautographic study of human pulmonary tissues. Thorax, 24, 61.

Lane, B.P., and Gordon, R.E., 1974. Regeneration of rat tracheal epithelium after mechnical injury. I The relationship between mitotic activity and cellular differentiation. Proc. Soc. Exp. Biol. & Med. 145, 1139.

Le Blond, C.P., 1964. Classification of cell populations on the basis of their proliferative behaviour. Natl. Cancer Inst., Monogr., 14, 119.

McDowell, E.M., Barrett, L.A., Glavin, F., Harris, C.C., and Trump, B.F. 1978. The respiratory epithelium. I. Human bronchus. J. Natl. Cancer Inst., 61, 539.

McDowell, E.M., Becci, P.J., Schurch, W., and Trump, B.F., 1979. The respiratory epithelium. VII Epidermoid metaplasia during regeneration following mechanical injury. J. Natl. Cancer Inst. 62, 995.

Messier, B., and Le Blond, C.P., 1960. Cell proliferation and migration as revealed by autoradiography after injection of thymidine-H^3 into male rats and mice. Am. J. Anat., 106, 247.

Meyer zum Gottesberg, A., and Koburg, E., 1963. Autoradiographische untersuchungen zur zellnenbildung im respirationstrakt, in der Tube, im mittelohr und ausseren Gehorgang. Acta. Otolaryngol (Stockholm), 56, 353.

Rhodin, J., and Dalhamn, T., 1956. Electron microscopy of the tracheal ciliated mucosa in rat. Z. zellforsch 44, 345.

Rhodin, J., 1959. Ultrastructure of the tracheal ciliated mucosa in rat and man. Ann. Otol. Rhinol. Laryngol., 68, 964.

Rhodin, J., 1963. An atlas of ultrastructure. W.B. Saunders & Co., Philadelphia and London, 82-86.

Rogers, A.W., 1979. Techniques of autoradiography, 3rd edit. Elsevier/North-Holland, Amsterdam, New York.

Shorter, R.G., Titus, J.L., and Divertie, M.B., 1964. Cell turnover in the respiratory tract. Dis. Chest., 46, 138.

Shorter, R.G., Titus, J.L., and Divertie, M.B., 1966. Cyto dynamics in the respiratory tract of the rat. Thorax, 21, 32.

Simmnett, J.D., and Hepplestone, A.G., 1966. Cell renewal in the mouse lung. The influence of sex, strain and age. Lab. Invest., 15, 1793.

Spencer, H., and Shorter, R.G., 1962. Cell turnover in pulmonary tissue. Nature. 194, 880.

Strelin, G.S., 1930. Uber in-vitro-kulturen der brochen des kaninehens mit besonderer berucksichtingung des epithels. Arch. Exp. Zellforsch, 9, 297.

Waller, C., and Bjorkman, G., 1882. Studien uber den bau der trachealschleimhaut mit besonderer beruchsichtigung des epithels. Biol. Untersuch., Stockholm, 2, 71.

Wells, A.B., 1970. The kinetics of cell proliferation in the tracheobronchial epithelia of rats with and without chronic respiratory disease. Cell Tissue Kinet., 3, 185.

                              DISCUSSION

LECTURER: Ayers   CHAIRMAN: Cumming

CUMMING: Thank you Margaret, this paper is now open for
                discussion.

HEATH: This may be a rather unfair question since you have
                not specifically spoken about Clara cells, but I
                wondered, do you have any information as to whether
                the kinetics of Clara cells are bound up at all
                with the system that you've been describing? Do
                they operate independently from the system of cells
                you've been talking about, or are they
                interrelated? Do you have any data on this at all?

AYERS: There are two pieces of information which I can
                give you, firstly, Peter and I have been working
                with Lynne Reed using sulphur dioxide as the
                irritant. We found that Clara cells in response to
                this irritation showed metaplasia and changed into
                serous cells, into secretory cells and into goblet
                cells. The observation was that a sequence of
                events could be seen changing from Clara, to
                serous, to goblet. There is also work by Evans
                using nitrogen dioxide as the irritant and he has
                concluded from his observations using pulse
                labelling in the same way, giving a pulse at the
                beginning of the observation and watching the cell
                changes over 14 days, he came to the conclusion
                that the Clara cell increased its rate of division
                in response to the irritant and was the progenitor
                for other cell types within the bronchiolar
                epithelium.

LAURENT: Margaret, it was nice to see your approach in
                trying to get some data on kinetics, and to do that
                you made a foray into the very difficult area of
                pulse labelling with radioactive compounds. Could
                I make a note of how careful one must be in the
                interpretation of this sort of data with labelled
                compounds and I'll do that in the first question
                and then pose a more general question after that.
                The problem of labelling with compounds is that
                after the initial pulse labelling, label can then
                leave the DNA in degradation and be reincorporated
                at a later time. This can cause quite drastic
                confusion. For example, one of your conclusions

with regard to the effect of smoking was that after the initial labelling there was a more rapid loss of labelled compound in the groups that have been exposed to smoke. The alternative is that as the label leaves the DNA in the smoking group smoking group it will be reincorporated, so that it's not just loss of labelling that you are measuring here, there is also the problem of reincorporation, which means that you have been measuring synthesis rate as well. So, loss of label not only means a lower rate of death, it also means an increased synthesis of new cells or new cell production.

CUMMING: May I ask you, Geoff, to let us know the half time of DNA decay?

LAURENT: With DNA one does not talk in terms of half life because current views are more of a life span, and the other thing is that the idea of life span of cells varies greatly, in other words one cell can have a completely different life span to another. But in some cases, with regard lung cells these times would be very short.

CUMMING: I'm sorry, I must insist: when you say very short the crucial question is can DNA be broken down and resynthesized in the time period of 14 days that Margaret is studying.

LAURENT: Without any doubt at all.

CUMMING: Thank you, I just wanted to be quite specific about that answer.

LAURENT: I'm going on too long. Finally, my general question is what do you think is the function of this very rapid turnover of cells which you related for some cells? Is it simply a matter of cells wearing out and dying or is there some other, perhaps more complex process involved.

AYERS: I take it you mean in the respiratory tract in particular, in this system, and I can only be very speculative. With that in mind I think within our system the function is a protective one, in the sense that the animals that we use are very pure in that they haven't had exposure to any pollutants and they have not been infected, so that they are as close to the fetal situation as is possible and there are correlates between the bronchial epithelium in these animals and the epithelium in human fetus, in that there are large numbers of serous cells. The response to a threat, to irritation, is the production of cells with lucent granules, which is supposed to produce mucus of a more viscous type, which you could speculate would be protective in the sense of protecting from the irritant and enabling the defences of the lung to get rid of the irritant more quickly.

CORRIN: I'm going to be very brave and challenge a chemist's concept of DNA breakdown. I think that Geoff's challenge of Margaret needs challenging. My concept with pulse labelling with thymidine is that DNA synthesizes from thymidine, but, Geoff, can you just tell us more details about the breakdown of DNA. I thought it wasn't released again, the label would not be released again as labelled thymidine, but more fundamental small molecular units would be released which could be taken up in other synthetic processes, maybe in the cytoplasm or maybe excreted as such. The idea of thymidine pulse labelling is that the thymidine is not taken up again by other cells.

LAURENT: Brian, the thymidine would certainly be incorporated by new cells in new synthesis, but I take your point, there are other pathways that thymidine and its degradation products would go through. But if you have a rapid synthesis or rapidly dividing and newly synthesized cells, quite a large proportion of the thymidine coming out of the DNA and breakdown would be reincorporated as thymidine.

AYERS: There is evidence of this within tumours, Steel has shown re-uptake in that the eventual labelling index is higher than expected as time progresses, because of cell breakdown and re-uptake. I think Geoff's point is valid and important in the interpretation of this phenomenon.

LAURENT: I'm talking of thymidine in nucleotides here.

JEFFERY: Geoff Laurent's point is absolutely right, I don't think there is any point in prolonging the discussion on this particular subject further. There is no doubt that tritiated thymidine can be re-utilized not only from lung but of course from the thymidine taken up by other tissues in the body and subsequently released. It is available for re-utilization and there are specific techniques which have to be employed to measure the extent of re-utilization. One such technique is the use of tritiated UDR, which we intended to do and needs to be done in this particular work. The other approach is to measure the amount of tritiated thymidine in the circulating pool in the serum, which is perhaps more difficult, but we are well aware of the need to do this.

CUMMING: Please expand UDR.

JEFFERY: The uridine components. Your second point was the interesting question as to what might be the stimulus for the increase in proliferation, it's a very interesting one and one of the suggestions that has been made by Lane and Gordon is that it is the increased loss of cells from the superficial compartment whmch is the stimulus for the proliferation of the stem cell compartment in the deeper layers. That would probably be the system which is working here because the irritant that was given was not noxious enough to cause frank ulceration and necrosis of the epithelium. So I think we are looking at sensitive changes in the system.

BIENNENSTOCK: I tried to introduce you to the immunologist's view of the world, which is slightly different from that which you suggested in terms of the purity of this population of animals, the specific pathogen-free animals, because in tobacco smoke there are large numbers of antigens which these animals could be responding to, and the response of the goblet cell population by proliferation might be due directly or indirectly to this immune response. This is not totally hypothetical because it has been shown that it is possible to transfer from one animal to another the goblet cell proliferation in the recipient animal. Presumably that is not the transfer of precursor cells, I think we would all agree, but it's likely to be a transfer of sensitized lymphocytes which on interacting with antigens release material which in turn can effect goblet cell and other cell proliferation.

KAY: Margaret, I would like to ask you another question about the mast cell. It follows up from the last comment. You may know that two or three years ago

David Lamb from Edinburgh and his colleagues reported some experiments in which they showed a proliferation of mast cells in the respiratory epithelium of human smokers. The data was preliminary but it was convincing and that fitted in nicely with studies of our own which showed that mast cell derived mediators are present in high concentrations in bronchial secretions of patients with chronic bronchitis and also symptomatic young cigarette smokers. That's my comment, my question is: did you have a look at mast cell proliferation in your experimental model, and if you didn't, would it be possible to go back to the blocks and do these types of count?

AYERS:

I haven't processed the quantitation of mast cells. First of all we have to differentiate between what we call globular leukocytes, intraepithelial granulated cells, and I observed in my counting several of these that were labelled, and those I counted. So, I have that material. I didn't particularly observe the submucosal mast cells, which are different morphologically. Presumably there is a relationship between them. But it certainly would be possible and would be worthwhile to go back because definitely they were in division. Whether there was a significant repsonse within our system of those cells I don't know, I'd have to go back and look at the data.

CUMMING:

Well, can I thank you Margaret, and draw the discussion to a close. We will now have a break for 30 minutes, which means we'll reassemble here at 11.35, and can I say that during the interval will you please collect your name badges, a list of participants and some touristic information. See you at 11.35.

CILIA

Bjorn A. Afzelius

Wenner-Gren Institute
University of Stockholm
Stockholm, Sweden

INTRODUCTION

The cilium and its longer variety, the flagellum, must be regarded as one of Nature's most successful inventions. The lucky event, when cilia came into being took place perhaps a thousand million years ago and in a eukaryotic cell, which according to Cavalier-Smith (1981) would be regarded as a fungal cell. The immediate precursor to the cilium might have been cytoplasmic microtubules such as those seen in the mitotic spindle. During subsequent evolution this ciliated cell gave rise not only to the fungal 'kingdom' but also to the protozoan, plant, and animal kingdoms. Cilia and flagella are common in all these kingdoms and have the same structure.

The main constituent of the cilium, the axoneme, appears identical in all eukaryotic organisms where cilia occur and there is also no structural difference between cilia and flagella, except for their length. I will hence use the term 'cilium' for both types of organelles throughout most of the text. There are on the other hand certain variations with respect to appendages attached to the outside of the flagellum (McQuade, 1977) and there are structural differences in the proximal region of the flagellum (Hibberd, 1979). These varieties have turned out to be useful characteristics in phylogenetic studies.

Not only are cilia widespread, particularly within the animal and protozoan kingdoms, each organism may also carry a great number of cilia (Purkinje and Valentin, 1835) (Fig. 1). Our own species is a good example of this. It can be estimated that we in our respiratory tracts have ciliated epithelia covering an area of 0.5

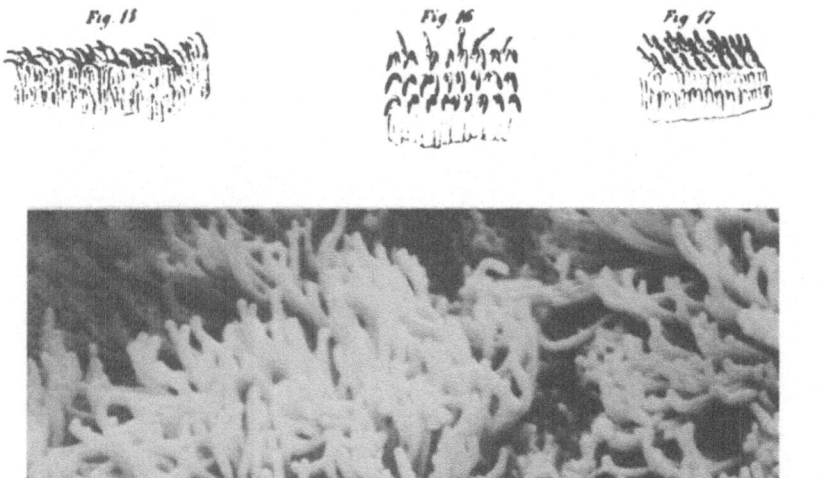

1

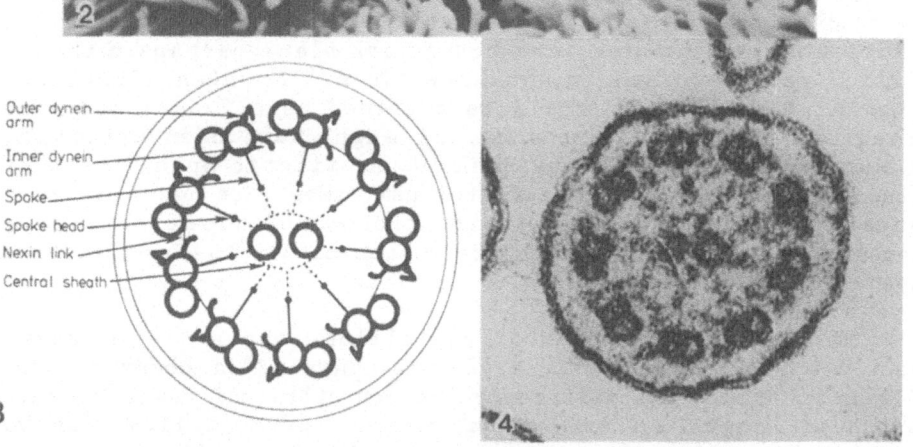

3

Fig.1. J E Purkinje was first to describe ciliated epithelia. In
this lithograph from 1835 cilia from the chicken oviduct (fig 15
and 16) and from the grass snake (fig 17) are shown. The ciliary
beat is correctly shown: A downstroke toward the spectator followed by a recovery stroke in which cilia return by bending toward
their left side.
Fig.2. Scanning electron micrograph from the ciliated epithelium
of the human nose. Micrograph by P Hörstedt, Dept of Pathology
Umeå University. Magnification x 6,000 times.
Figs 3 and 4. Diagram and cross-sectioned cilium. x 180 000 times.

m^2. As the density of the cilia is 6 per μm^2, the total number of cilia in the human body is in the order of 3.10^{12}.

The purpose of this communication is to review what is known about the fine structure and functions of cilia and of the role of cilia in human health.

CILIARY STRUCTURE

Evidence has been given above for the belief that the cilium is the most common motor in the world. (A motor is in Webster's dictionary defined as a small, compact engine, a definition which superbly suits the cilium). It would be natural to assume that it has a simple construction and a simple mode of operation. This is not so. The number of components within the motor is unknown, but Huang et al. (1979) have given evidence for their opinion that over 100 different polypeptides compose the axonemal motor. We do not know the localization and function of more than a few of these components.

The axoneme has a characteristic configuration: nine microtubular doublets in a ring around two central microtubules (or singlets). This is the famous 9 + 2 pattern (Figs. 3 and 4). Three types of bonds keep the 9 + 2 microtubules together: (i) the dynein arms which are extensions from one side of the doublets toward the neighbouring doublet, (ii) the nexin links which are slender bonds connecting neighbouring doublets, and (iii) the spokes which extend from the doublets toward the central microtubules. Most spokes have a thickened end portion, the spoke head, that lies close to a structure called the central sheath. The central sheath surrounds the two central microtubules and may be filamentous projections from them.

The dynein arms and spokes are attached to one of the two Siamese twin-tubules making up a doublet. This doublet is called the A-tubule and it has been shown by Gibbons (1961 a) that the A-tubules all have a clockwise direction if the cilium is seen from the interior of the cell toward its tip. All the 9 + 2 microtubules are assumed to run straight along the length of a cilium, which is 5 μm in human ciliated epithelia. If a field of cross-sectioned human cilia is examined, it can usually be seen that the orientation of the cilia is rather fixed (Fig. 5). A line through the two central microtubules in one cilium is fairly parallel to a line through the central microtubules in neighbouring cilia. Deviations usually are less than 30°. A line perpendicular to that line will hit an outer doublet, which by definition is called number 1. The direction of the effective beat in beating cilia has been shown to be away from doublet number 1 and thus toward doublets number 5 and 6 (Gibbons, 1961 b).

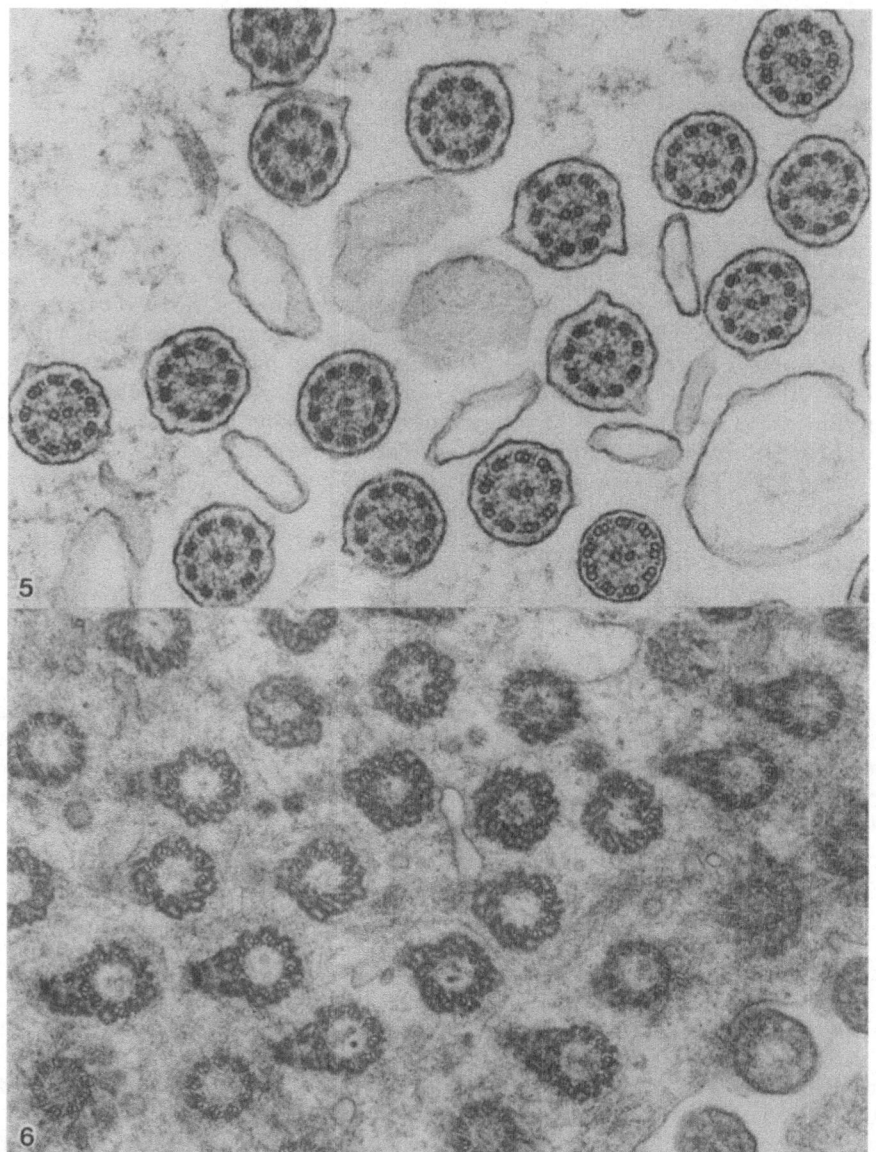

Fig. 5. Cross-sectioned human cilia. Note that the cilia have a
fairly fixed orientation. Magnification 55,000 times.
Fig. 6. Cross-sectioned basal bodies from the same biopsy as in
fig. 5. Note that the triangular basal foot processes all point
in the same direction; this would be the direction of the effective
stroke of the cilia. Magnification   45,000 times.

Cilia are outgrowths from intracytoplasmic organelles called basal bodies. These are structurally indistinguishable from the centrioles at the poles of mitotic spindles. There is also a fixed polarity in the orientation of the basal bodies, best seen in the foot process (also called the basal foot), which projects from one side of the basal body (Fig. 6). The basal foot can be regarded as a convenient marker of the direction of the effective beat; it points in the same direction as do doublets number 5 and 6.

The transitional region between the basal body and the free cilium has several characteristics. It is a constricted region which contains Y-shaped connections between the doublets and the plasma membrane, there is a so-called ciliary necklace in the plasma membrane (Gilula and Satir, 1972) and the two central microtubules originate in this region.

Close to the ciliary tip 9 + 2 single microtubules are seen. These are the nine A-tubules and the two central ones. The A-tubules do not carry dynein arms at this level. There may also be claw-like projections at the tip (Jeffery and Reid, 1975), sometimes called the ciliary crown.

CILIARY FUNCTIONS

The cilia of a paramecium or a ctenophore enable the animal to swim around, propelled by the ciliary beats. The cilia of a sessile clam drive a water-current over the gills of the animal. The cilia in our own respiratory tracts do neither but rather act by pushing a mucous layer along the ciliated epithelum. The frequency of the beating in the tracheobronchial tree is around 15 Hertz in the young and middle-aged and perhaps somewhat less in old persons (Yager et al. 1980). Two phases can be distinguished in the ciliary movement: (i) an effective (or power) stroke in which the cilium swings forward about 110° by a bend at the base; the cilium behaves as a stiff rod in this phase and the stroke is almost planar. (ii) a recovery (or preparatory) stroke in which a bend is propagated up the cilium; it is a three-dimensional counter-clockwise rotation as viewed from above. The details of these strokes were first studied on instantaneously fixed paramecia by Parducz (1967) and later in great detail by Omoto and Kung (1980) and, on tracheal cilia by Sanderson and Sleigh (1981). The mucous blanket is pushed forward during the power stroke but remains stationary when the cilium returns in the preparatory stroke.

It is believed that cilia move by the work exerted by the dynein arms and that the work is a sliding of the doublets relative to each other. The elastic nexin links permit a sliding only to a certain degree (maximally less than 0.1 μm) and this restriction transforms the would-be sliding of the doublets into a bending and thus of the entire cilium. The spokes may also have proteins of the dynein type which would allow them to climb on the central sheath. Another function of the spokes is to prevent the axoneme from collapsing when the cilium bends.

The two central microtubules apparently rotate counter-clockwise in beating cilia, $360°$ with each beat cycle (Omoto and Kung, 1980). It has been assumed that this motion provides the trigger by which the nine doublets are initiated to slide. Because the central pair of singlets is twisted around its axis, the rotation will cause a trigger of different doublets at different levels along the cilium. Evidence for this hypothesis has been obtained from work on some protozoa (mainly Paramecium) and may be valid also for other types of cilia. So far, however, there are no data to show that a twisting and rotation of the central microtubules indeed occurs in cilia from ciliated epithelia; on the contrary the central tubules have a fixed position as far as can be seen with techniques employed so far.

The basal bodies are generally assumed to be the sites at which the ciliary beats originate. If a flagellum is severed, only the proximal fragment will continue beating. An understanding of the basal bodies is hence of importance for an understanding of the ciliary work. It has recently been shown that actin, myosin, tropomyosin all are located within the region of the basal bodies and in particular in the basal feet (Gordon et al. 1980). We still do not know what function these contractile proteins have in the basal body, and whether they participate in giving the basal feet their proper orientation. The functioning of the ciliary motor is certainly imperfectly understood.

ROLE OF CILIA IN THE HUMAN BODY

Even if the intricacies of the ciliary motor remain incomprehensible, it is possible to estimate the importance of the work performed by this motor. This is due to the recent discovery of a human disorder, which is characterized by the cilia being either completely immotile or else so poorly motile that the cilia cannot transport the mucous layer in the respiratory tract (Afzelius, 1976, 1979). Cilia from patients having this disease or from controls can be observed in the living state and also seen in thin sections by electron microscopy. Spermatozoa are easily obtainable and both vital microscopy and electron microscopy are routine techniques in many laboratories. By such simple methods it has been possible to conclude that affected persons have cilia and spermatozoa that are structurally defective and incapable of motility. The disorder has been called the immotile-cilia syndrome.

It is highly likely that the immotile-cilia syndrome is an

inborn disease of genetic origin. The following features are in
favour of this: The symptoms can be seen in the newborn. The
syndrome is often found among sibs. The parents of affected
persons are close relatives in a greater than average percentage.
There are affected and unaffected sibs in a family but no
intermediates. The cilia from the affected persons are
structurally defective with a pattern that resembles the paralyzed
mutant strains of the flagellate alga Chlamydomonas.

The syndrome is a heterogeneous one in that several subgroups
can be characterized based on the ultrastructural appearance of the
cilium. Interestingly enough the clinical appearance of these
different subgroups do not differ. The most common subgroup is one
where the dynein arms are strongly reduced in number or even are
completely absent (Fig. 7). In another subgroup the spokes and
probably the nexin links are defective, and as a result the axoneme
is disorganized (Fig. 8) (Sturgess et al., 1979). Still other
subgroups have been described (Afzelius and Eliasson, 1979;
Jahrsdoerfer et al., 1979; Sturgess et al., 1980). In most cases
the cilia have a random orientation in these various forms of the
immotile-cilia syndrome; this indicates that the cilia would
counteract if they were motile.

It is not surprising that persons with the immotile-cilia
syndrome have clinical symptoms from their respiratory tracts; all
having chronic bronchitis from early infancy. It is more
surprising that persons with a very slow or non-existing mucociliary transport still survive and even can live an active life
and have a relatively normal life expectancy.

Patients with the immotile-cilia syndrome often acquire
obstructive lung disease. Emphysema may be seen even in young
patients and bronchiectasis is common. Most patients also have
chronic sinusitis, rhinitis, and otitis. Some slight atrophic
changes in the brain ventricles have been recorded in some
patients, possibly as a result of the ependymal cilia being
immotile (Afzelius, 1979). The oviduct cilia do not seem to have
an essential function, since women with the syndrome may become
mothers. It is not known whether there is some reduction in female
fertility. Males are infertile due to an inability of the
spermatozoon to move or to show progressive motility. The sperm
tail is a flagellum with the same type of axoneme as that of the
cilium. It has been estimated that the prevalence of the immotile-
cilia syndrome is about 1:20,000. It is hence a rare but not
exceedingly rare inborn disorder.

Fifty percent of the affected persons have situs inversus
totalis; this seems to be the case regardless of whether the defect
is one of the dynein arms, or the spokes or some other component.
It has been suggested that this particular feature of the immotile-

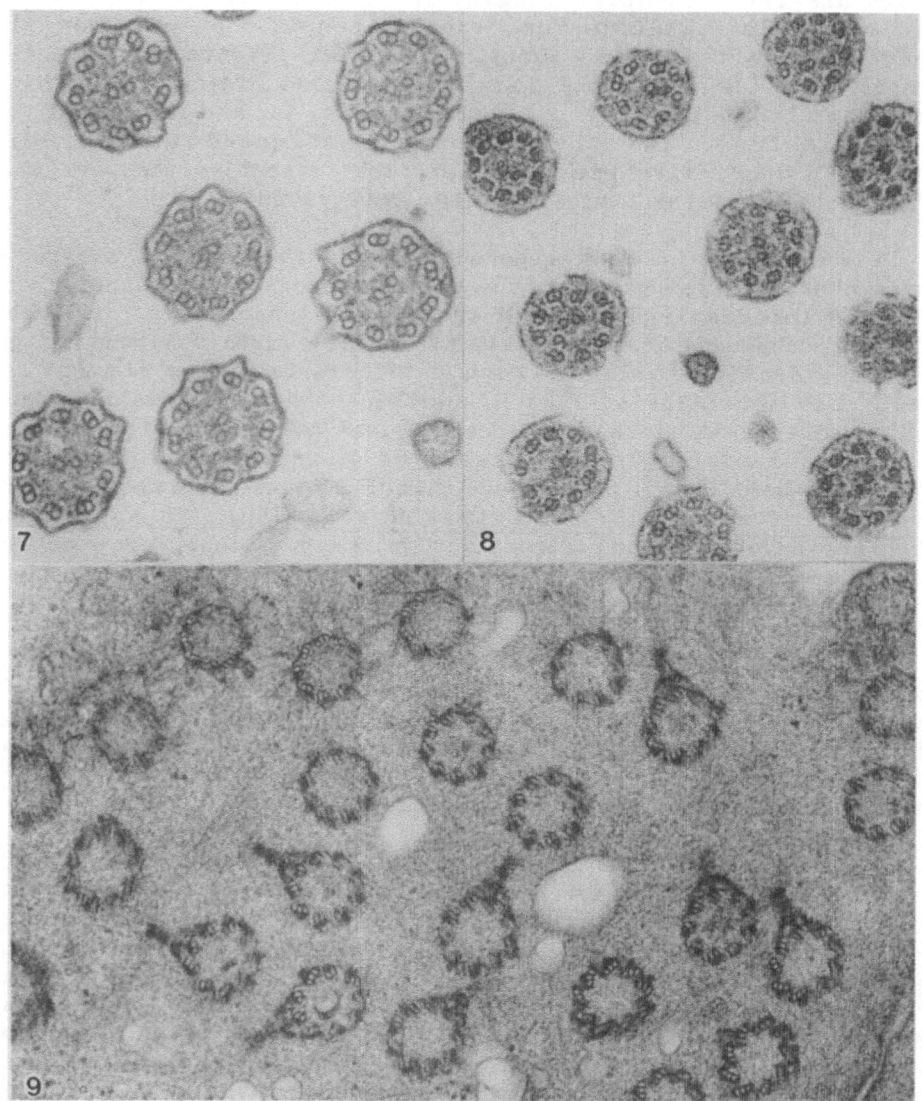

Fig. 7. Cross-sectioned bronchial cilia from a patient with the immotile-cilia syndrome. Note that most dynein arms are missing and that the cilia have a random orientation. x 75,000 times.
Fig. 8. Cross-sectioned cilia from another patient with the immotile-cilia syndrome. The outer dynein arms are present but the central part of the axoneme is disorganized. x 50,000 times.
Fig. 9. Cross-sectioned basal bodies from the same biopsy as in fig. 7. Note that the basal feet point in random directions. Magnification x 50,000 times.

cilia syndrome is caused by the visceral laterality being randomly determined rather than determined by the work of embryonic cilia (Afzelius, 1976).

SUMMARY

The cilium of one of Nature's oldest and most successful inventions. Although it has been assumed to be a derivative of the cytoplasmic microtubules it has a considerably more intricate structure. The mode of operation of the ciliary motor is not understood although it is clear that a sliding of the microtubular doublets relative to each other is part of the work. The trigger and regulatory mechanisms are largely unknown. The ciliary motor is defective in certain mutaed experimental organisms and in some human subjects. This disorder in man has been called 'the immotile-cilia syndrome'. A study of patients with this syndrome has shown that ciliary dysfunctions often lead to obstructive lung disease.

REFERENCES

Afzelius, B.A., 1976, A human syndrome caused by immotile cilia, Science, 193:317.

Afzelius, B.A., 1979, The immotile-cilia syndrome and other ciliary diseases, Intern. Rev. Exp. Pathol., 19:1.

Afzelius, B.A. and Eliasson, R., 1979, Flagellar mutants in man. J. Ultrastr. Res., 69:43.

Cavalier-Smith, T., 1981, The evolutionary origin and phylogeny of eukaryote flagella. Soc. Exp. Biol. Symp. 35 (in press).

Gibbons, I.R., 1961 a, Structural asymmetry in cilia and flagella, Nature, 190:1128.

Gibbons, I.R., 1961 b, The relationship between the fine structure and direction of beat in gill cilia of a lamellibranch mollusc, J. Biophys. Biochem. Cytol., 11:179.

Gilula, N. and Satir, P., 1972, The ciliary necklace, J. Cell Biol. 53:494.

Gordon, R.E., Lane, B.P., and Miller, F., 1980, Identification of contractile proteins in basal bodies of ciliated tracheal epithelial cells, J. Histochem. Cytochem. 28:1189.

Hibberd, D.J., 1979, The structure and phylogenetic significance of the flagellar transition region. BioSystems, 11:243.

Huang, B., Piperno, G., and Luck, D.J.L., 1979, Paralysed flagella mutants of Chlamydomonas reinhardtii, J. Biol. Chem. 254:3091.

Jahrsdoerfer, R., Feldman, P.S., Rubel, E.W., Guerrant, J.L., Eggleston, P.A. and Selden, R.F., 1979, Otitis media and the immotile cilia syndrome, Laryngoscope, 89:769.

Jeffery, P.K. and Reid, L., 1975, New observations on rat airway epithelium, J. Anat. 120:295.

McQuade, A.B., 1977, Origins of the nucleate organisms, Quart. Rev. Biol. 52:249.

Omoto, C.K. and Kung, C., 1980, Rotation and twist of the central pair microtubules in Paramecium. J. Cell. Biol. 87:33.

Parducz, B., 1967, Ciliary movement and coordination in ciliates, Intern. Rev. Cytol. 21:91.

Purkinje, J.E. and Valentin, G., 1835, Commentaţio physiologica de phenomeno motus vibratorii continui, Wratislaw.

Sanderson, M.J. and Sleigh, M.A., 1981, Ciliary activity of cultured rabbit tracheal epithelium, J. Cell. Sci. 47:331.

Scheeberger, E.E., McCormack, J., Issenberg, H.J., Schuster, S.R. and Gerald, P.S., 1980, Heterogeneity of ciliary morphology in the immotile-cilia syndrome in man. J. Ultrastr. Res. 73:34.

Sturgess, J.M., Chao, J., and Turner, J.A.P., 1980, Transposition of ciliary microtubules. New Engl. J. Med. 303:318.

Sturgess, J.M., Chao, J., Wong, J., Aspin, N., and Turner, J.A.P., 1979, Cilia with defective radial spokes. ibid. 300:53.

Yager, J.A., Ellman, H., and Dulfano, H.J., 1980, Human ciliary beat frequence. Amer. Rev. Resp. Dis. 121:661.

DISCUSSION

LECTURER: Afzelius   CHAIRMAN: Cumming

CUMMING: This paper is now open for discussion.

SOUROUR: I have in Alexandria two families in which three
              brothers in one family have Kartagener's syndrome
              and in the other family a brother and a sister show
              the same syndrome. I was discussing these matters
              with Professor Bryan Corrin two years ago and he
              told me about another family which he knew with
              Kartagener's syndrome. I wonder if there is any
              genetic relationship between the integrity of the
              dynin arms and ciliary function.

AFZELIUS: I was glad that you mentioned that the name of this
              syndrome when the heart is on the right side.
              Kartagener's syndrome was described in 1935 and I
              believe that all cases of Kartagener's syndrome are
              due to the cilia being immotile, and apart from
              Kartagener's syndrome there is an equally big group
              where the heart is on the eft side, and they have
              exactly the same thing. If you have a family where
              there is a person with Kartagener's syndrome and
              you examine the sibs, you find that 25% have the
              same syndrome with regard to the cilia — they have
              chronic bronchitis and sinusitis. Half of them
              have also situs inversus, so there we have peculiar
              inheritance, clearly it's not 1:3 but 1:7 with
              regard to the localisation of the heart. There are
              reasons to believe that it's a genetic disease: it
              occurs in families, there are sibs which are
              healthy and those which are sick and no
              intermediates, and they are more common in places
              where there is a greater degree of inbreeding.
              This syndrome in man resembles very strongly the
              same thing in chlamidomonas, a unicellular alga,
              except that there it's called paralysed flagella
              mutants.

CUMMING: Thank you. I think before asking Brian Corrin to
              ask the next question we should perhaps distinguish
              between situs inversus and dextrocardia. The heart
              being on the right side is dextrocardia, whereas I
              understand that situs inversus relates to the
              dextroposition of the abdominal viscera as well as
              the heart, so there are two conditions I think we
              should distinguish there.

CORRIN: I was very interested in the situs inversus
               syndrome in mice which I was previously unaware of.
               Do these mice have immotile sperm tails, do they
               suffer bronchiectasis and sinusitis? If not, you
               are postulating genetic defects of their embryonic
               cilia without the adult cilia being affected.

AFZELIUS: You are correct, they don't have immotile cilia and
               immotile spermatozoa, they are healthy by and
               large. Initially, when this mutant was isolated a
               great number of them died from lung defects, but
               nowadays the strain is pure and does not have a
               high degree of lung defects. I presume in the case
               of the mice only embryonic cilia are affected, I
               believe, or else there is something else. Whether
               these situs inversus mice are a good model for
               Kartagener's syndrome I don't know.

CUMMING: Perhaps I can raise one problem while the next
               person thinks up his question. There is a general
               prejudice in biology that the conversion of
               chemical energy to mechanical movement is via the
               medium of the actinomyosin system. Now it's
               suggested that the microtubules have no
               actinomyosin and yet these are transducers of
               mechanical movement. Can you give us any
               information about this entirely different area of
               mechanical transduction.

AFZELIUS: Yes, I think it's a completely different area and I
               think we have by and large in the cell at least two
               different motor systems or contractile systems.
               One is the actinomyosin system and the other is the
               dynine tubular system, and there is a great
               resemblance between the myosin-actin system and the
               tubulin dynine system, but one possibility does not
               exclude the other, that is to say that in ciliated
               epithelia we may have both, and in the axoneme most
               of the work for sure is done by the dynine tubulin,
               but in the basal bodies, particularly in the basal
               body region we have actin, and myosin. So, at
               least down at the basement membrane we also have
               the actin myosin system and maybe they have
               something to do with the orientation of the basal
               body, I don't know.

CUMMING: Thank you.

JEFFERY: I wonder if in your studies you have looked at the
               correlation between the structure and the function.

On the basis of structure there are basically three abnormalities recognized in cilia, namely deficiency in the dynine arms, transposition of tubules and deficiency in the radial spokes. Have you in your studies looked at the functional correlation between cilia which beat or do not beat and any of these three abnormalities? My question is: if you have cilia dynine arms deficiency do the cilia still beat, but perhaps irregularly, or if there is tubule transposition do the cilia still beat? The second point is when you show and talk about primary cilia with a deficiency in the central tubules, would you expect the cilia to beat or would they be non-functional.

AFZELIUS: As for your first question, last September I met together Patricia Sturgess from Toronto and the group from Paris and we all had examined about 40 patients each. So when we pooled the material we had together a fairly large sum of people who have the immotile cilia syndrome but since we had divided it up slightly differently we couldn't strictly compare it. But is turns out that whether the defect was in the spoke system or the dynine system or in some other system the clinical picture was the same. So it seems that from the clinical point of view it doesn't matter if one thing or the other is broken, we have motor defect. Now, it's true that in some cases with broken dynine arms, cilia do not beat at all, they are completely immotile, as the name implies. And I think when the dynine arms are completely gone then we have completely immotile cilia. In other cases cilia beat back and forth like a pendulum rather than making the intelligent movement they should and these cases have very little spoke defects, but transposition defect and so on. One can distinguish that with the light microscope but not clinically, as far as I know. As for your other question about the primary cilia, the ones which I showed in the last slide, all evidence seems to be that they don't beat: I want them to beat, but I do not know whether the beat as I wish.

BAUM: I wonder if you could tell us what is the latest data about the cystic fibrosis factor.

AFZELIUS: The last slide I have, which I didn't show, was cilia from a man having cystic fibrosis from which he died. These cilia look perfectly normal. The cystic fibrosis factor seems to have many effects. One of them being that it increased the mucus transport and these patients have an enormous amount of mucus, and cilia by themselves are active but they are overwhelmed by the work, they cannot work because of too much mucus to transport, so they die.

BAUM: But I understood that there is a factor that could be isolated from the serum of those patients which would stop the beating of cilia in bivalves.

AFZELIUS: They stop the beating only because of mechanical obstruction, the mucus is too much and too tough, but if you wash away the mucus they beat. So the cilia by themselves can beat, but they are not permitted to beat because of the tough mucus.

CUMMING: Are you saying that there is no serum factor? Because the implication that it is a purely mechanical limitation means it cannot be a circulating factor. Harold Baum seems to think there may be a circulating factor and you are I think denying it, is that correct?

AFZELIUS: I say there is a serum cystic fibrosis factor, but that this factor acts on the goblet cells or something else, not on the cilia directly.

HEATH: You were saying that the cilia in the effector stroke all beat in the same direction. I wonder if you would like to make some comments as to the method of control so as the cilia don't beat towards the head. What is the control mechanism so that the cilia don't beat the mucus down into the alveolar spaces? And also, as an extension of that question, do you know anything about a similar control of the direction of the effector beat of the cilia or the flagella in lower forms of life?

AFZELIUS: The first part of the question I wish I had an answer, but I know of no study which shows what the control mechanism is. I think I remember some experimental work where someone took a piece of trachea and turned it upside down. The first effect would of course be that the mucus stopped at a certain level, but whether nature by some way overcomes this obstacle or not I don't remember. Maybe somebody else can answer, do you remember?

JEFFERY: They beat in the wrong direction and it is thought that the direction is genetically determined.

AFZELIUS: Yes, in my lung they beat up to the pharynx, in my nose they beat down to the pharynx, in my ears they beat centrally to the pharynx.

CUMMING: The use of the words up and down in the lung is rather difficult, as in the upper lobes it beats downwards and in the lower lobe it beats upwards. One may get some information from the fact that the homotransplanted lung, completely denervated, still has normal cilial function, so it's unlikely to be neurologically directed.

JEFFERY: Just to come back to the point raised by Harold Baum, last year we had a presentation on serum factors and the effect on cilia and ciliary dyskinesia and I think Michael Sanderson from the Sleigh group in Southampton clearly showed that sera not only from cystic fibrosis patients but indeed normal sera did cause ciliary dyskinesia and they put forward on good evidence that this was due to IgM agglutinating the cilia by heterophil antigen.

LAURENT: Just one quick one. I like you, Gordon, am fascinated by the analogy between the actinomyosin system which was proposed by Huxley 20 years ago now. What would your comments be on the teleology of these two systems and have any sequence studies been done on this, aminoacid sequence, on these proteins and has any analogy been shown between these two systems in molecular terms?

AFZELIUS: If I first take the simplest thing, molecular weight, tubulin and actin have about the same molecular weight and both form chains. The dynine and the myosin being the two longest polypeptide chains formed on one gene. So thus far the analogy is pretty nice. For the aminoacid, not sequence but proportions, Hido Mori from Tokyo has made some comparisons and found that there is not a great deal of similarity in the aminoacid composition of actin and tubulin or dynine or myosin. It seems that the two systems have evolved separately and one could speculate why. For instance each cell has both systems operating, actin-myosin for some purposes and tubulin-dynine for other purposes, such as using the actin-myosin for contraction and the other for relaxation. This is one theory, so, like we have in our bodies benders and stretchers, so the cell has also benders and stretchers, benders being actin-myosin and stretchers tubulin dynine.

KUHN: To return to this question of the control of direction in paramecium there is a phenomenon of ciliary reversal, where the cilia change and beat in the opposite direction, so the paramecium can back away from the noxious agent. I was wondering if in that situation the basal foot rotates and turns to point in the opposite direction. The paramecium does not have the basic foot, so it rotates in a symmetric body of the paramecium but in man and high organisms I don't think that such means of changing that direction exists; we are less flexible, more rigid.

FABBRI: I'd like to know whether in patients with fixed cilia syndrome the physicochemical features of the mucus are perfectly normal.

ROSSI: There is no reason to believe that the properties of mucus are different, but I don't know, I have no answer on this question, it has not been examined.

CUMMING: I think I should terminate the discussion at that point, ladies and gentlemen. I have sensed a feeling amongst the audience that maybe the temperature in this room is rather low for comfort. Would that be the general view? I will do my best over the lunch interval to have that rectified. I hope we have got the audiosystem now working properly, but if anybody else has a problem about the mechanics of communication, please see me during the interval. Thank you very much.

LAVAGE EOSINOPHILS AND HISTAMINE

P.L. Haslam, A. Dewar and M. Turner-Warwick

Departments of Medicine and Electron Microscopy
Cardiothoracic Institute, London SW3 6HP U.K.

In recent years, development of the technique of
bronchoalveolar lavage has provided a simple and safe way of
obtaining samples of inflammatory cells and other components which
accumulate in the air spaces of the lungs as a consequence of
disease. These components can be examined free from the tissues,
and are providing a great deal more information about inflammatory
processes in different lung disorders. In particular,
bronchoalveolar lavage has been used to study interstitial
disorders of the lung including granulomatous lung disorders,
namely sarcoidosis and extrinsic allergic alveolitis and
interstitial fibrosing lung disorders, including asbestosis and
cryptogenic fibrosing alveolitis (synonym: idiopathic pulmonary
fibrosis). One of the main findings in these samples is that while
lavage lymphocyte increases are a striking feature of the
granulomatous lung disorders, by contrast lavage neutrophil
increases are the most striking feature of interstitial fibrosis[1].

Recent observations at the Cardiothoracic Institute and
Brompton Hospital, London, however, suggest that eosinophils as
well as neutrophils, may be an important component of inflammation
in some patients with interstitial fibrosing lung disorders.

INCREASES IN LAVAGE EOSINOPHILS IN INTERSTITIAL FIBROSING LUNG
DISORDERS

The techniques for light microscopical examination and
differential counting of cell in bronchoalveolar lavage fluids have
been previously described[2]. In brief, the cells are spun onto
microscope slides using low speed cytocentrifugation, followed by
staining with an appropriate differential white cell stain, our

preference being for May Grunwald-Giemsa stain. Macrophages are the predominant cell in control lavages of patients with radiographically normal lungs (the contralateral lungs of patients with confirmed or suspected bronchial carcinoma). Lymphocytes, neutrophils and eosinophils can also occur, but very rarely exceed 11%, 4% or 3% respectively of the total cells.

While in agreement with others that the main inflammatory cell increased in lavage samples in granulomatous lung disorders is the lymphocyte, and the main cell increased in fibrosing lung disorders is the neutrophil, we also find notable and statistically significant increases in lavage eosinophils in many of our patients with cryptogenic fibrosing alveolitis (CFA) and asbestosis. Moreover, eosinophils provide an even clearer demarkation between the fibrotic and granulomatous groups than neutrophils.

The significant increases in lavage eosinophils in CFA and asbestosis are demonstrable by absolute number counts of the cells, as well as by differential counts. Thus, although there are no significant increases in blood eosinophils, nor any other signs of atopy in CFA and asbestosis, it appears that eosinophil accumulation can be a feature of inflammation in the lungs. This conclusion is not without precedent in CFA when we look at the historical literature, where "Peabody and co-workers (1950, 1964) were the first to suggest the possibility of an 'allergic' aetiology for the condition based on the presence of eosinophils in the interstitial tissues"[3]. Indeed, eosinophils are present in small numbers in the interstitium and within alveolar spaces in lung biopsies from many of our patients with CFA.

CLINICAL CORRELATES SUGGESTING ASSOCIATION WITH DISEASE PROGRESSION AND FAILURE OF RESPONSE TO CORTICOSTEROIDS

The importance of the eosinophil as a component of inflammation in interstitial fibrosing lung disorders has been substantiated by certain clinical correlates which have emerged in our patients with CFA[4]. There is a trend indicating that increases in eosinophils in lavage samples are especially associated with progression of disease. Increased numbers of eosinophils before treatment reflect progression as assessed by subjective deterioration in breathlessness over the 12 months preceding lavage. In addition, the increases show a significant association with post-lavage deterioration assessed from physiological follow-up over at least 12 months. Lavage neutrophil counts also tend to be higher in progressors than in those with stable disease but this is not statistically significant.

There is also a significant association of increased lavage eosinophils with failure to respond to corticosteroids in CFA as assessed by changes in lung function tests over at least 12 months[4].   Neutrophils are also slightly higher in the non-responders, but again, as with progression, this is not statistically significant.

The correlations between increased eosinophils and adverse clinical response in CFA indicate that although they are less frequently increased in this disorder than neutrophils, nevertheless they appear to mark an amplification of inflammation which has especially harmful consequences.

POSSIBLE REASONS FOR ACCUMULATION

It now becomes important to consider more fully the nature and role of eosinophils in CFA. A first question to consider is why eosinophils should accumulate in some of these patients? This could be no more than a non-specific response to local inflammation, but another possibility is that it is a selective response to a local chemotactic stimuli. Further studies are needed to answer these questions.

Our current observations show that eosinophil increases invariably occur in association with increases in neutrophils. Neutrophil increases, however, frequently occur alone. This suggests that eosinophils may accumulate as a consequence of a secondary circuit of inflammation superimposed upon the primary inflammatory circuit which effects influx of neutrophils. In asbestosis, lavage eosinophils are also invariably associated with neutrophils suggesting that similar primary and secondary inflammatory circuits may also operate in this disorder.

Neutrophils can produce chemotactic factors for eosinophils when appropriately stimulated[5]. The possibility that neutrophils may directly contribute in the secondary circuits leading to eosinophil influx in these interstitial fibrosing lung disorders should, therefore, now be fully explored.

There are a number of other possible sources of eosinophil chemotactic factors in CFA lungs. One is that complement activation may occur, giving rise to activation fragments such as C5a and C3a which are well known to exert chemotactic attraction for eosinophils as well as neutrophils. Evidence in support of local complement activation has been obtained from the demonstration in our laboratory of C3 conversion products using Laurell crossed immunoelectrophoresis. The presence of soluble immune complexes in the lavage fluids of some of these patients also provides a possible explanation for local complement activation.   In asbestosis, on the other hand, there is experimental evidence suggesting that asbestos fibres may directly activate the alternative pathway of complement under some circumstance[6].

Another possible source of eosinophil chemotactic factors in the lungs of these patients is the mast cell. Although rarely seen in lavage or in alveolar spaces, these cells can be readily demonstrated in lung tissue from our CFA patients using electron microscopy and toluidine blue-stained 1μ sections[7]. They are significantly more common in areas of dense fibrosis than in areas with minimal alveolar wall thickening, and reviewing the literature it appears that they are increased in numbers compared with normal lungs. Moreover, high power detail of the granules in the electron microscope has demonstrated evidence of some granules with a fine granular matrix contrasting with the normal scroll form appearance. This fine granular appearance is a degranulation change which some workers have ascribed to the possible consequence of a process of slow degranulation differing from that of classical anaphylaxis. This morphological evidence of degranulation thus suggests the possibility that mast cells may provide either a direct source of, or perhaps an indirect stimulus for, eosinophil chemotactic factor release in CFA lungs (as well as, of course, other mediators of inflammation).

Further evidence suggesting a role for mast cells in CFA has come from our assays of histamine in lavage fluids which have demonstrated significant increases in CFA compared with controls or patients with sarcoidosis[7]. The lavage fluid histamine increases show a significant correlation not only with the numbers of eosinophils in the lavage fluids of patients with CFA, but also with the numbers of neutrophils. Since mast cells are the likely source of histamine, this provides further evidence that mast cells and neutrophils may in some way be linked in the inflammatory circuit leading to eosinophil influx in CFA.

The reasons for mast cell degranulation in CFA lungs are unknown but a number of possibilities arise from other observations in this disorder. If, as discussed, complement activation is occuring, then complement-derived anaphylatoxins may provide one possible explanation. The evidence of local production of IgG in CFA lungs[1] suggests that cytophilic IgG may also be worthy of consideration. There is preferential accumulation of lymphocytes as well as mast cells in interstitial sites in CFA lungs. little is known of the nature of these interstitial lymphocytes, but circulating T-lymphocytes sensitised to type I collagen and single-stranded DNA have been demonstrated in some patients[1]. Recent experimental data indicates a possible interaction of sensitised T-lypmphocytes and/or their products with mast cells[8]. If confirmed in humans, this may provide yet another possible explanation for mast cell degranulation in CFA lungs.

In summary, eosinophils may accumulate in the lungs of patients with CFA as a consequence of a secondary circuit of inflammation superimposed upon the primary inflammatory circuit leading to neutrophil influx. Possible components of this secondary circuit may include complement activation fragments, and products derived from neutrophils and mast cell.

MORPHOLOGICAL FEATURES SUGGESTING LONGEVITY AND DEGRANULATION

The lavage eosinophils in CFA and asbestosis resemble tissue rather than blood eosinophils in showing a number of changes suggestive of maturation. These include nuclear hypersegmentation suggesting longevity, and cytoplasmic vacuolation and granule alterations, suggesting loss of release of granule contents.

Nuclear hypersegmentation can be seen clearly in cryocentrifuge preparations of the cells, and three, four or even five-lobed nuclei are often present. This has long been recognised as a feature of granulocyte maturation. Cytoplasmic vacuolation is also shown clearly in cytocentrifuge preparations, occurring in up to 50% of lavage eosinophils from CFA, and in up to 40% from asbestosis. Similar changes have been reported in blood eosinophils in various conditions associated with hypereosinophilia[9].

Cytocentrifugation gives a distorted view of cells and may induce artefactual vacuolation. Electron microscopic studies have, however, confirmed that lavage eosinophils from CFA and asbestosis show features consistent with complete or partial loss of granule contents. While there are usually a few normal granules, the majority show alterations. These include clearing of the granule matrix fragmentation of the central "crystalloid" core, membranous material in the matrix or complete clearing and vacuolation. These changes are recognised to occur when eosinophils interact with membrane-stimulating agents[10], and in the blood in situations of prolonged eosinophilia, for example in Loffler's cardiomyopathy syndrome[11], and also in eosinophils infiltrating tissues as, for example, in lymph nodes from patients with Hodgkin's lymphoma[12].

Electron microscopic studies of eosinophils in lung biopsies from patients with CFA show that eosinophils in blood capillaries are practically normal, with little or no evidence of granule alteration. Eosinophils which have infiltrated to the interstitial connective tissues, on the other hand, still have many normal granules but granule alteration has quite clearly commenced. However, by the time eosinophils have reached the alveolar spaces, granule alteration is a very striking feature comparable with the findings in lavage.

Lung biopsy observations thus confirm the lavage observations indicating that eosinophils which have infiltrated the lung tissues of patients with CFA show striking changes suggesting release or loss of granule contents.

THE POSSIBILITY OF ACTIVATION

The reasons for loss or release of eosinophil granule contents in the lungs of patients with interstitial fibrosing lung disorders are unknown. One possibility is that this may be a process of normal degeneration within the tissues. Little is known about normal eosinophil clearance in man, but self-autolysis in tissue sites has been one proposal.

There are, however, in the lungs of patients with CFA a number of agents which might enhance the potential of eosinophils to participate in inflammatory reactions. These include immune complexes, lysosomal enzymes, neutral proteases and mast cell products which have been demonstrated in studies of cell-free lavage fluids[1]. The evidence to support the potential role of such agents in enhancing eosinophil binding capacity is substantial[13,14]. Immunofluorescence studies of lavage eosinophils from our patients with CFA have provided evidence of immune complexes interacting with some of these cells. Brightly stained, homogeneous, IgG-containing inclusions within lavage eosinophils have been observed in eight of our patients with CFA and in one of our patients with asbestosis. The staining is specific and quite distinct from non-specific appearances due to charge interaction of fluorescent dye with eosinophil granules. Inclusions of similar appearance can be induced in vitro by incubation of eosinophils with immune complexes or immunoglobulin aggregates but not native immunoglobulins[15]. It is thus probable that the IgG inclusions within lavage eosinophils represent the uptake of IgG immune complexes or aggregates by these cells.

Electron microscopy has confirmed the presence of inclusions in cases with positive immunofluorescence. Inclusions of finely electron dense material have been observed in some of the cells. These again resemble appearances which can be obtained by in vitro incubation of eosinophils with immune complexes or immunoglobulin aggregates[16].

Thus, evidence has been obtained of lavage eosinophils interacting with at least one agent capable of stimulating degranulation. Confirmation of eosinophil activation in the lungs of these patients must, however, now be obtained by, for example, studying Fc and C3b surface receptor enhancement, peroxidase release, oxygen consumption and by examing the capacity of lavage fluid components to activate normal eosinophils.

SPECULATION ON ROLE IN TISSUE DAMAGE

Increases in lavage eosinophils in patients with CFA have invariably been reconfirmed where follow-up lavage information is available. This suggests that once initiated, the eosinophilic

component of inflammation is maintained. It thus appears that we
are dealing with a disorder where prolonged release and
extracellular accumulation of eosinophilgranule contents can occur.
This conclusion may be relevant to the apparently harmful clinical
effects of eosinophils in the lungs of these patients. It is now
apparent that a number of eosinophil granule components are
potentially harmful. Two forms of cationic proteins from the

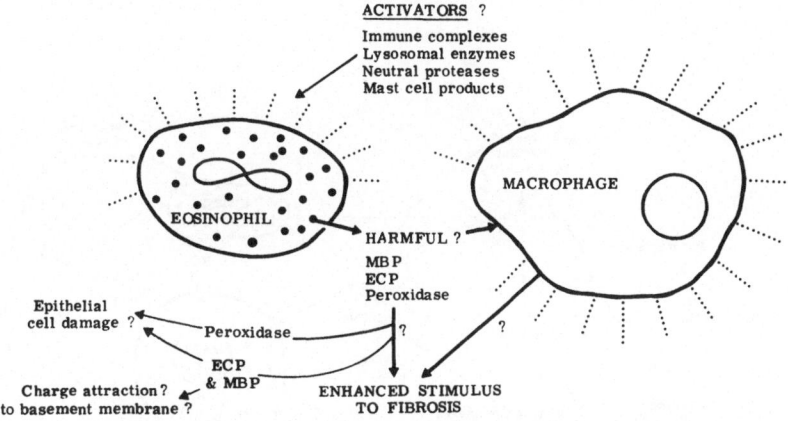

Figure 1.   Theory on harmful effects of eosinophils in
             interstitial fibrosing lung disorders.

granule "crystalloid" core have been isolated, the Major Basic
Protein[17] and the Eosinophil Cationic Protein[18]. Both have been
shown to have toxic effects on cells under experimental conditions.
The Major Basic Protein has been demonstrated to exert cytotoxic
effects on guinea pig intestinal and splenic cells, porcine

endothelial cells, human blood mononuclear cells ane epidermal cells in vitro, and to impede tracheal ciliary function[19].

The Eosinophil Cationic Protein has been demonstrated to have a toxic effect on neuronal cells in vivo[20]. In addition, peroxidase released from eosinophil granules might cause damage to cells.

Thus it is reasonable to spectulate as has been proposed in Loffler's cardiomyopathy syndrome[11] that the harmful effects of eosinophils in CFA may be due to maintained release and accumulation of harmful granule products. the biopsy observations in our patients with CFA suggest that the main sites of degranulation are the air spaces. Toxic products might, therefore, exert their main effects by acting to maintain damage to the alveolar epithelium. Epithelial damage is possibly initiated by neutral proteases secreted from macrophages and neutrophils (Figure 2).

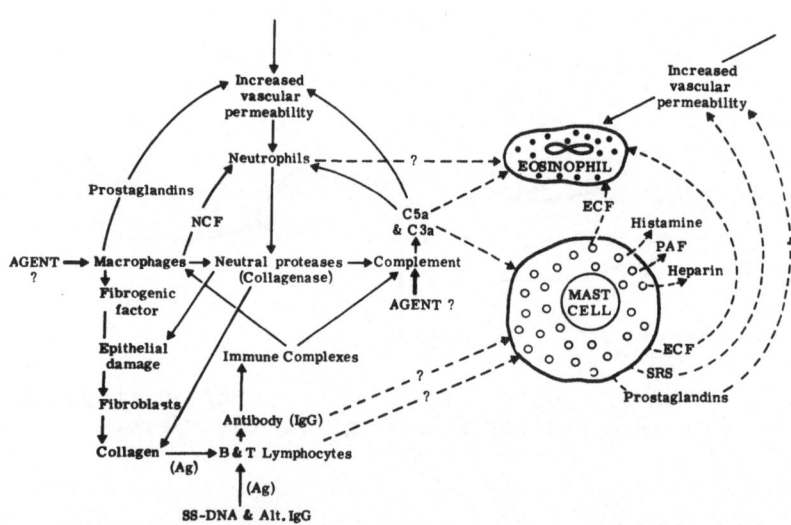

Figure 2.   Theory on inflammatory circuits leading to granulocyte influx in interstitial fibrosing lung disorders.

Toxic eosinophil products may exert a direct toxic effect on alveolar epithelial lining cells. It is also theoretically possible that charge attraction binding of the highly cationic "crystalloid" proteins to negatively charged exposed epithelial basement membranes could maintain epithelial damage by preventing attachment of regenerating epithelial cells.

Another possible way in which eosinophils might act to enhance inflammation in CFA and asbestosis may be through an ability to enhance alveolar macrophage activation. This has been suggested from studies using ability to spread on glass in vitro as a marker of alveolar macrophage activation, where we have found a significant association of spreading with the presence of increased lavage eosinophils. Since activated alveolar macrophages probably play an important primary role in inflammation both in CFA and asbestosis, enhancement of their activation by eosinophils or their products might prove especially harmful. Electron microscopy has demonstrated phagocytosed eosinophil granules still containing "crystalloid" material within occasional alveolar macrophages, and suggests a way in which a stimulus to enhanced activation might occur. This is obviously an important area for further study.

The speculations on harmful effects of eosinophils in interstitial fibrosing lung disorders are summarised in Figure 1. To provide a balanced perspective, a theory summarising our current views on the multiple facets of primary and secondary inflammatory circuits leading to granulocyte influx in these disorders is also given in Figure 2.

In conclusion, studies in patients with CFA and asbestosis emphasise the importance of eosinophils as well as neutrophils in inflammation in interstitial fibrosing lung disorders. They also emphasise the importance of taking eosinophils into account when devising new approaches to the therapy of patients who fail to respont to corticosteroids.

ACKNOWLEDGEMENTS

We are most grateful to the Medical Research Council, The Tobacco Research Council and The Chest, Heart and Stroke Association for their support. We would also like to thank our many colleagues who have helped in collection of lavage samples.

REFERENCES

1.   Turner-Warwick M., Haslam P.L., Lujoszek A.m Townsend P., Allan F., du Bois R.M., Turton C.W.G. and Collins J.V. (1981). Cells, enzymes and interstitial lung disease. Journal of the Royal College of Physicians of London; 15, 5-16.
2.   Haslam P.L., Turton C.W.G., Lukoszek A., Salsbury A.J., Dewar A., Collins J.V. and Turner-Warwick M. (1980). Bronchoalveolar lavage fluid cell counts in cryptogenic fibrosing alveolitis and their relation to therapy. Thorax; 35, 328-339.
3.   Read J. (1958). The pathogenesis of the Hamman-Rich syndrome:

a review from the standpoint of possible allergic aetiology. American Reviews of Tuberculosis; 78, 353-367.

4.   Rudd R.M., Haslam P.L. and Turner-Warwick M. (1981). Cryptogenic fibrosing alveolitis: relationships of pulmonary physiology and bronchoalveolar lavage to respond to treatment and prognosis. American Review of Respiratory Disease (in press).

5.   Koing W., Czarnetzki B.M. and Lichtenstein L.M. (1976). Eosinophil chemotactic factor (ECF) II. Release from human polymorphonuclear leucocytes during phagocytosis. Journal of Immunology; 117, 235-241.

6.   Sain-Remy J-M.R. and Cole P.J. (1980). Interactions of crysotile asbestos fibres with the complement system. Immunology; 41, 431-437.

7.   Haslam P.L., Cromwell O., Dewar A. and Turner-Warwick M. (1981). Evidence of increased histamine levels in lung lavage fluids from patients with cryptogenic fibrosing alveolitis. Clinical and Experimental Immunology; 44, (in press).

8.   Askenase P.W. (1980). Effector cells in late and delayed hypersensitivity reactions that are dependent of antibodies or T cells. In "Fourth International Congress of Immunology, Progress in Immunology IV". Editied by M. Fougereau and J. Dausset, p. 829-845.

9.   Connell J.T. (1968). Morphological changes in eosinophils in allergic disease. The Journal of Allergy; 41, 1-9.

10. Zucker-Franklin D. (1968). Electron microscopic studies of human granulocytes: structural variations related to function. Seminars in Haematology; 5, 109-133.

11. Spry C.K.F and Tai P.C. (1976). Studies on blood eosinophils II. Patients with Loffler's cardiomyopathy. Clinical and Experimental Immunology; 24, 423-434.

12. Parmley R.T. and Spicer S.S. (1975). Altered tissue eosinophils in Hodgkin's disease. Experimental and Molecular Pathology; 23, 70-82.

13. Tai P.C. and Spry C.J.F. (1980). Enzymes altering the binding capacity of human blood eosinophils for IgG antibody coated erythrocytes (EA). Clinical and Experimental Immunology; 40, 207-219.

14. Anwar A.R.E. and Kay A.B. (1978). Enhancement of human eosinophil complement receptors by pharmacologic mediators. Journal of Immunology; 121, 1245-1250.

15. Litt M. (1964). Studies in experimental eosiniphilia VI. Uptake of immune complexes by eosinophils. Journal of Cell Biology; 23, 355-361.

16. Takenaka T., Okuda M., Dawabori S. and Kubo K. (1977). Extracellular release of peroxidase from eosinophils be interaction with immune complexes. Clinical and Experimental Immunology; 28, 56-60.

17. Gleich G.J., Loegering D.A., Mann K.G. and Maldonado J.E. (1976). Comparative properties of the Charcot-Leyden crystal

protein and the major basic protein from human eosinophils. Journal of Clinical Investigation; 57, 633–640.

18. Olsson I. and Venge P. (1974). Cationic proteins of human granulocytes II. Separation of the cationic proteins of the granules of leukemic myeloid cells. Blood; 44, 235–246.

19. Gleich G.J., Frigas E., Loegering D.A., Wassom D.L. and Steinmuller D. (1979). Cytotoxic properties of the eosinophil major basic protein. Journal of Immunology; 123, 2925–2927.

20. Olsson I., Olofsson T., Venge P. and Winquist I. (1980). The eosinophil cationic protein and the eosinophil in inflammator reactions. Transactions of the Royal Society of Tropical Medicine and Hygiene; 74 (suppl.), 7–10.

DISCUSSION

LECTURER: Haslam   CHAIRMAN: Cumming

CUMMING·   Dr. Rossi will open the discussion

ROSSI: Do you find any correlation between the level of immune complex and the number of eosinophils in the lavage fluid of your patient?

HASLAM: No, we don't. We have only used the Cl_q binding test to detect immune complexes. What we find, stangely enough, is that the complex we can detect with the Cl_q binding test shows a significant association with the lymphocytes in the fluid of our cryptogenic fibrosing alveolitis patients and a significant lack of association with neutrophils and eosinophils. What I think I can conclude from this is that the method that we were using to detect those kinds of immune complexes is not appropriate for the detection of immune complexes which may be particularly important to the neutrophil-eosinophil inflammatory circuit, since I think that our observations on the fluorescence of eosinophils suggest that the immune complexes are there.

ROSSI: We have the same finding that immune complexes are there, but we have found a positive correlation, as we will say later on, between the number of immune complexes and the percentage of neutrophils.

HASLAM: Did you use a different test for the immune complex estimation?

ROSSI: No, the Cl_q binding and then we use also haemolytic blocking agent to test the actual number of immunoglobulin secreting cells. The other thing is, looking at your slide on the fluorescence one wonders whether you washed the cells before making the slide or not? Because it is very usual for us to find a specific immune fluorescence on macrophages also.

HASLAM: That's interesting because the reason for undertaking the lavage observation was to look for immune complex association with the surface of macrophages. I wonder whether your series is quite comparable with ours because there are a number of technical differences between your series and ours.

ROSSI: For us is quite common to find immune complex on the surface of the macrophages also if we incubate the cells after washing for two hours just to be sure that we don't get any non-specific immune complex binding.

HASLAM: Why I think the observations in our preparations may be valid is because we have found, in the same study, a significant increase in the amount of IgG associated with the surface of macrophages from patients with sarcoidosis. The point is that we are able to demonstrate something when it is there.

ROSSI: The last question and then I stop. The presence of eosinophils in lavage fluid is constant in the same patient when you follow it up or is it changing?

HASLAM: We haven't analysed our follow-up lavage information in detail yet, but from what I have seen at a glance, when we have done repeated lavages on patients who have not been treated, the eosinophil counts are very comparable at the second lavage. When we do repeated lavages on people who have received corticosteroids, their eosinophils levels still remain constant, but occasionally if one uses cyclophosphamide instead of prednisone, that the eosinophil counts can fall; we have had successful response in patients with eosinophils using cyclophosphamide when we have not had a good response with prednisone.

ROSSI: I ask another question. We have a similar problem, when you say that the presence of eosinophils is bad for the prognosis of the patient, what do you mean exactly? Because in our series at least 40 to 50% of patients with a high number of immune factor cells in the alveoli do not respond to the therapy. They all respond by evaluation of broncho-alveoli lavage and deterioration of lung function in a short time like one year or six months, and they respond to cyclophosphomide.

HASLAM: Of course this presentation was based on results obtained to the time before cyclophosphamide was really thought of as useful in this disease. I must say the reason why we became interested in cyclophosphamide at the Brompton is interesting; when we were first analysing our initial lavage results we were looking at the relationship between lavage count and response to treatment. We noticed that the few patients that responded to prednisone happened to have been tried on cyclophosphamide as a new sort of measure and that these were showing response, suggesting that we should try a controlled trial of cyclophosphamide but I can't speak about the results of that yet.

BIGNON: I confirm that in bronchial lavage in asbestosis patients we have found some cases with eosinophils in the lavage fluid; but only very few cases, in one series of cases we found a high level of lymphocytes, about 20% in the broncho-alveoli lavage in asbestosis. My question is: how can we explain the increase of eosinophils in lavage fluid with asbestos fibres?

HASLAM: I have given a lot of thought to this. One must think of the reason why immune complexes might be present in cryptogenic fibrosing alveolitis and whether immune complexes can possibly be present in patients with asbestosis; I am very interested in the immune overtones in both these disorders because I think it's possible that there is a secondary consequence of the inflammatory process, if collagenases produced by activation of macrophages or by neutrophils enter the lungs in both these situations. The collagenase then influences the collagen so that it becomes antigenic and anticollagen antibodies would be found in asbestosis as in cryptogenic fibrosing alveolitis. I am wondering whether immune complexes involving anticollagen antibodies may be particularly important in this disorder and immune complexes activate complements which would set off the secondary circuit.

BIGNON: There is some strain of guinea pig with normal lung containing a lot of eosinophils, about 15%. How can we explain these findings in normal guinea pigs?

HASLAM: I am not very sure, I don't know very much about inflammatory cells in experimental animals but I do know that when you lavage normal dog lung you get a tremendous number of neutrophils and I wonder whether this might apply in experimental animals.

WILLIAMS: Thank you for your interesting talk. What I would like to ask is how reproduceable is this eosinophilic increase, because in my experience of doing some of this in sarcoidosis and some on CFA, if you do it at a different time, you get a completely different answer and aso if you do it in a different part of the lung you get a different answer. So could you please tell us how reproduceable it is?

HASLAM: I think it is very important that one stays with the selected site because we ourselves haven't done very much work on different areas of the lung; but we know that Dr. Crystal's group has; I don't know whether Dr. Rossi would like to comment. The question of how reproduceable is the lavage sampling in different sites in the lung, we haven't done very much work on that, but I know you have.

ROSSI: In the last two years since we needed a lot of cells to make all studies we have performed on patients with interstitial lung disease and normal controls, we have been using a three-site lavage and we actually lavaged the lingula, the left lower lobe and the right middle lobe. We compared the results in terms of cell recovery, fluid recovery, and percentage cells, immunoglobulin production and so on. The only difference you can find between different lobes in interstitial diffuse lung disease is the amount of fluid recovered because for anatomic reasons of one site being representative of what's going on in the ret of the lung at least for interstitial lung disease as IPF and sarcoidosis there is no doubt.

CUMMING: The other question is: what is the reproducibility in temporal terms, if you did it three weeks later.

ROSSI: There is some work done on dogs showing that if you lavage one site of the lung and then you relavage it after one day, you get an increased number of neutrophils in that site. That, I think, is related to the trauma, there was a paper published, I think six or seven months ago, on the American Review of Respiratory Diseases on reproducibility, since in the normal there is no question that you get the same amount. In diseases like IPF and sarcoidosis, in which you expect a peak of the disease, the disease spontaneously regresses without treatment. That's what you see normally in sarcoid if you evaluate for example the enlargement of lymph nodes and things like that. It's not unusual to see a difference in the activity of the prcess; anyhow, I don't have so much experience in that because if we see patients with active interstitial lung disease we treat them, we don't expect to follow them to know what is the natural history of the disease by bronchoalveoli lavage. We know they'll deteriorate and we try to stop it.

CUMMING: Dr. Roussouw, may I ask you to speak rather slowly? We are having a little difficulty with the translation.

ROUSSOUW: I just want to make one very short pregmatic clinical remark on Dr. Haslem's paper. In our series of 400 consecutive broncho-alveoli lavage procedures in patients with sarcoidosis and CFA and a number of fibrosing alveolitis patients, we found only one patient with an eosinophil count of more than 4% and this was 29%. That was done a long time ago, in September, 1977; and in January 1978, July 1978 and January 1980 the oesinophil dropped from 29% to 11% to 1% to 0%, and that was 100% correlated with the radiological and clinical improvement of this patient.

CUMMING: May I make an intervention there, Dr. Roussouw? Are you from Tygerberg?

ROUSSOUW: Yes, sir.

CUMMING: Were the patients black, white or coloured.

ROUSSOUW: This was a coloured patient with hardly any sign of parasite infestation.

CUMMING: That's for one coloured patient, the remaining 400 or so lavages in which you found very few eosinophils, were they in general a black population, a coloured population or a white population?

ROUSSOUW: I think our patients are more or less evenly distributed between the races: Europeans and coloured patients.

CUMMING: Would you like to comment on that picture?

HASLAM: I have been thinking about our particularly striking increases in eosinophils, because in our series these increases are more striking than have been published by other groups. They are more

striking than yours in South Africa; yours reached
10% but ours reached 20%, so they are more striking
and although high levels were seen in the Paris
series, (they were higher than Dr. Crystal's
group), but not reaching quite the levels we see. I
was wondering whether this may be a reflection of
the stage of the disorder at the time of the
lavage, because I don't know whether, when you did
your lavages, you would think that the patients
were at an early first referral point, because at
the Brompton Hospital its a secondary referral
centre. I think it is possible what we have seen
in our patients slightly later than you and I
wonder whether this is why we have seen more
eosinophils.

CUMMING: Thank you Pat. May I have your name, please?

CAPELLI: Although our experience with lavages dates back to
                a very recent time, we are able, however, to lend
                support to your view concerning an eosinophilic
                increase at least in allergic alveolitis cases.
                Similarly, we have found no increased eosinophil
                levels in sarcoidosis and in controls.   Now,
                however, you have claimed that using histamine as
                an index it clearly emerged that both in
                sarcoidosis and primitive fibrosing alveolitis,
                histamine levels increase, which might count as a
                sign of mast cells degranulation;   on the contrary
                there is no eosinophil increase in sarcoidosis.
                How could you explain this?

HASLAM: The question was how can we explain why in
                sarcoidosis there is an increase in histamine but
                no increase in eosinophils.   Well, in fact,
                although occasionally some of our patients in our
                sarcoidosis series have perhaps slightly higher
                histamine level than in the normals, that was not a
                very strong trend in the series and it was not a
                significant trend.   In fact there was a significant
                increase in histamine in cryptogenic fibrosing
                alveolitis compared to either sarcoidosis or
                control.   So I think that's really why we do not
                see too many eosinophils.   Interestingly enough,
                the histamine levels in our sarcoid group showed an
                association with fibrosis, there can be a slight
                increase of neutrophils in that group in our
                experience and it is interesting that the
                neutrophils showed a correlation with the histamine
                and moreover all showed a correlation with evidence

of contraction on the radiographs of those patients. So it seems to be the ones who got the fibrosis that were getting the neutrophils in their histamines, which I think is very interesting. There were no eosinophils at that stage.

ROSSI: You mentioned that a patient with IPF with eosinophils has an increased number of lymphocytes. Did you find any sub-group difference in the lymphocytes, is it an increase in the T cells, in the B cells or in the newer cells?

HASLAM: I am afraid we were not very clever at this, and we had trouble. The lymphocytes increases in our fibrosing alveolites patients are very moderate. This means that there are not really enough cells for us to be able to obtain a sufficiently good separation of the lymphocytes from the total lavage population to do what I would consider satisfactory studies. We can certainly demonstrate T cells and what seem to be B cells, but as to the relative proportion of one to the other I wouldn't like to comment on that.

CUMMING: Thank you very much for your presentation. It's very clear that you have identified sufficient differences of view that will keep you occupied happily for several years to come. We will pursue the problem of eosinophils and neutrophils in Kay's presentation. I wonder if I may clear up a word before we proceed. You have used several times in your talk the word crystalloid; that to me has a specific meaning, but I think our meanings must differ, because on one slide you have shown that the crystalloid contains a high degree of protein and these two things seem to be mutually exclusive. Can you clarify it?

KAY: It refers to the large granules of the eosinophils because the central core looks crystalline under the electron microscope and has a beautiful lattice structure.

NEUTROPHILS AND EOSINOPHILS IN IgE-DEPENDENT

ALLERGIC TISSUE REACTIONS

A.B. Kay

Department of Clinical Immunology
Cardiothoracic Institute
Fulham Road, London SW3 6HP

INTRODUCTION

It is well recognized that allergic tissue reactions involving mast cell degranulation are accompanied by the infiltration of inflammatory cells, including neutrophils and eosinophils. Experimentally, it can be shown that following the interaction of specific antigen with mast cell-bound IgE (or equivalent homocytotropic antibody such as guinea pig IgG_1 or rat IgG_{2a}) a mechanism is inititated which results in the local accumulation of various cell types. For instance, when IgG_1 was introduced into the skin of a normal guinea pig and, after a latent period, blue dye and specific antigen was injected intravenously (as in the usual passive cutaneous anaphylatic reaction) cell accumulation could be observed at the site of the blueing reaction when the tissue was excised at time intervals after antigen challenge[1]. The first cells observed were neutrophils and these were present in large numbers at 4 hr. Eosinophils appeared in appreciable numbers at 8 hr after challenge and were present in largest numbers at 12 hr. After this time macrophages were the prominant cells and they continued to accumulate for up to 72 hr (at which time the experiment was terminated). Similar histological findings have been observed in the skin of non-human primates (marmosets) using human IgE and pollen allergen[2]. Thus homocytotropic antibody (i.e. antibody which has the capacity to passively sensitize tissue for the subsequent antigen-induced release of pharmacological mediators of anaphylaxis) together with specific allergen has the capacity to initiate, in normal recipients, the events leading to infiltration of neutrophils and eosinophils, and later the macrophage. In atopic humans using the "skin window" technique a similar series of events have been described following the application of the appropriate allergen to abraded skin sites[3].

95

These experiments on the kinetics of cell infiltration into the site of certain allergic tissue reactions raise a number of closely related questions. Confining attention to the granulocytes such questions might be as follows. What is the mechanism by which neutrophils and eosinophils are attracted to the reaction site? Are chemotactic factors released from mast cells, and/or other cells, in vitro and do these preferentially attract certain cell types? Can chemotactice factors be identified in vivo following antigen challenge? What is the role of eosinophils and neutrophils in allergic tissue reactions? Do they, for instance, have a "dampening" or homeostatic effect on mast cells by inactivating mast cell mediators or do they contribute to the inflammatory reaction as a result of release of their lysosomal enzyme? Finally, what is the fundamental biological role of mast cell/IgE interactions and their ensuing granulocyte infiltration? Are they redundant mechanisms which cause "allergies" or are IgE-mediated reactions of biological advantage in certain types of adaptive immunity? These questions will be discussed below.

MAST CELL-DERIVED CHEMOTACTIC FACTORS

(a) Studies in vitro

Products of the anaphylactic reaction can be assessed for their capacity to attract neutrophils and eosinophils in vitro using techniques such as the Boyden micropore assay. Using this system it was shown that "anaphylactic" diffusates prepared from human lung fragments preferentially attracted eosinophils from a mixed leucocyte population[4,5]. The fragments were passively sensitized with serum from atopic individuals and challenged with specific antigen. The eosinophil chemotactic activity eluted from Sephadex G-25 with molecules having a molecular weight between approximately 400 and 1000. This low molecular weight activity was referred to as the "eosinophil chemotactic factor of anaphylaxis (ECF-A)"[4,5]. It is now appreciated that ECF-A consists of many substances with diverse chemical structures, a few of which have been chemically characterized. They include (a) amines such as histamine and a major catabolite, immidazole acetic acid[6]; (b) mast cell-derived peptides - those recognized at present are (i) the tetrapeptides Val/Ala-Gly-Ser-Glu[7], (ii) a heterogenous group of 600-1000 molecular weight peptides[8] and (iii) peptides of higher molecular weight (1500-2500) released by immunological stimulation of rat mast cells[9]. More recently, lipid chemotactic factors, made available by the release of free arachidonic acid have been described. Both cyclo-oxygenase and lipoxygenase products of arachidonic acid have been shown to posses neutrophil and eosinpophil chemotactic activity. The active agents include 5-HETE (and other HETE's), 5,12-di-HETE (leukotriene B_4), 5-HPETE and

HHT[9,10]. At low concentrations PGD_2 and PGE_2 are preferentially chemokinetic for eosinophils. In general, it appears that the lower molecular weight mast cell-derived peptides preferentially attract eosinophils from a mixed leucocyte population whereas the complex fatty acids derived on both neutrophils and eosinophils. It should be emphasized that histamine and the lower molecular weight eosinophilotactic peptides are preformed in the mast cell granule from which they can be readily extracted by non-specific manoeuvres such as multiple freezing and thawing. On the other hand, arachidonic acid metabolites are released, probably from secondary cells such as neutrophils and mononuclear phagocytes, which in turn are triggered as a result of primary mast cell activation.

Using single mast cell preparations from the rat it has been shown that the majority of labelled arachidonic acid offered is metabolized to PGD_2 (and to a lesser extent to PGI_2)[11]. Therefore, mast cells cannot account for the release of all the chemotactic factors mentioned above and are presumably not the only sources of the lipidopeptides, i.e. LTC and LTD which comprise the slow reacting substance of anaphylaxis (SRS-A)[12]. Thus it seems likely that SRS-A, and the chemotactic derivatives of eicosatetraenoic acid, are produced from neutrophils or monocytes by an as yet unidentified signal from mast cells containing bound IgE. This has important implications since it can no longer be assumed that mast cells are the sole source of eosinophilotactic agents. The initial neutrophil infiltration observed histologically following IgE/mast cell interaction (described above) could account for the subsequent eosinophil infiltration. That is to say that neutrophils might be a potent source of eosinophil chemotactic factors and in this respect the neutrophil-derived ECF, shown to be a small molecular weight fatty acid which is probably derived from arachidonic acid, may be relevant to eosinophil accumulation in vivo.

So far, attempts to reproduce the potency and the selectrive eosinophil-attracting activity of ECF-A using synthetic, or highly purified, mediators, either alone or combination, have been unsuccessful.

Activities with predominant neutrophil-attracting activity have been recognized both from whole lung fragments and isolated basophil leukaemia cells following IgE stimulation. This "neutrophil chemotactic factor of anaphylaxis (NCF-A)" has a molecular size of approximately 750,000. This material is predominantly chemotactic for neutrophils and is probably performed in the mast cell granule although this fact is yet to be ascertained with certainty.

(b)   Studies in vivo

In general there is fairly good correlation between the chemotactic factors recognizd from the in vitro techniques described above and those which can be detected in the circulation or extracted from the tissue sites following an IgE (or equivalent)/antigen interaction in vivo.

For instance, the site of guinea pig active cutaneous anaphylactic reactions was examined and the skin area excised at intervals and material extracted for subsequent assay in a Boyden chamber technique. Two factors were recognized, one had a high molecular weight and was heat liable whereas the other was heat stable and had many of the characteristics of ECF-A[13]. As mentioned above, histamine has chemotactic properties for eosinophils in vitro and is also released in vivo following anaphylaxis in experimental animals as well as being demonstrable in the circulation following bronchial challenge by specific antigen[14], or following exercise-induced challenge[15].

In man, IgE-dependent release of chemotactic factors into the circulation in vivo following the appropriate manipulations has been observed in two general situations. The first is cold urticaria where thermal shock is thought to reorientate the IgE molecule (in the absence of allergen) leading to histamine release from mast cells. When a limb was immersed in ice-cold water, NCF-A and ECF-A appeared in the venous blood within minutes whereas these mediators were not detectable in the contralateral limb kept at ambient temperature. A macromolecule with neutrophil chemoattractant properties (and, therefore, thought to be analogous to NCF-A release from single cells), can also be detected in the circulation in asthmatics following bronchial challenge. Exercise[15], but not isocapnic hyperventilation, also leads to the elaboration of a high molecular weight NCF. NCF released by antigen or exercise was inhibited by prior administration of disodium cromoglycate. These experiments indicate that NCF may be mast cell-derived and has the potential for being a useful "marker" in bronchial asthma. It has yet to be ascertained whether NCF is a single substance and whether it is released into the circulation in sufficient amounts to enable its full chemical characterization.

ROLE OF NEUTROPHILS AND EOSINOPHILS IN ALLERGIC TISSUE REACTIONS

(a)   Eosinophils and neutrophils as regulators of mast cell
      function

It has been proposed that eosinophils serve as modulators of the anaphylactic response. For instance, the high content of eosinophil histaminase, arylsylphatase and phospholipase D are thought to equip this cell for a special role in "dampening" the

allergic response by inactivating histamine, SRS-A and platelet
activating factor (PAF) respectively[16]. However, it is recognized
that there is also a high content of diamine oxidase ("true
histaminase") in neutrophils and that histamine N-methyl
transferase which converts histamine to 1-methyl-4-immidazole
acetic acid (via 1, 4-methyl histamine), is present predominantly
in mononuclear phagocytes. For these reasons it is difficult to
attribute a special role to eosinophils in histamine inactivation
since there are many other potential sources of histamine
inactivating enzymes. Similarly, the chemical characterization and
total synthesis of SRS-A as leukotrienes C and D (LTC and LTD) also
raises doubt as to the role of arylsulphatase in inactivating these
agents[12]. LTC and LTD differ only in the terminal amino acid of
the glutathione moiety of their lipidopeptide structure; the
cysteine at C-6 having a thiol ester linkage. Thus these
leukotrienes do not contain a free sulphate ester which are normal
substrates for arylsulphatases. Similarly, the presence of
phospholipase D is surprising as this enzyme was previously
localized only within plant tissues. The recent demonstration that
1-0-alkyl-2-acetyl-sn-glycero-3-phosphorylcholine possesses
platelet activating activity holds promise for a true evaluation of
this mediator, and the enzyme(s) which inhibit it, in acute
allergic reactions[17].

For all these reasons it would now seem reasonable to re-
evaluate the role of granulocytes, especially the eosinophil, in
terms of its role in modulating anaphylactic reactions.

(b) Neutrophil and eosinophil derived enzymes and tissue damage

It is well recognized that neutrophil-derived lysosomal
enzymes, and other granule constituents such as cationic proteins,
probably contribute to tissue damage when there are large
infiltrations of neutrophil leucocytes, i.e. in Arthus (type III)
tissue reactions. Similarly, eosinophil-derived granular
constituents, especially the major basic protein (or eosinophil
cationic protein) have potential for tissue destruction. There is
much evidence accumulating at the present time to suggest that
these highly basic proteins have a deleterious action on various
organs. For instance, eosinophils and their products are known to
be toxic for central and peripheral nerves, to lyse endothelial and
tumour cell lines, to damage cardiac muscle, as in the
cardiomyopathy associated with hypereosinophilic syndrome and to
destroy tracheal cilia and denude tracheal epithelium. These
latter findings are reminiscent of the changes observed
histologically in severe bronchial asthma. Evidence that
eosinophil-derived granular material may contribute to tissue
damage has also been shown in Crohn's disease and ulcerative
colitis. For these reasons it seems reasonable to suggest that
granulocytes, particularly eosinophils, may actually contribute to

tissue damage in IgE-mediated reactions in which they are known to accumulate in large numbers. It is well recognized that infiltration by eosinophils is a prominent feature in the lungs of patients who have died from acute severe asthma. If eosinophils are being inappropriately recruited in certain IgE/mast cell reactions, why have mast cells been retained in phylogeny and what is the biological significance of the IgE response? The notion that these mechanisms may have been retained in adaptive immunity against helminthic larvae is discussed below.

GRANULOCYTES, ALLERGY AND METAZOAN PARASITES

The reader is referred to previous articles in which the association between eosinophils, neutrophils, IgG, IgE, mast cells, complement, allergy and immunity to helminths has been discussed in detail[18,19]. The hypothesis, which is based largely on in vitro experiments using an assay system based on the killing of schistosomula of Schistosoma mansoni, is summarized as follows. Depending on the experimental conditions, eosinophils and neutrophils both kill appropriately opsonized schistosomula. Although there is some disagreement as to whether complement or antibody-coated schistosomula are more susceptible to granulocyte damage most investigators agree that the combination of antibody and complement, i.e. fresh serum from an individual with schistosomiasis, renders schistosomula particularly susceptible to damage by eosinophils, and to a lesser extent, by neutrophils. The helminthocidal property of eosinophils in these systems can be markedly increased by prior incubation with mast cell-derived meditors such as the ECF-A tetrapeptides or histamine[18]. Since these are known to be mast cell-derived it is suggested that the IgE-mediated release of pharmacological mediators act both to locally recruit eosinophils and to arm them for subsequent damage or death of helminthic larvae. Allergens, possibly because of their similar chemical configuration to certain larvae, may inappropriately trigger the same series of events thereby leading to the release of pharmacological mediators with the ensuing clinical syndromes, i.e. hay fever and allergic asthma. Thus "allergies" may be a relatively trivial accompaniment of an otherwise useful biological mechanism.

SUMMARY

The role of eosinophils and neutrophils in IgE-dependent allergic tissue reactions has been discussed. The kinetics of neutrophil and eosinophil infiltration in vivo have been described as have numerous chemotactic factors released from sensitized mast cells challenged with specific antigen in vitro. Some of the in vivo observations have been confirmed in vivo by detecting the appropriate chemotactic factors after manipulations such as bronchial challenge in antigen sensitive asthmatics. The role of neutrophils and eosinophils in allergic tissue reactions were considered and doubts expressed as to whether these cells have a fundamental function in dampening the acute allergic response. It seems likely that in some situations eosinophils contribute to the inflammatory response as a result of the releaseof potent lysosomal enzymes – in much the same way as neutrophils contribute to tissue damage in immune complex disease. The notion that eosinophils, and to a much lesser extent neutrophils, combine with mast cell products to provide an effective defence against helminthic larvae was proposed indicating that an important role for mast cells is as repositories of factors which recruit and arm eosinophils for helminth destruction.

REFERENCES

1. A.B. Kay, Studies on eosinophil leucocyte migration I. Eosinophil and neutrophil accumulation following antigen-antibody reactions in guinea pig skin, Clin. exp. Immunol. 6:75 (1970).
2. L.W. Turnbull, D.P. Evans, and A.B. Kay, Human eosinophils, acidic tetrapeptides (ECF-A) and histamine. Interactions in vitro and in vivo, Immunology 32:57 (1977).
3. D.H. Bryant, and A.B. Kay, Cutaneous eosinophil accumulation in atopic and non-atopic individuals. The effect of an ECF-A tetrapeptide and histamine, Clin. Allergy 7:211 (1977).
4. A.B. Kay, D.J. Stechschulte, and K.F. Austen, An eosinophil leukocyte chemotactic factor of anaphylaxis, J. Exp. Med. 133:602 (1971).
5. A.B. Kay, and K.F. Austen, The IgE-mediated release of an eosinophil leukocyte chemotactic factor from human lung, J. Immunol. 107:899 (1971).
6. L.W. Turnbull, and A.B. Kay, Eosinophils and mediators of anaphylaxis. Histamine and imidazole acetic acid as chemotactic agents for human eosinophil leucocytes, Immunology 31:797 (1976).
7. E.J. Goetzl, and K.F. Austen, Purification and synthesis of eosinophilotactic tetrapeptides of human lung tissue: ientification as eosinophil chemotactic factor of anaphylaxis, Proc. Nat. Acad. Sci. USA 72:4123 (1975).
8. E.J. Goetzl, and K.F. Austen, Natural eosinophilotactic peptides: evidence of heterogeneity and studies of structure and function, Chapter 9, in: The Eosinophil in Health and Disease, A.A.F. Mahmoud, and K.F. Austen, eds., Grune & Stratton, New York (1980).
9. R.N. Boswell, K.F. Austen, and E.J. Goetzl, Intermediate molecular weight eosinophil chemotactic factors in rat peritoneal mast cells: immunologic release, granule association, and demonstration of structural heterogeneity, J. Immunol. 120:15 (1978).

10. L. Nagy, T.H. Lee, E.J. Goetzl, and A.B. Kay, Complement receptor enhancement and chemotaxis of human neutrophils and eosinophils by leukotrienes and other lipoxygenase products, Submitted for publication.

11. R.A. Lewis, S.T. Holgate, L.J. Roberts II, J.A. Oates, and K.F. Austen, Preferential generation of prostaglandin D_2 by rat and human mast cells, in: Proceedings of the Fourth International Symposium on the Biochemistry of the Acute Allergic Reactions, in press.

12. B. Samuelsson, Oxidative products of arachidonate: Leukotrienes, a new group of compounds including SRS-A, in: Proceedings of the Fourth International Symposium on the Biochemistry of the Acute Allergic Reactions, in press.

13. M. Hirashima, and H. Hayashi, The mediation of tissue eosinophilia in hypersensitivity reaction I. Isolation of two different chemotactic factors from DNP-ascaris extract-induced skin lesion in guinea pig, Immunology 30:203 (1976).

14. P.C. Atkins, M. Norman, H. Weiner, and B. Zweiman, Release of neutrophil chemotactic activity during immediate hypersensitivity reactions in humans, Ann. Int. Med. 86:415 (1977).

15. T.H. Lee, L. Nagy, M.J. Walport, and A.B. Kay, Exercise-induced release of a neutrophil chemotactic factor in bronchial asthma, Submitted for publication.

16. E.J. Goetzl, Modulation of human eosinophil poly-morphonuclear leukocyte migration and function, Am. J. Pathol. 85:419 (1976).

17. D.J. Hanahan, C.A. Demopoulos, J. Liehr, and R.N. Pinckard, Identification of platelet activating factor from rabbit basophils as acetyl glyceryl ether phosphorylcholine (AGEPC), J. Biol. Chem. 255:5514 (1980).

18. A.B. Kay, The role of the eosinophil, J. Allergy clin. Immunol. 64:90 (1979).

19. A.B. Kay, The role of the eosinophil in physiological and pathological processes, in: Recent Advances in Clinical Immunology, Chapter 5, R.A. Thompson, ed., Churchill Livingstone, Edinburgh (1980).

DISCUSSION

LECTURER: Kay   CHAIRMAN: Cumming

SOLIMAN: Being Egyptian I have a great interest in the
             schistosoma infestation and I would like to ask and
             discuss several points. Why did you choose mansoni
             and not Haematobrum? Is there any difference?

CUMMING: If you are going to ask several questions would you
             take one at a time?

KAY: We chose Mansoni because most people had been
             working in this field and we felt it was important
             to compare our results, and to be sure that any
             discrepancy was not a species difference. That't
             the first reason; the second reason is that Mansoni
             is in our hands, rather easier, people who advised
             us told us that Mansoni was much easier to keep
             than Tripolitan or Haematobium in a cycle with mice
             and snails.

SOLIMAN: The second point about this local action: the
             larvae will be surrounded by this complex and mast
             cells and eosinophilic enhancement. Does it apply
             to the endemic areas or just for the first
             infestation?

KAY: Time didn't allow me to talk about primary
             infection in schistosomiasis but very briefly what
             I believe happens is that complement activation is
             the important recruiter of cells and almost
             certainly the neutrophils play their role there.
             Obviously in the primary infection there will be no
             IgE or IgG, it is very likely that complement has
             an essential role in primary infections.

SOLIMAN: The next point: what about the role of the lungs
             from an immunological point of view? Especially in
             cases of cor pulmonale, we have a lot of cases of
             schistosoma cor pulmonale. What is the role of the
             lungs in this.

KAY: There's not much point in getting into great detail
             about the life cycle of schistosomiasis. It has a
             life phase of course, we don't have to talk about
             the complex host-parasite relationship or the fact

that by the time it gets to the lung it has escaped from the mechanism I'll be talking about, I should think it is possibly something we should talk about privately during the week.

BIENENSTOCK: Would you like to comment on the possibilities that exist for local eosinophils being, either as a result of their locl environment, different from those recruited from the blood or at least being activated relative to being inactivated and what the effect of this is on the killing that you talked about.

KAY: I hope it was clear from the slides that when these cells get into the tissues they are subjected to the influence I have been describing such as mast cell products and in that respect they would be different from circulating eosinophils. If you are asking: have we looked at other body fluids apart from blood as a source of eosinophils for what we were describing, the answer is no. This is in humans, because in the human situation, as you know, it is not all that easy getting eosinophils from other sources. Now, having said that, complement receptors in cerebrospinal fluid have been looked at from patients with a disease that i forget now, that was associated with schistosoma infection. Anyway, he found that there was a marked increase in C3 receptors in this situation, so that certainly fits with the sort of hypothesis that I have been talking about. We have tried to get eosinophils from the sputum but there are formidable difficulties in getting them from the sputum in a form that we can do the sort of functional experiments that we have been talking about. We may be luck, I suppose, occasionally and get some from pleural aspirate, but again there are obvious difficulties.

JEFFERY: What is the relationship, if any, between the leucotactic and the chemotactic factors; if there isn't any, what's the role of the leukocytes.

KAY: This is a question I have difficulties in giving a short answer to, but very briefly, as you know, there are families of molecules which are the result of the action of lipoxegenase activity. Some of these molecules have peptides on them and are what we used to call slowly-acting substance and in that respect the pathological role of amines is suspected in asthma and the like conditions. Where they fit in the scheme I have been describing is not clear. They probably act like other mediators in killing. I have been trying to do that for the last few weeks, but it hasn't worked and i don't know why, which is slightly embarassing. I think that if we do it right, they will really have similar effects in potentiating eosinophil and neutrophil mediated damage. That's a short answer to your question, maybe it's enough.

CUMMING: Thank you very much, Barry.

COHEN: In your slide of the killing of the schistosomes you never reached 100%. I suppose the maximum was 70 or 80%. Isn't there any basis for the resistance of those refractory schistosomes?

KAY: The reason for this is purely the condition of the "in vitro" assay. There has to be a limit on the affected cell ratio, because if we have too many cells it just makes it very difficult to read. If we were to go up to double the number of eosinophils present we would get 100% killing and if we use a chromium release I think we could. The visual assay we use is more reliable than chromium release assay in which the dead organisms take up the methylene blue.

LAURENT: I have a very short question. The word cathodic protein gives nothing away, is there anything known about what sort of protein it is? What effect has it on cells?

KAY: Well, it is cathodic because it is very basic. The evidence is that it probably does not have enzymatic activity and that its action is purely a charge effect.

CUMMING: Thank you very much Barry for that clear presentation. We pass onto the next paper. I should perhaps preface it with a comment that we invited Ronald Crystal to come to this symposium and he was able to accept, but at the last minute he had some personal difficulties in the U.S.A. which prevented him attending with us today; he very kindly arranged for Dr. Rossi, who has come to us from Genova, to deliver the paper on his behalf and we are very grateful to Dr. Rossi for taking this onerous task at such short notice. He also

agreed, although he is Italian of course, to
present the paper in English because it ismy
judgement that the bulk of the audience are not
native Italian speakers and therefore it would be
easier if he does as he agreed.

EVALUATION OF INFLAMMATORY AND IMMUNE PROCESSES IN THE INTERSTITIAL

LUNG DISORDERS: USE OF BRONCHOALVEOLAR LAVAGE

Giovanni A. Rossi,[*] Gary W. Hunninghake and
Ronald G. Crystal
Pulmonary Branch
National Heart, Lung, and Blood Institute
Bethesda, MD 20205

The interstitial lung diseases are a group of chronic, usually fatal disorders that diffusively involve the alveolar structures.[1-9] Although there are many types of interstitial diseases, all are characterized by two general features: (1) "alveolitis" – a chronic accumulation of inflammatory and immune effector cells within the alveolar structures; and (2) "fibrosis" – a derangement of the alveolar structures including changes in the number and type of lung parenchymal cells accompanied by alterations of the extracellular matrix, particularly collagen.[1-9] Importantly, while fibrosis of the alveolar walls is usually considered to be the hallmark of these disorders, it is now recognized that the key to the pathogenesis of the interstitial diseases is the alveolitis.[7-9] Not only does the alveolitis precede the other changes of the alveolar structures, but by virtue of a potent armamentarium of mediators, the cells comprising the alveolitis mediate the cellular and extracellular derangements that lead to progressive loss of function.[7-9] In this context, it is critical to define and characterize the alveolitis in order to understand the pathogenesis of the interstitial disorders and to rationally stage and treat these patients.

Classically, the alveolitis of interstitial disease has been evaluated by open lung biopsy.[1-6] However, this is an invasive procedure performed only once during the course of the disease and thus is not a useful approach to monitoring disease activity. In addition, while morphologic methods are very useful in diagnosis, they give little insight into the function and state of activation of the cells comprising the alveolitis. This is important, as it is the relative activity of the alveolitis that defines the rate and extent of injury to the alveolar structures.

The new approach to the study of the alveolitis of interstitial disease is the adaptation of the fiberoptic bronchoscope to "lavage", or wash out, inflammatory and immune effector cells and mediators present in the epithelial fluid of the alveolar structures. These components can then be evaluated by morphologic and functional methods to define the alveolitis.[7-12] Most importantly, since this procedure is rapid, safe, and readily accepted by patients, it can be used repeatedly to characterize the alveolitis to stage patients and evaluate their response to therapy.[8-13]

Technique of Bronchoalveolar Lavage

Following premedication with Valium (2-10 mg, IV) and atropine (0.8 mg, IM), local anesthesia is administered with a 2 percent lidocaine spray. A flexible fiberoptic bronchoscope (model BF-B2, Olympus Corp. of America, New Hyde Park, N.Y.) is then inserted through the nose or the mouth into the bronchial tree. After routine evaluation of the airways, the tip of the instrument is wedged in a subsegmental bronchus. Any lobe can be used, but the lingula or right middle lobe are most convenient.[12] Lavage is performed with a total amount of 100 ml sterile saline in five, 20 ml aliquots. After each aliquot is inserted, the fluid is immediately aspirated under 50 to 100 mm Hg negative pressure (e.g., utilizing a usual clinical suction apparatus such as the Ohio Intermittent Suction Unit, Ohio Medical Products, Madison, Wis.) and collected in a suitable container (e.g., 50 ml specimen trap, Cheesebrough-Ponds, Inc., Greenwich, Conn.). The return fluid is then placed on ice and immediately taken to the laboratory for analysis.

The fluid is filtered through two layers of surgical gauze and the volume measured. The cells are separated from the fluid by centrifugation (500 g, 5 min.) and the total number of cells counted (Coulter counter model FN, Coulter Electronics, Hialeah, Fla.). The cells are then resuspended in Hank's balanced salt solution (without Ca^{++} or Mg^{++}) at the desired cell density (10^7 cell/ml is convenient). The bronchoalveolar lavage cells can then be evaluated for viability and differential count,[8,11-12] lymphocyte subpopulations,[8,14-16] immunoglobulin in secreting cells,[17-19] and spontaneous release of mediators,[15,20-21] as previously described.

The cellular fluid can be frozen and stored; it is most convenient to first divide the sample into aliquots to obviate the need for repeated freezing and thawing in the future.[8] A variety of analyses can then be carried out to evaluate inflammatory and immune mediators in the fluid. If conventional analyses are used (e.g., Ouchterlony agar diffusion), then the fluid must be concentrated approximately 50-fold using a method such as Amicon filtration (Diaflo chamber, Amicon Corp., Lexington, Mass.).[11-12] A newer approach to lavage fluid analysis is to utilize an enzyme-linked immuno-absorbant (ELISA) test.[22] This can be carried out on unconcentrated fluid and can be adapted to quantify immunoglobulins, complement components, and other mediators.

Inflammatory Immune Effector Cells and Mediators in Normal Lung

Lavage of a normal adult human with 100 ml saline yields 52 $\pm$ 8 ml of fluid containing 5-10 x 10^6 cells.[8] The differential count reveals that of these cells, 93 $\pm$ 3% are macrophages, 7 $\pm$ 1% are lymphocytes, and less than 1% are neutrophils, eosinophils, and basophils (Table I). Of the total lymphocytes present, 73 $\pm$ 4% are T-lymphocytes and 8 $\pm$ 3% are B-lymphocytes;[8] of the T-lymphocytes, 46 $\pm$ 3% are helper cells and 25 $\pm$ 2% are suppressor cells.[23-24]

In individuals who smoke cigarettes, the number of broncho-alveolar cells recovered by lavage is increased (10-20 x 10^6 cells recovered per 100 ml lavage) and neutrophils are present (4.4 $\pm$ 1.2% of the total cells recovered). The proportions of the other cell types, however, are similar to that of nonsmokers.[8,25]

In normal nonsmoking individuals the inflammatory and immune effector cells in the lower respiratory tract are in a resting state and produce little, if any, mediators[8,26] (Table II). For example, although they can be stimulated in vitro to do so, alveolar macrophages recovered from normals are not producing the chemotactic factor for neutrophils,[25] growth factor for fibroblasts,[21] or lymphocyte activating factor.[24] They do produce fibronectin, an adhesive glycoprotein that attracts fibroblasts and attaches them to the extracellular matrix, but at a very low level.[22] T-lymphocytes recovered from the normal lung are capable of being induced to produce mediators such as migration inhibitory factor, leukocyte inhibitory factor, monocyte chemotactic factor, or "help" for B-cell immunoglobulin production, but these cells do not spontaneously release significant amounts of such mediators.[8,15,19]. Evaluation of B-lymphocytes of the normal lung show that, like normal blood B-cells, a small proportion (<1%) are spontaneously producing immunoglobulins.[8]

* All data is presented as mean $^+$ standard error of the mean.

In general, the function of the effector cells recovered from the lungs of cigarette smokers are similar to that of nonsmokers. The major exception is the alveolar macrophage; in smokers these cells are releasing the chemotactic factor for neutrophils, the likely mediator that attracts neutrophils to the lower respiratory tract of these individuals.[8,25]

Lavage fluid of normal individuals contains various proteins

TABLE I

PROPORTIONS OF INFLAMMATORY AND IMMUNE EFFECTOR CELLS
PRESENT IN BRONCHOALVEOLAR LAVAGE FLUID OF PATIENTS
WITH SARCOIDOSIS AND IDIOPATHIC PULMONARY FIBROSIS
THAT HAVE ACTIVE ("HIGH INTENSITY") ALVEOLITIS[a]

		Condition	
Cell	Normal[b]	Sarcoidosis	Idiopathic Pulmonary Fibrosis
Macrophage	$93 \pm 3\%$	decreased	decreased
Lymphocyte	7 ± 1	markedly increased	normal
T-cell	73 ± 4^c	markedly increased	normal
Helper-T	46 ± 3^d	markedly increased	normal
Suppressor-T	25 ± 2^d	markedly decreased	normal
B-cell	8 ± 3	decreased	normal
Neutrophil	$<1^e$	normal[f]	markedly increased
Eosinophil	<1	$\pm$ increased	increased
Basophil	<1	$\pm$ increased	$\pm$ increased

[a] See references 8, 11-12 for details.

[b] Data presented as the proportion of all bronchoalveolar lavage cells recovered in non-smokers.

[c] Data presented as the proportion of lymphocytes recovered.

[d] Data presented as the proportion of T-lymphocytes recovered.

[e] For cigarette smokers, neutrophils comprise 3-5% of all cells recovered.

[f] Patients in the late stages of sarcoidosis can have neutrophils present (34).

TABLE II

FUNCTION OF INFLAMMATORY AND IMMUNE EFFECTOR CELLS IN THE LUNGS
OF PATIENTS WITH SARCOIDOSIS AND IDIOPATHIC PULMONARY FIBROSIS
THAT HAVE ACTIVE ("HIGH INTENSITY") ALVEOLITIS[a]

Cell	Mediator Release	Condition		
		Normal	Sarcoidosis	Idiopathic Pulmonary Fibrosis
Macrophage	Chemotactic factor for neutrophils	no[b]	no	yes
	Growth factor for fibroblasts	no[b]	yes	yes
	Lymphocyte Activating factor	no[b]	yes	no
	Fibronectin	low levels	increased	increased
T-lymphocytes	Monocyte chemotactic factor	no[c]	yes	no
	Migration inhibitory factor	no[c]	yes	no
	Leukocyte inhibitory factor	no[c]	yes	no
	"Help" for B-cell Ig production[d]	no[c]	yes	no
B-lymphocytes	IgG[e]	$0.23 \pm 0.04\%$	increased	increased
	IgM[e]	$0.10 \pm 0.02\%$	± increased	increased
	IgA[e]	$0.31 \pm 0.06\%$	± increased	± increased

[a] See references 8, 15, 19, 21-22, 24-26, 30 for details.

[b] Normal macrophages can be stimulated *in vitro* to produce these mediators.

[c] Normal T-lymphocytes can be stimulated *in vitro* to produce these mediators.

[d] Specific "help" mediator(s) produced by human lung T-cells have not been identified.

[e] Data for immunoglobulin production presented as percentage of B-lymphocytes spontaneously producing immunoglobulin.

which are important components of the inflammatory and immune processes in the lung (Table III). These include immunoglobulins IgG, IgA, and IgE. However, IgM is rarely detected in normal individuals using conventional radial immunodiffusion plates to evaluate lavage fluid concentrated 30-50 fold.[8,10-11] All pathways of the complement system are present, including the classical, alternative, and common pathways.[8,10-11,27] The major antiprotease of the lower respiratory tract is α 1-antitrypsin, a serum protein that diffuses through the alveolar structures to provide antielastase protection.[8,10] α 2-macroglobulin can be detected in lavage fluid of normals, but it is present in very low amounts and is not likely an important antiprotease for the lower respiratory tract.[8,10] The major connective tissue proteases, elastase and collagenase, are not found in normal lavage fluid.[8,28-29] Fibronectin is present, but in relatively small amounts.[22]

Inflammatory and Immune Effector Processes in the Pathogenesis of Interstitial Lung Disease

Although the inflammatory and immune effector cells and mediators present in the normal alveolar structures have the potential to injure the lung parenchyma using a variety of mechanisms (Tables II, III), no damage is normally apparent. There are several reasons for this. Firstly, the release of mediators by effector cells generally occurs only when such cells are activated, but in the normal lung, only a small proportion of the total effector cell population is activated.[8,21-26] Secondly, while immunoglobulins are present in the normal lung, they are likely not directed against lung constituents.[8,30] Likewise, while the classical, alternative, and common complement pathways are present, all available evidence suggests that in the normal lung the complement system is turning over very slowly.[8,10-11,27] Thirdly, the alveolar structures possess a variety of defense and repair processes (e.g., antiproteases, antioxidants) that seem to effectively balance any low level of injury that does occur in the normal lung.[8,10]

In contrast, the interstitial diseases are characterized by increased numbers of effector cells in the alveolar structures, changes in the relative proportion and state of activation of the various cell types, and alterations in the quantity and type of mediators present. The result of this alveolitis is the derangement of lung parenchyma.[1-12]

The adaptation of bronchoalveolar lavage techniques to the study of interstitial disease has led to the important observation that the alveolitis of the various diseases can be remarkably different, i.e., while many interstitial diseases have common features such as fibrosis, the inflammatory and immune effector processes comprising the alveolitis may be quite disparate.[8,11-

TABLE III

INFLAMMATORY AND IMMUNE MEDIATORS IN THE LUNGS OF PATIENTS WITH SARCOIDOSIS AND IDIOPATHIC PULMONARY FIBROSIS THAT HAVE ACTIVE ("HIGH INTENSITY") ALVEOLITIS[a]

Component	Normal	Condition	
		Sarcoidosis	Idiopathic Pulmonary Fibrosis
Immunoglobulins			
IgG	0.31 ± 0.03[b]	increased	increased
IgM	<0.01[b]	rarely detected	not detected
IgA	0.72 ± 0.02[b]	$\pm$ increased	$\pm$ increased
Complement			
C4	1.7 ± 0.2[c]	?[d]	normal
C6	4.5 ± 0.9[c]	?[d]	normal
C3	12 ± 3[e]	increased in a few patients	increased in a few patients
C5	<1	not detected	not detected
factor B	1.8 ± 0.7[e]	?	increased
Antiproteases			
α1-antitrypsin	50 ± 15[f]	normal	normal
α2-macroglobulin	<5[g]	normal	normal
Proteases			
elastase	none	none	none
collagenase	none	none	present
Other mediators			
fibronectin	1.17 ± 0.27[h]	increased	increased

[a] See references 8, 10-12, 22, 27-28 for details.

[b] Data presented as amounts of immunoglobulin (mg/mg albumin).

[c] Data expressed as hemolytic activity/mg albumin.

[d] ? = unknown.

[e] Data expressed as amounts of each complement component (μg/mg albumin).

[f] Data expressed as amount of α1-antitrypsin (μg/mg albumin).

[g] Data expressed as amount of α2-macroglobulin (μg/mg albumin).

[h] Data expressed as amount of fibronectin (μg/mg albumin).

[12,31] In addition, comparison of the findings of bronchoalveolar lavage to that in blood has demonstrated that while the interstitial diseases are characterized by active inflammatory processes in the alveolar structures, this is usually not reflected by inflammatory and immune processes elsewhere.[8,12,15-16,19,23]

Two interstitial disorders, sarcoidosis and idiopathic pulmonary fibrosis (IPF), serve as prototypes that illustrate these concepts and point out the valuable use of bronchoalveolar lavage to help in understanding pathogenesis, staging patients, and making therapy decisions. Sarcoidosis and IPF have been studied extensively by bronchoalveolar lavage over the past several years and a large body of information is now available relating to the amount, type and state of activation of inflammatory and immune processes comprising the alveolitis of both disorders.

Sarcoidosis

The characteristic feature of the alveolitis of sarcoidosis is the presence of increased proportions of lymphocytes in the alveolar structures (Table I).[8-9,12] Not only are there increases in the proportion of total effector cells that are lymphocytes, but there is a marked increase in the proportions of T-lymphocytes and in the proportions of T-lymphocytes that are "helper" cells.[15-16,23-24,32] As in normals, neutrophils are rarely found in the lung parenchyma of patients with sarcoidosis. The exception to this statement is in two situations: the sarcoid patient that smokes cigarettes (3-5% neutrophils are generally present) and the sarcoid patient in the progressive, late stages of disease.[8-9,25,33-34] A few eosinophils and basophils can be found in the sarcoid lung, but these cells represent only 2-3% of all effector cells comprising the alveolitis.[8,15]

As in most interstitial diseases, sarcoidosis is characterized by a reduction in the proportions of alveolar macrophages recovered by lavage. However, this does not imply that the total numbers of macrophages in the sarcoid lung is reduced. In fact, the opposite is true - sarcoidosis is characterized by an increase in the total number of macrophages.[8,12,15,31-32] Rather, lavage tells us that sarcoidosis is a disease in which the alveolitis is characterized by a shift in the types of effector cells present so that lymphocytes become a relatively much more important component of the effector cell populations of the alveolar structures.

Evaluation of the function of the effector cells present in the sarcoid lung demonstrates that not only are there changes in the proportion of cells present, but also in their state of activation (Table II). Most importantly, bronchoalveolar lavage analysis has demonstrated that the T-lymphocyte populations in the sarcoid lung are activated and producing a variety of mediators.[15]

One of these T-cell products is monocyte chemotactic factor, a 10-15,000 dalton protein that attracts monocytes to the alveolar structures. This mediator is important to the pathogenesis of sarcoidosis since monocytes are the building blocks of granuloma formation. The activated T-cells in the sarcoid lung also release migration inhibitory factor and leukocyte inhibitory factor, lymphokines that likely play a role in helping the newly recruited monocytes to differentiate into macrophages, epithelial cells, and giant cells - the cells that comprise the granulomata of this disease.[15,32-33,35]

The presence of activated T-lymphocytes in the sarcoid lung also explains why these patients have high levels of circulating immunoglobulins. Among the mediators released by these activated T-cells are "helper" factors that stimulate lung B-cells to produce immunoglobulins in a nonspecific, polyclonal fashion. Thus, the sarcoid lung is characterized by an increased proportion of B-cells that are producing immunoglobulins; these immunoglobulins spill over into blood causing the hyperglobulinemia that is characteristic of this disease.[19]

Comparison of blood T-lymphocytes to the T-lymphocytes recovered by lavage demonstrated a marked partitioning of immune effector cells in sarcoidosis. At the site of disease (e.g., the alveolar structures) there are increased proportions of activated T-cells producing a variety of mediators that are responsible for at least some of the derangements of the lung parenchyma characteristic of the disease. In contrast, blood T-cells are reduced in number and are not in a state of activation.[15] This dichotomy between lung and blood in sarcoidosis is an excellent example of the usefulness of bronchoalveolar lavage in directly evaluating the alveolitis of interstitial disease, i.e., evaluation of blood not only fails to indicate the status of lung inflammatory and immune processes, but can also give misleading concepts concerning lung effector processes.

The reason why the lung T-lymphocytes are increased in number and are altered to include higher proportions of helper cells and lymphokine producing cells is not clear. One possible explanation is that the T-cell populations are influenced by mediators produced by activated alveolar macrophages in the local milieu. In this context, evaluation of macrophages recovered by lavage of sarcoid patients with active disease demonstrates that these cells are releasing lymphocyte activating factor (Table II), a mediator that induces T-cell proliferation.[24,36]

The alveolar macrophage also plays an important role in mediating the fibrosis that occurs in many patients with sarcoidosis. By producing fibronectin (which attracts the fibroblasts and attaches them to the extracellular matrix)[22] and growth factor for fibroblast (which stimulates the recruited cells to replicate),[21] the sarcoid macrophage increases the number of fibroblasts in its local environs (Table II). Since dibroblasts produce collagen, the result is an increase local collagen production, i.e., fibrosis. Consistent with this concept, fibronectin levels are known to be increased in sarcoid lavage fluid (Table III).[26].

The role of complement in the pathogenesis of sarcoidosis is unclear (Table III). Furthermore, proteases and antiproteases do not seem to be central to the pathogenesis of sarcoidosis. The antiprotease screen of the sarcoid lung is normal and the connective tissue specific proteases elastase and collagenase are not present in sarcoid lavage fluid (Table III).[8,28]

Idiopathic Pulmonary Fibrosis

The characteristic feature of the alveolitis of IPF is the presence of neutrophils in the alveolar structures (Table I).[3,6-9] In contrast to the normal lung that contains few neutrophils and the lungs of cigarette smokers that have 3-5% neutrophils, the lung of a patient with active IPF is characterized by neutrophils representing up to 50% of all effector cells present.[11-12,37] The neutrophil carries such a potent array of mediators that the presence of so many neutrophils overwhelms the local defenses with results that are devastating to the alveolar structures.[8,38-40] For example, the neutrophil also releases a variety of oxidants that can mediate damage to lung parenchymal cells.[41] Consistent with the presence of large numbers of neutrophils in the IPF lung, evaluation of lavage fluid of these patients has demonstrated the presence of neutrophil collagenase, an enzyme that specifically cleaves type I collagen, the major connective tissue component of the extracellular matrix of the alveolar structures (Table III).[28] Collagenase is particulary effective in deranging the alveolar structures since the lung has no effective anticollagenase to inhibit it. In contrast, while the neutrophil carries elastase, this enzyme is prevented from acting in the IPF lung because of the presence of ample amounts of α 1-antitrypsin, an effective antielastase (Table III).[10]

While analyses of IPF lavage demonstrate a reduction in the proportions of alveolar macrophages (Table I), this is simply a reflection of the shift in the IPF alveolitis toward neutrophils. Like other interstitial diseases, the total number of macrophages in the IPF lung is clearly increased.[3,8-9,11,42] In addition, analysis of the macrophages recovered from the IPF lung by lavage reveals that these cells are activated in such a manner that they release a number of mediators that help to explain the pathogenesis of this disease (Table II). For example, IPF macrophages are releasing neutrophil chemotactic factor, thus explaining the presence of neutrophils in the alveolar structures of these patients.[41] Like the sarcoidosis macrophages, the IPF macrophages are releasing fibronectin and growth factor, thus accounting for the fibrosis of the disease.[21-22] Thus, activation of the alveolar macrophages is central to the pathogenesis of IPF.

Evaluation of the lymphocyte populations recovered by lavage of IPF patients has shown that while the T-cell population appears normal, the B-cell population is producing large amounts of immunoglobulins, particularly IgG.[30] In addition, evaluation of lavage fluid of these patients not only shows increased amounts of IgG, but the presence of immune complexes as well. These immune complexes activate the alveolar macrophage population by binding to the macrophages through surface receptors.[30,43]

Neutrophils are not the only polymorphonuclear leukocytes found in the IPF lung. Lavage analyses also demonstrate eosinophils (3-5%) and small numbers of basophils.[8,11-12,37] The role played by these cells in the pathogenesis of IPF is unclear, but eosinophils do carry a collagenase that can attack the major collagens of the alveolar structures.[44]

The role of complement in the pathogenesis of IPF is unclear. Lavage fluid analyses in IPF have shown normal levels of C4 and C6 and undetectable C5.[11] However, some patients have increased C3 levels and most have increased factor B levels.[27]

In summary, evaluation of IPF by bronchoalveolar lavage has demonstrated it to be an immune complex disease mediated by the local production of immune complexes that in turn stimulate alveolar macrophages to recruit and activate neutrophils and fibroblasts, thus explaining a great deal of the abnormalities characterizing this disease.[42]

Importantly, bronchoalveolar lavage has clearly shown that the alveolitis of IPF is markedly different than that of sarcoidosis. Yet like sarcoidosis, the inflammatory and immune effector processes of IPF are local to the lung, i.e., they are not necessarily reflected in blood.[3,8-9,31,42,45]

Use of Bronchoalveolar Lavage in Making Staging and Therapy Decisions

Since the alveolitis of the interstitial disorders precedes and mediates the derangements of the alveolar structures, and since bronchoalveolar lavage accurately portrays both the character and intensity of the alveolitis, it is reasonable to hypothesize that this technique is useful in staging and making therapy decisions concerning these patients.

In sarcoidosis, an active alveolitis (so-called "high intensity alveolitis") is characterized by an increased proportion of activated T-lymphocytes and increased numbers of activated alveolar macrophages. In this context, prospective studies by Keogh and colleagues have demonstrated that 63% of untreated patients with sarcoidosis with high intensity alveolitis will deteriorate in at least one lung function parameter over a 6 month period.[44] In contrast, those with low intensity alveolitis will remain stable or improve over the same study period. Importantly, these studies have shown that the prognostic value of evaluating the alveolitis is independent of the chest film or pulmonary function status at the time of evaluation. In other words, while the status of the alveolitis can predict deterioration in sarcoidosis, conventional parameters such as chest roentgenograms and pulmonary function testing indicate only the damage that has already occurred, i.e., they cannot be used to accurately stage the activity of the disease.[9,46]

Prospective studies of the use of alveolitis estimates to guide therapy decisions concerning sarcoidosis are still ongoing. However, preliminary evaluation of this data suggest what might be expected: patients with high intensity alveolitis treated with corticosteroids will improve. Thus techniques such as bronchoalveolar lavage may enable clinicians to establish uniform guidelines for making therapy decisions in sarcoidosis, i.e., those with high intensity alveolitis should be treated and those with low intensity alveolitis observed. However, since sarcoidosis can spontaneously go from active to inactive (or vice versa), it is necessary to periodically revaluate the alveolitis every 3 to 12 months depending on the activity of the disease.[46]

In IPF, "high intensity" alveolitis is characterized by the presence of significant proportions of neutrophils in the bronchoalveolar lavage analysis. Studies by Keogh et al.[47] and by Rudd et al.[48] have shown that the patients who deteriorate are those in whom neutrophils are a significant component of the alveolitis. Keogh et al. have added an estimate of alveolar macrophage activity to this analysis: those with 10% neutrophils and large numbers of active macrophages are those that deteriorate.[47] In this context, both staging and therapy decisions can be made with bronchoalveolar lavage. For example, if high intensity alveolitis persists in an individual with IPF being treated with corticosteroids, the likelihood is that this individual will continue to deteriorate. In addition, like in sarcoidosis, there is clear evidence that conventional methods such as chest X-ray and lung function tests can not be used to accurately assess the alveolitis of IPF.[46] However, with evaluation by bronchoalveolar lavage every 3 to 12 months, rational staging and therapy decision can be made.[47-48]

Validity and Potential Problems in Utilizing Bronchoalveolar Lavage in Assessing Inflammatory and Immune Cells and Mediators in the Lower Respiratory Tract

There are several questions that are commonly asked in relation to the validity of bronchoalveolar lavage as a means to assess inflammatory and immune processes in the lower respiratory tract.

(1) What is the relative contribution of airways versus alveoli to the cells and mediators recovered by lavage? The fiberoptic bronchoscope is 5.9 mm in diameter; when it is wedged for the lavage procedure it is usually at the fourth to fifth order bronchus. Hence, the lavage fluid must pass 11 to 12 orders of bronchi to reach the alveoli and again pass the same bronchial before recovery.[8] However, the surface area of the bronchial tree through which the fluid travels is small in comparison to the alveolar epithelial surface that is washed out, so that effector cells present in the airway epithelial fluid contribute little to the overall analysis. For example, if 7×10^6 alveoli are lavaged, the maximum alveolar surface area sampled would be 17,000 cm^2 compared to a maximum bronchial surface area of 172 cm^2.[49-50] Even so, it is critical that at least 60 ml of fluid (total) be used for lavage. Sequential studies have shown that the first 20 ml reflect primarily airways. However, if at least 60 ml fluid is used, the airway contribution to the total lavage analysis becomes a minor component since the number of cells and mediators contained in the second and third 20 ml aliquots not only reflect the alveolar contribution but are many fold greater in quantity than those in the first 20 ml.[51] In this context, direct comparison of the effector cells contained in lavage analyses have demonstrated a remarkable similarity to the effector cell populations present in open lung biopsies; this is true for normals, cigarette smokers, and patients with interstitial disease.[8,31]

(2) Is bronchoalveolar lavage valid in the presence of inflammatory airway disease? To date, detailed analyses of lavage fluid have been carried out only in individuals with no inflammatory airway disease (i.e., individuals with airways that have a normal appearance and contain no purulent secretions). The mucus associated with inflammed airways contains large numbers of neutrophils;[8] this may confuse analyses of cells and mediators recovered by lavage. It is possible that evaluation of macrophages and/or lymphocytes has validity in the presence of airway disease, but this has not been examined closely.

(3) What is the best way to present analyses of the cells recovered by lavage? Three methods have been used, including: (a) as percentage, i.e., as in a cell differential of blood; (b) as the total numbers of each cell type; and (c) as the numbers of each

cell type per volume of fluid recovered, e.g., numbers of neutrophils/ml fluid returned. Although the volume of fluid recovered is usually 50 to 60% of that infused, there is a broad range, even with normals. This fact, together with the concept that effector cells may be "stickier" in some patients compared to other, argues strongly for the concept that lavage cell data should only be expressed as a proportion of the total cells recovered.[8] The opposing argument is that while percentage data gives an accurate view of the relative distribution of cell types, it is also worthwhile to consider the data as total numbers or numbers/ml, approaches which theoretically give insight into effector cell density in the lower respiratory tract. However, only the percent method of analysis has been validated, and, at least for the present, it remains the method of choice.

(4) What is the best way to present analysis of the mediators (and other noncellular components) of recovered lavage? This is a major problem. What one wants is an estimate of the concentration of the mediator in the epithelial fluid, but the lavage fluid returned is a variable mixture of epithelial fluid and saline.[8] Thus, expression of the data as amount (or activity) per ml of fluid has little meaning. To overcome this problem, investigators use albumin as a standard by which to relate noncellular lavage fluid constituents.[8,10,12] The concept underlying this approach is that albumin is produced external to lung, is not metabolized by lung, and, by virtue of its moderate molecular weight (69,000 daltons), diffuses freely through the alveolar structures. Since many lavage fluid components are also present in serum (e.g., immunoglobulins, complement, antiproteases), comparison of the component in lavage as amount (or activity)/mg albumin can be easily related to serum. If the level compared to albumin is increased, this strongly suggests local production (or decreased removal) in the alveolar structures. However, if the level/mg albumin is the same in lavage fluid and serum, it is likely that the component reached the alveolar structures by diffusion from blood. The argument against this is that if there are "leaky" alveolar capillary units (as occurs in acute disease such as adult respiratory distress syndrome), the denominator (i.e., albumin) is increased; this may mask an increase in the numerator (i.e., the component being analyzed), and thus mask increased production.[8,10,12]

(5) Is bronchoalveolar lavage dangerous? No, as long as reasonable criteria are used for patient selection and reasonable precautions are taken. Specifically, we use the following criteria for patient selection: (a) FEV_1 1 litre; (b) PaO_2 75 mm Hg (with or without supplemental oxygen); (c) no CO_2 retention; and (d) no significant cardiac disease. As a precaution during the procedure, all subjects are monitored by continuous ECG and ear oximeter, an intravenous line is in place, and resusitation

equipment is readily available. With this approach, the
complication rate is 5%, with all complications minor and
equivalent to that expected from fiberoptic bronchsocopy itself.[13]

Future Prospects

The adaptation of the fiberoptic bronchoscope as a tool to
sample inflammatory and immune processes in the lower respiratory
tract to patients with interstitial disease has "opened" the
alveolar structures of these patients to rational assessment of
disease activity in a quantitative, repetitive manner. The
procedure is safe and readily acceptable to patients; some
individuals in our study population have been lavaged more than 20
times over a 5 year period. Together with the realization that it
is the alveolitis of these disorders that mediates tissue injury
and hence functional deterioration, bronchoalveolar lavage
represents a clinical tool that permits the clinician to directly
assess disease activity. While routine chest roentgenograms and
lung function tests maintain an important place in evaluating these
disorders, they are more useful as a means of assessing what has
happened to the patient. In contrast, bronchoalveolar lavage tells
the clinician what is actively occurring, and thus, what the future
holds.

REFERENCES

1. H. Spencer, Interstitial Pneumonia. <u>Annu. Rev. Med</u>. 18:423
    (1967).

2. J.G. Scadding and K.F.W. Hinson, Diffuse fibrosing alveolitis
    (diffuse interstitial fibrosis of the lungs): correlation with
    histology at biopsy with prognosis. <u>Thorax</u> 22:291 (1967)

3. R.G. Crystal, J.D. Fulmer, W.C. Roberts, M.L. Moss, B.R. Line,
    and H.Y. Reynolds, Idiopathic pulmonary fibrosis: clinical,
    histologic, radiographic, physiologic, scintigraphic,
    cytologic and biochemical aspects. <u>Ann. Intern. Med</u>. 85:769
    (1976).

4. J.D. Fulmer and R.G. Crystal, The biochemical basis of
    pulmonary function. <u>in</u>: The Biochemical Basis of Pulmonary
    Function", R.G. Crystal, ed., Marcel Dekker, New York, (1976).

5. F. Basset, P. Soler, and J.F. Bernandin, Contribution of
    electron microscopy to the study of interstitial pneumonias.
    <u>Prog. Resp. Dis</u>. 8:45 (1975).

6. J.D. Fulmer and R.G. Crystal, Interstitial lung disease. <u>in</u>:
    "Current Pulmonology", D.H. Simmons, ed., Houghton Mifflin,
    Boston (1979).

7. B.A. Keogh and R.G. Crystal, Chronic interstitial lung
    disease. <u>in</u>: "Current Pulmonology", D.H. Simmons, ed., Wiley,
    New York (1981), Vol 3: pp. 237-340.

8. G.W. Hunninghake, J.E. Gadek, O. Kanawani, V.J. Ferrans, and
    R.G. Crystal, Inflammatory and immune processes in the human
    lung in health and disease: evaluation by bronchoalveolar
    lavage. <u>Am. J. Pathol</u>. 97:149 (1979).

9. R.G. Crystal, J.E. Gadek, V.J. Ferrans, J.D. Fulmer, B.R.
    Line, and G.W. Hunninghake, Interstitial lung disease: current
    concepts of pathogenesis, staging, and therapy. <u>Am. J. Med</u>.
    70:542 (1981).

10. H.Y. Reynolds and H.H. Newball, Analysis of proteins and
    respiratory cells obtained from human lungs by bronchial
    lavage. <u>J. Lab. Clin. Med</u>. 84:559 (1974).

11. H.Y. Reynolds, J.D. Fulmer, J.A. Kazmierowski, W.C. Roberts,
    M.M. Frank, and R.G. Crystal, Analysis of bronchoalveolar
    lavage fluid from patients with idiopathic pulmonary fibrosis
    and chronic hypersensitivity pneumonitis. <u>J. Clin. Invest</u>.
    59:165 (1977).

12. S.E. Weinberger, J.A. Kelman, N.A. Elson, R.C. Young, H.Y. Reynolds, J.D. Fulmer and R.G. Crystal, Bronchoalveolar lavage in interstitial lung disease. Ann. Intern. Med. 89:459 (1978).

13. I.J. Strumpf, M.K. Feld, M.J. Corneilus, B.A. Keogh, and R.G. Crystal, Safety of fiberoptic bronchoalveolar lavage in the evaluation of interstitial lung diseases. Chest 80:268 (1981).

14. M.S. Weiner, C. Bianco, V. Nussenzweig, Enhanced binding of neuraminidase-treated sheep erythrocytes in human T-lymphocytes. Blood 42:939 (1976).

15. G.W. Hunninghake, J.D. Fulmer, R.C. Young, J.E. Gadek, and R.G. Crystal, Localization of the immune response in sarcoidosis. Am. Rev. Resp. Dis. 120:49 (1979).

16. G.W. Hunninghake, J.D. Fulmer, R.C. Young, and R.G. Crystal, Comparison of lung and blood lymphocyte subpopulations in pulmonary sarcoidosis. in: "Proceedings of the Eighth International Congress on Sarcoidosis and Other Granulomatous Disease", W.J. Williams and B.H. Davies, eds. Alpha Omega Publishing, London. p. 426 (1980).

17. E. Gronowicz, A. Coutinko, and F. Melcher, A plaque assay for all cells secreting Ig of a given type or class. Eur. J. Immunol. 6:588 (1976).

18. E.C. Lawrence, R.M. Blaese, R.R. Martin, P.M. Stevensen, Immunoglobulin secreting cells in normal human bronchial lavage fluid. J. Clin. Invest. 62:832 (1978).

19. G.E. Hunninghake and R.G. Crystal, Mechanisms of hypergamma-globulinemia in pulmonary sarcoidosis: site of increased antibody production and role of T-lymphocytes. J. Clin. Invest. 67:86 (1981).

20. L.E. Spitler, and C. Von Muller, Leukocyte migration inhibition in agarose. in "In Vitro Methods in Cell Mediated and Tumor Immunity", B.R. Bloom, J.R. David, eds., Academic Press, New York (1976).

21. P. Bitterman and R.G. Crystal, Pulmonary alveolar macrophages release a factor that stimulates human lung fibroblasts to replicate. Am. Rev. Resp. Dis. 121:58A (1980).

22. S. Rennard, P. Bitterman, G. Hunninghake, J. Gadek and R. Crystal, Alveolar macrophage fibronectin: a possible mediator of tissue remodelling in fibrotic lung disease. Clin. Res. 29:374 (1981).

23. G.W. Hunninghake and R.G. Crystal, Pulmonary sarcoidosis: a disorder mediated by excess helper T-lymphocyte activity at sites of disease activity. N. Engl. J. Med. (in press) (1981).

24. G.W. Hunninghake, P. Broska, R. Haber, B.A. Keogh, B.R. Line and R.G. Crystal, Correlation of lung T-cell and macrophage function with disease activity in pulmonary sarcoid. Clin. Res. 29:550A (1981).

25. G.W. Hunninghake, J.E. Gadek, and R.G. Crystal, Mechanisms by which cigarette smoke attracts polymorphonuclear leukocytes to the lung. Chest (Suppl) 77:273 (1980).

26. G.W. Hunninghake, J.E. Gadek, S.V. Szapiel, I.J. Strumpf, V.J. Ferrans, B.A. Keogh and R.G. Crystal, The human alveolar macrophage. in: "Methods and Perspectives in Cell Biology: Cultured Human Cells and Tissues in Biomedical Research", C.C. Harris, B.F. Trump and G.D. Stoner, eds., Academic Press, New York (1980).

27. R.A. Robbins, J.E. Gadek, and R.G. Crystal, Potential role of the complement system in propagating the alveolitis of idiopathic pulmonary fibrosis. Am. Rev. Resp. Dis. 123:50A (1981).

28. J.E. Gadek, J.A. Kelman, S.E. Weinberger, A.L. Horwitz, H.Y. Reynolds, J.D. Fulmer, and R.G. Crystal, Colagenase in the lower respiratory tract of patients with idiopathic pulmonary fibrosis. N. Engl. J. Med. 301:737 (1979).

29. A. Mayem, A. Scharfman, A. Laine, J. Lebas, and A. Kopaczewski, Etude de la compontion proteique et enzymatique des liquides obetnus a differents miveau du tractus respiratoire de subjects temorius. Bul. Eur. Physiopathol. Respir. 15:17P (1979).

30. G.W. Hunninghake and R.G. Crystal, Localization of immunoglobulin production to sites of disease activity in idiopathic pulmonary fibrosis. (submitted).

31. G.W. Hunninghake, O. Kawanami, V.J. Ferrans, R.C. Young, W.C. Roberts, and R.G. Crystal, Characterization of the inflammatory and immune effector cells in the lung parenchyma of patients with interstitial lung disease. Am. Rev. Resp. Dis. 132:407 (1981).

32. R.G. Crystal, W.C. Roberts, G.W. Hunninghake, J.E. Gadek, J.D. Fulmer, and B.R. Line, Pulmonary sarcoidosis: a disease characterized and perpetuated by activated lung T-lymphocytes. Ann. Int. Med. 94:73 (1981).

33. G.W. Hunninghake, J.E. Gadek, R.C. Young, O. Kawanami, V.J. Ferrans, and R.G. Crystal, Maintenance of granuloma formation in pulmonary sarcoidosis by T-lymphocytes within the lung. N. Engl. J. Med. 302:594 (1980).

34. C. Roth, G.J. Juchon, A. Aronoux, G. Stanislas-Leguern, J.H. Marsac and J. Chretien, Bronchoalveolar cells in advanced pulmonary sarcoidosis. Am. Rev. Resp. Dis. 124:9 (1981).

35. G.W. Hunninghake, B.A. Keogh, B.R. Line, J.E. Gadek, O. Kawanami, V.J. Ferrans, and R.G. Crystal, Pulmonary sarcoidosis: pathogenesis and therapy. in: "Basic and Clinical Aspects of Granulomatous Diseases", D. Boros and T. Yoshida, ed., Elsevier-North Holland, New York, (1980).

36. E.R. Unanue, The regulation of lymphocytic functions by macrophage. Immunol. Rev. 40:227 (1978).

37. P.L. Haslam, C.W.G. Turton, B. Heard, A. Lukoszek, J.V. Collins, A.J. Salsbury and M. Turner-Warwick, Bronchoalveolar lavage and pulmonary fibrosis: comparison of cells obtained with lung biopsy and clinical features. Thorax 35:9 (1980).

38. J.E. Gadek, J. Fells, G.W. Hunninghake, R. Zimmerman and R.G. Crystal, Alveolar macrophage-neutrophil interaction: a role for inflammatory cell cooperation in the disruption of lung connective tissue. Clin. Res. 27:397A (1979).

39. J.E. Gadek, J. Fells, G.W. Hunninghake, and R.G. Crystal, Interaction of alveolar macrophage (AM) and the circulating neutrophil: AM-induced neutrophil activation. Am. Rev. Resp. Dis. 119:66A (1979).

40. G.W. Hunninghake, S.V. Szapiel, J.D. Fulmer, B.A. Keogh, and R.G. Crystal, Neutrophil-mediated fibroblast destruction in idiopathic pulmonary fibrosis (IPF). Am. Rev. Resp. Dis. 119:72A (1979).

41. W.J. Martin II, J.E. Gadek, G.W. Hunninghake, and R.G.
     Crystal, Oxidant injury of lung parenchymal cells.
     J. Clin. Invest. (in press).

42. G.W. Hunninghake, B.A. Keogh, J.E. Gadek, P.B. Bitterman,
     S.E. Rennard, and R.G. Crystal, Inflammatory and immune
     characteristics of idiopathic pulmonary fibrosis. in:
     "Clinical Immunology Update", Vol III, R.G. Buckley, D.
     Boniach, J.L. Elsevier-North Holland, (in press).

43. G.W. Hunninghake, J.E. Gadek, T.J. Lawley, and R.G. Crystal,
     Mechanisms of neutrophil accumulation in the lungs of patients
     with idiopathic pulmonary fibrosis. J. Clin. Invest. (in
     press) (1981).

44. W.B. Davis, J.E. Gadek, G.A. Fells, and R.G. Crystal, Role of
     eosinophils in connective tissue destruction
     Am. Rev. Resp. Dis. 123:55A (1981).

45. G.W. Hunninghake and R.G. Crystal, Inflammatory and immune
     mechanisms in chronic diseases of the lung parenchyma. in:
     "The Proceedings of the 6th Irwin Strasburger Memorial Seminar
     on Immunology", G.W. Siskind, ed. (in press).

46. B.A. Keogh and R.G. Crystal, Pulmonary function testing in
     interstitial pulmonary disease. What does it tell us? Chest
     78:856 (1980).

47. B. Keogh, B. Cline, M. Rust, G. Hunninghake, J. Meier-Sydow,
     and R. Crystal, Clinical staging of patients with idiopathic
     pulmonary fibrosis. Am. Rev. Resp. Dis. 123:89A (1981).

48. R.M. Rudd, P.L. Haslam, and M. Turner-Warwick, Cryptogenic
     fibrosing alveolitis. Relationships of pulmonary physiology
     and bronchoalveolar lavage to response to treatment and
     prognosis. Am. Rev. Resp. Dis. 124:1 (1981).

49. E.R. Weibel, "Morphometry of the Human Lung", Academic Press,
     New York (1963).

50. G. Cumming and L.B. Huntg, "Form and Function of the Human
     Lung", Williams and Wilkins C., Baltimore (1968).

51. G.W. Hunninghake, G.A. Rossi and R.G. Crystal, unpublished
     observation.

DISCUSSION

LECTURER: Rossi   CHAIRMAN: Cumming

CUMMING: Thank you very much Dr. Rossi for that elegant
                presentation.   Might I ask a question of
                clarification to begin with? In the patients who
                were treated with steroids 40 to 50% deteriorated
                after one year and yet in your last slide neither
                the IPF nor sarcoidosis did so.

ROSSI: The IPF deteriorate, the sarcoidosis not. Most of
                them, the slide says almost everybody, but there
                are a few patients with sarcoidosis who do not
                respond to steroids.

BONSIGNORE: What is the relationship between the functional
                test and broncho-alveoli lavage in the long-term
                monitoring of fibrosis. If I remember rightly I
                have read one paper published in the Journal of
                Clinical Investigation which indicates a strict
                relationship, it also indicates the best test to
                monitor long-term fibrosis. What is your
                experience about the best test?

ROSSI: Of tests we used to follow patients with
                interstitial lung disease, the most useful was
                DL_{co}. In interstitial lung disease the lung
                volumes are also important, because there is a
                deterioration of lung volumes, but the most
                sensitive tests to the varition of the activity of
                the disease and the ones which can detect the small
                differences in terms of deterioration are DL_{co} and
                arterial blood gases on exercise with a fall in
                PO_2. We see that patients with sarcoidosis with a
                completely clear chest x-ray but with lymph node
                involvement without tracheal or bronchial
                compression have quite normal physiology except for
                the fact that they drop the PO_2 on exercise.

WILLIAMS: Thank you, I enjoyed your presentation and I know
                the nice work you and Dr. Crystal are doing. I was
                a little perturbed that your histology slide of
                sarcoidosis in that you showed granuloma and
                commented on the degree of interstitial
                inflammation. This is not a common feature of
                sarcoidosis, it is a common feature of extrinsic
                allergic alveolitis such as farmers lung, it is a

common feature of Beryllium disease but it is interesting that it is a feature that can be used to distinguish the ordinary case of sarcoid from those diseases with circulating immunoglobulin immune complexes. This might mean that your cases where you coloured a number of cells in your lavage fluid that these cell groups have got superinfected or altered in some other way. What would you like to comment on this?

ROSSI: I didn't get the beginning of your question. You said that the interstitial infiltration of cells is peculiar to other diseases like hypersensitivity pneumonitis in which there are immune complexes but not in sarcoid, where the immune complexes are not present, that't right. I can answer this question in several ways. The simplest is that as far as I know the infiltration of the interstitium in sarcoidosis is a typical pattern which was described for the first time in a paper in 1976, which was describing the presence of early granuloma and interstitial infiltration of mono-nuclear cells in the lung of patients with stage I sarcoidosis. In all patients with active disease we see infiltration of mono-nuclear cells and collection of lymphocytes in the alveoli structures. That's the first part of the question. The second question you asked me is to compare the pathology and pathogenesis of other interstitial lung diseases as hypersensitivity pneumonitis and sarcoidosis. In hypersensitivity pneumonitis there are two major mechanisms of the disease and one is in the acute stage. There is a reaction with immune complex deposition along the capillary in the lung, clearly documented by experimental models on animals and a few pathological case reports of patients who died of acute exposure and in those cases alveolitis is characterised by neutrophils. In a second time there is a cell mediating immune response with production of granuloma. So I think there are two different patterns in hypersensitivity pneumonitis which clearly are mediated by different types of reaction.

CUMMING: Are you happy with that, Dr. Williams?

SOLIMAN: I would like to ask about other cells which might desquamate and may be found in the bronchial lavage. How could you show these in the lavage samples?

ROSSI: You are speaking about type I, type II and
              epithelial cells. Normally we do not see any type
              I or type II;   I don't have any experience of
              broncho-alveoli lavage performed on patients with
              specific desquamation of type I or type II cells,
              as in acute oxygen intoxication or acute radiation
              of the lung. I know that in England they have
              performed experiments on mice exposing them to
              radiation in the lung and they got cells which were
              thought to be alveolar macrophages but light
              microscopy did not permit their proper
              characterisation. About epithelial cells or
              bronchial cells I can tell you that it is a
              question of technique; if the patient is not very
              well anaesthetised and he is coughing a lot, you
              get a lot of bronchial cells and very very few
              alveolar macrophages. In those cases we don't
              consider alveoli lavage valid.

SOLIMAN: The point I'd concentrate on is this:- you know
              that the granular pneumocytes of type II cells
              should increase at one stage of the disease and
              this has another reflection on the physiopathology
              of the disease from the ventilation-perfusion
              imbalance point of view. So I would expect that
              you might see these cells in the lavage especially
              the type II cells, not particularly the type I.

ROSSI: We made for two years a study on every broncho-
              alveoli lavage and we couldn't detect
              any significant presence of those kinds of cells
              indicating a stage of disese or a peculiar pattern
              of the disease.

SOLIMAN: But you detected them, but not significantly.

ROSSI: Well, I am not sure about that, I don't know it,
              but not at a level to be interesting.

SPENCER: I have been listening this afternoon to a great
              deal about broncho-lavage and I am going to put to
              you a very simple question.

ROSSI: I am afraid of that.

SPENCER: Where do the lung macrophages come from? I have
              looked at literally 100,000 sections of lungs both
              at light and electron microscopic level, I have
              never yet seen an alveolar macrophage half way in

and half way out of the alveolar wall. Where do they come from?

CUMMING: Before you go on Herbert, have you seen them actually in the alveoli?

SPENCER: Oh, plenty in the alveolar lumen and in the interstitial tissues, but how do they get from one to the other?

CUMMING: It is a transport problem.

ROSSI: It is a problem of transportation and, I don't know, I think they can migrate directly but for sure they are seen in both places.

CUMMING: I have a long list of questioners, I have them all in order. Don't get excited, you'll get your turn.

KAY: Can I just see that I have got the facts right; you took what you call a neotrophil-rich preparation and showed that this would increase chromium release from fibroblasts and you compared these with the neutrophil-poor preparations. Now I noticed the neutrophil-poor preparations and the number of macrophages in the nuetrophil-rich was about 90% and in the neutrophil-poor was about 70%. My question is, and I am sure that time would have allowed you to have said this, I wonder whether the incubation could influence that.

ROSSI: We didn't incubate with them, we just centrifuged them.

KAY: Exactly, but did you centrifuge your neutrophil-rich ones because you said "unfractionated".

ROSSI: May I have the slide, it is one of the green ones. I don't know, can you just show me on the slide?

KAY: I think the point is a simple on,e in the unfractionated ones you didn't. Could it be that the solution you used to fractionate damaged the macrophages and it is really macrophages and not neutrophils?

ROSSI: O.K. I get the problem. I think that one way to explain the cell damaged fibroblast is through the release of proteolitic enzymes and other mediators. If we compare the amount of collagenases in the last stage secreted by two types of cells, you notice that neutrophils release more collagenases and elastases than macrophages. I can't prove that it is not damaging the alveolar macrophages. I know that this technique for sure didn't alter the properties of alveolar macrophages in all other experiments and it is not damaging lymphocytes, so it would surprise me if that happened, but I can't answer your question in the negative.

??   Dr. Rossi, I would like to ask you a question about your scheme for the sequence of events that occurs in IPF. On the left hand side of the scheme you showed a sequence of events suggesting that a growth factor was released from the alveolar macrophage which stimulates the fibroblast and there is quite a lot of evidence that such a sequence can occur. You also showed some experiments which suggested that alveolar macrophages can release a factor which causes increased numbers of fibroblasts. Several years ago one of Crystal's group published a paper suggesting that the synthesis rate of collagen doesn't change in patients with idiopathic pulmonary fibrosis. Do you have some sort of reconciliation of these observations?

ROSSI: That is a very interesting question. The amount of collagen secreted in vitro is enhanced by the fact that there are more fibroblasts and that's one mechanism by which the actual synthesis of collagen in the lung can be enhanced, but we don't have any data about the degradation which could be enhanced too. So I don't think we can consider that this experiment destroys what we have said for years that the amount of collagen in the lung in those diseases is not increased. Its a dynamic process. Are you happy with that or not?

HUTCHISON: Would you mind if I return to Professor Spencer's simple question and ask him to set it in perspective by telling us how often pathologists see cells that are known to migrate, half way in and half way out in the alveolar structure.

ROSSI: Frequently, We have several examples of migration of cells. I thought he spoke about normal people or normal animals. In pathology there is no question that there is a lot of migration from the capillary to the interstitium. He was speaking about normals and I don't have any picture in normals showing that.

CUMMING: Herbert, can I ask you, are you referring solely to normals or is your comment applying to all the normals and abnormals you have looked at?

SPENCER: What I have said applies equally to the normals and the abnormals. You find the alveoli filling macrohages, they are one of the most common cells in lung lavage. You see them in the interstitial tissues of the lung but how do they get from one to the other. I have never seen one half way in or half way out.

CUMMING: That's clearly an unresolved question.

HASLEM: I was going to comment on some remarks but I have so many questions! It is very interesting that one might expect not only to find a decreased proportion of macrophages in sarcoid alveoli, but also a decreased number as if they were for some reason retarded in the interstitial spaces, but there is not a decreased number the number is as you would find in a normal situation.

CUMMING: It is a dilution phenomenon really that you are suggesting.

HASLEM: No, there is no reduction in the number of macrophages in the alveolar spaces, so there again we come back to the original question: where the macrophages in the alveolar spaces come from, and I wonder what you think. Do you think there is any possibility that things which would be described as lymphocytes could in fact be monocyte precursors and this might particularly be so in sarcoidosis because of the so called transform cells which we don't yet know the origin of?

SPENCER: May I add a sort of addendum? Some years ago a colleague said to me: "I have put some lymphocytes in the fridge and they were at about $4^{\circ}C$. I came back and collected them 24 or 48 hours later, a lot of them were macrophages.

CUMMING: Brian Corrin is the next on the list.

CORRIN: Two papers this afternoon which dealt with broncho-alveoli lavage have shown very clearly the value of this technique with regard to diagnosis and also prognosis, but also also we have had a lot of speculation about pathogenesis that goes along with Dr. Rossi's idea about pathogenesis of sarcoidosis; but coming to the pathogenesis of intersititial fibrosis and cells that are washed out from the lung, they are really coming from the wrong place, the disease is in the interstitium of the lung. I wonder if I might consider some diseases of the lung which are marked by a tremendous increase in the number of the cells in question in the alveoli lumen for example an eosinophilic pneumonia which might get the drug reaction migrating to the lung or more often an idiopathic eosinophilic pneumonia. This is not marked by an interstitial fibrosis; some of them are bacterial pneumonias, we have an increase in the number of the neutrophils to a large extent but excluded from the alveoli, a very marked increase. Occasionally fibrosis developed but it is a very different pattern of fibrosis, interstitial fibrosis that we were considering. Tremendous numbers of macrophages are found in smokers and a minority of idiopathic interstitial pneumonias which are called desquamative interstitial pneumonias. It is the one variety where fibrosis is least evident.

ROSSI: I think you posed several questions. The first one is: have we had any evidence that broncho-alveoli lavage is reflecting what's going on in the interstitium in idiopathic pulmonary fibrosis for example? The answer is yes. We have published a paper comparing the results of the percentage of cells and immunoglobulin secretion properties of macrophages in lung biopsies and broncho-alveoli lavage. I have a slide showing that the number of immunoglobulin secreting cells were increased both in BAL and in biopsy. So we think, at least in the disease we have evaluated, that broncho-alveoli lavage is reflecting events in the interstitium. The second question was what's happening to smokers with many alveolar macrophages but who do not develop interstitial fibrosis. Smoke particles are ingested by the macrophage which becomes altered in many ways; but one of the results is the secretion of the chemotactic factor for neutrophils and there are many neutrophils in the alveoli of young smokers, shown by BAL and biopsy. So the question is why neutrophils in IPF patients are producing fibrosis, and are producing emphysema in smokers.

One possible explanation for that is that in smokers the level of antitrypsin is normal, but the activity of this enzyme is half or less than half than normal. So the elastase released by the neutrophils is not inactivated by antitrypsin and is fated to derange elastic fibres and produce emphysema.

HEATH: Mr. Chairman, I cannot but think that a mystery has emerged this afternoon because I am sure that pathologists spend a long time looking at lung diseases and realised just how common hyperplasias of type II pneumocytes are in conditions associated with fibrosis of the alveolar walls and especially in conditions leading to that. In patients we have been waiting this afternoon for the granular type II pneumocytes to appear, but nobody has mentioned it, Dr. Rossi actually says that in his, what is obviously extensive experience with this technique, he has never seen one. This is extraordinary to me because it seems that there is total disagreement between the results you get from this technique and what we know to be the pathology of the disease on classical histopathology and electron microscopy. What is wrong? Why are the cells not appearing in these results? There is something that is not right somewhere and you know I am suggesting that the results of this technique are in some way not accurately reflecting the pathology and I think it is a very important criticism of the technique as presented today.

CUMMING: Would you like to comment on that, Dr. Rossi? it seems to me relevant; Dr. Soliman mentioned type II and it is rarely apparent in your results and in the literature I gathered. Is one fixed to another mobile or how do you explain this?

ROSSI: I didn't get the question, excuse me.

CUMMING: The question was: in conventional pathology it's usual to see type II granulocytes and your answer to Dr. Soliman is that you very rarely see them in BAL, but Professor Heath's question was why is it that in the experience of pathologists it is common to see the cells and in the experience of lavagists it is very rare to see the cells and why are the two techniques giving different results.

ROSSI: I don't know how to answer this question. Can you answer?

CORRIN: These cells are altered considerably by the
           technique, the monocytes are activated and a lot of
           cells are in the English idiom "bogged up". I can
           recognise granular pneumocytes in broncho-alveoli
           lavage fluid, because I know from straight
           microscopy what they look like in the normal state,
           and I haven't seen them often, but what I believe
           are partly degenerated granular pneumotyces. I
           believe Professor Heath was referring to cells
           fixed to the basement membrane, this is very common
           and indicates to me that he is not making remarks
           about the alveolar walls, but is referring to cells
           seen in the alveolar lumen. There have been a
           series of papers on electron microscopy of tissue
           sections of the so called desquamative interstitial
           pneumonia. The results are various but taken
           collectively it would be accurate to say that the
           great majority see either only macrophages free in
           the lumen or a majority of macrophages and a
           minority of truly desquamative epithelial cells. I
           think it is true to say, if you survey the
           literature, that the great majority of such cases
           reflect an exudation of macrophages rather than a
           true desquamation of epithelial cells. it may be a
           physical phenomenon of course and one patient may
           on one occasion desquamate epithelial cells and on
           another exude macrophages, which are believed to
           come from the interstitium. We do have in our
           literature a case of electromicrograph showing a
           macrophage in transition from the interstitium and
           the alveolus.

CUMMING: I still want to pursue Professor Heath's point
           which doesn't seem yet to me to be adequately
           answered. May I ask Dr. Rossi then: when you look
           at your cells, are there any groups of cells which
           go unrecognised and which you cannot attribute to
           each of your categories?

ROSSI: If the preparation is carried properly, no. Also
           because we have the support of electron microscopy
           in this case.

HEATH: You see, if you take Professor Corrin's statement
           at face value, he says desquamative granular
           pneumocytes are not very common. Even if you
           believe that, then they would be expected to appear
           somewhere, there would be some, but the fact is
           that this class of cells doesn't appear at all in

any of this afternoon's papers. You know, of course, from your own published work, I think I have quoted this against you before, but when you are doing your work, you pointed out in that paper how jolly difficult it is to make certain that you are not looking at granular pneumocytes or macrophages which have ingested phospholipids and you have to go to great lengths to distinguish these two classes of cells. Now what I am wondering about is when people carry out BAL and they identify cells as macrophages just how careful are their criteria; I am sure they are fairly careful, but I don't know why a cell which we know to be involved in fibrosing diseases of the lung doesn't appear anywhere.

CUMMING: We should let Pat Haslem answer this question. I should point out that Pat Haslem works with Margaret Turner-Warwick and this is an identical rerun of the discussion we had here one year ago and Margaret gave an answer which I now predict Pat Haslem will also give.

HASLEM: Well, I was brought up in an atmosphere of the conventional views about desquamative interstitial pneumonia, so when I started to work on lavage I couldn't understand why I was not finding desquamative type II cells and then I looked at some of the stained conventional biopsies and I really couldn't distinguish between alveolar macrophages and the pneumocytes, I couldn't do it. So I thought we might do it with other methods, and we turned to electron micriscopy and in our series it does seem just as Bryan Corrin said, that there are very few of the cells either in our cystic fibrosis biopsies or in our lavage material. There are some but they are a minority; so it is insufficient because of the possibilities of phagocytosis to exclude type II pneumocytes, so we went to yet another method; we started to play with non-specific esterase staining and looking at the biopsy material non-specific esterases positively stain cells and show them to be present in the alveoli; but the lining cells of the alveoli do not stain very well with non specific esterases and all the lavage cells we see were very strongly stained with non-specific esterases, so I think one way we can do it is to use esterase staining.

CUMMING: A little more work has obviously gone on from last year, because we got an extended answer.

KUHN: I have several comments that I could make but I think the most pertinent one probably is that I think BAL, especially as carried out on patients, is probably a very inefficient procedure and you are probably only sampling a tiny fraction of the most loosely attached cells, leaving behind a lot of macrophages; in order to even approach 50% efficiency in an experimental animal you have to add metal ions, which you could not give to human subjects obviously. So I think in order to get both type II cells you really have to be quite traumatic. Beyond that I think one very rarely sees detached type II cells, looking at electron micrographs or even disease tissues, they are always bound to the basement membrane. I would echo Dr. Haslem's comment that histochemical methods would very clearly demonstrate that type II cells are stuck on the alveolar walls and do not desquamate. More specific type II cells do have non specific esterases so they are making a quantitative judgement.

CUMMING: Thank you for a partial solution to that difficulty.

SOLIMAN: I just have a comment on macrophages in smokers. You find a lot of macrophage cells in the broncho-alveoli lavage of smokers but I think that in the macrophage of a smoker there are crystalloid inclusions, so that one can differentiate between the macrophages of smokers and macrophages due to other diseases.

ROSSI: The question was why cells are damaging the lung in different ways, not whether macrophages from smokers were different. Bronchial lavage from a smoker is brown instead of white as in a normal person; it's very easy to see if somebody is smoking or not by the lavage. I can end by saying to Jones-Williams I am sure that "bogged up" is a Welsh expression and not an English one.

CUMMING: Do you have any views on what makes a predominant phagocytic macrophage turn into a non phagocytic epithelial cell, and could it be the epithelial cell which is the source of the growth factor?

ROSSI: That's an interesting point. The macrophages we
              tested did not come from biopsy but from BAL and
              from the cytological point of view they were
              typical big round cells with a lot of cytoplasm, so
              I think the cells secreting the growth factor are
              alveolar macrophages. To extend this hypothesis if
              you take alveolar macrophages from normal
              individuals and assimilate them with different
              stimuli like immune complex, or if you make them
              phagocytise particles they produce growth factor,
              so I think it is a function of the alveolar
              macrophage. I don't know if anybody has tested
              blood monocytes to see if it is possible to
              transform them in growth factor secreting cells.
              That part of the question I can't answer.

CUMMING: Thank you Dr. Rossi.

JANSSEN: You mentioned the existence of immune complexes in
              the lavage fluids and stated that most of them
              refer C1q binding SS. You mentioned also your
              problems with immunofluorescence studies on the
              cells and the binding of ITT on the surface of the
              macrphages. We do have the same experience in
              those latter problems, but have you ever seen any
              phagocytised immunoglobulin inclusions in the
              cells, e.g. any macrophages or any neutrophils,any
              immunoglobilin inclusions in immunofluorescence
              studies inside the cells.

ROSSI: Not that I know.

CUMMING: Let us make it more specific. You have done
              immunofluorescence studies. You haven't seen
              things that Pat Haslem told us about earlier.

ROSSI: On the surface, yes.

CUMMING: No, within the cell.

ROSSI: You ask me if I saw concentrations of
              immunofluorescent material inside the cells like
              she showed in the eosinophils? No I didn't.

CUMMING: I repeat that you said you saw them in the
              neutrophils and in the macrophages as well, is that
              right.

ROSSI: Not in the macrophages.

CUMMING: In the neutrophils, but not in the macrophages,
                right.   Any further questions, ladies and
                gentlemen? No, I think we have uncovered some
                useful areas of incompatibility today which are
                generally the source of further advance and I am
                sure that will take it away and think about it
                carefully and then when next we talk about alveolar
                lavage we will have some more answers. So thank you
                very much and "Buon Appetito"!

THE DEVELOPMENT OF LUNG CANCER

Herbert Spencer

Emeritus Professor of Morbid Anatomy
University of London

The development of lung cancer should be considered in the context of cancer development in all sites in the body. In many organs the neoplastic process often starts as a metaplasia of the surface epithelium which may sometimes proceed to dysplasia if the initiating cause is not removed. Later the stage of carcinoma in situ may occur. At any of these intermediate steps the process may be reversed but this is most likely to occur during the initial stages. The sequence of changes described above is found mostly in the large central bronchi and is particularly associated with heavy cigarette smoking. The changes are usually multifocal and involve a field of epithelial cells. Carcinoma in situ changes in the bronchial epithelium not only involve the surface cells but also the lining cells of the bronchial mucous gland ducts and the glandular epithelium itself. The resultant neoplasm usually present as plaque-like growths but can also produce polypoidal lesions. Polypoidal bronchial epithelial tumours present a spectrum of lesions varying from the well differentiated benign squamous cell papilloma of the bronchus through a less well differentiated transitional type of tumour to the undoubted papillary bronchial carcinoma in situ. The usual signs of hyperchromatism and dedifferentiation characterise the tumours at the malignant end of the spectrum.

It is uncertain how often a carcinoma in situ in the bronchus becomes an invasive tumour but if the initiating cause, cigarette smoking, continues unabated, it seems probable that ultimate invasive change commonly results. It is however, now recognised that if a lifelong smoker abandons the habit the likelihood of the development of invasive bronchial cancer recedes with each year that smoking is discontinued and hence it is probable that even established carcinoma in situ which is commonly found in the bronchial tree of heavy cigarette smokers can undergo reversal.

The importance of the multifocal nature of carcinoma in situ is that if often accounts for the appearance of further independent tumours in either lung following an apparent successful removal of the first tumour. Such recurrent tumours may be wrongly attributed to secondary spread from the initial tumour.

Cancer frequently arises in many organs as a consequence of continued simple reparative hyperplasia of an epithelium which ultimately becomes an uncontrolled cell growth or carcinoma. Examples of such cancers include malignant transformation of a chronic dental ulcer, of chronic ulcerative colitis and schistosomal ulceration in the urinary bladder. Lung damage mainly affects lung tissue distal to the terminal bronchioles and both bronchiolar and alveolar epithelium possess very considerable regenerative capacity. Total alveolar epithelial destruction often results in bronchiolar epithelial proliferation to reline the damaged alveolar wall. Incomplete alveolar epithelial damage results in a reparative proliferation of type 2 pneumocytes and this may very occasionally progress to neoplastic proliferation. Because interstitial alveolar fibrosis results in type 2 metaplasia of the lining epithelium lung cancer is particularly associated with this type of lung damage. Interstitial alveolar fibrosis is in turn a very common response to lung damage due to a wide variety of causes and often complicates other forms of tissue response to lung damage. When malignant change supervenes in a lung which is the seat of alveolar interstitial fibrosis, the cancer cells not infrequently line the internal surfaces of the damaged alveolar walls and result in a gland-like structure which is mistakenly referred to as an adenocarcinoma although it has no relation whatsoever to any gland structure. The neoplastic cells can furthermore also produce islands of squamous celled cancer of solid nests of undifferentiated cells. The resulting scar cancer may spread at an early stage to the hilar lymph glands from whence spread to the wall of an adjacent bronchus can occur simulating a primary cancer arising in that situation. Both radiological and pathological confirmation of this sequence of events exists. Unfortunately by the time most peripheral lung cancers become clinically detectable the growth is already large by pathological as opposed to clinical standards.

The third variety of lung cancer, the origin of which is now more clearly understood, is the small or oat-celled lung cancer. With the gradual realisation of the importance and ubiquity of the APUD (parocryne) system of cells which are found throughout the embryological entodermal structures, it was found that Kulschitsky cells exist both in the bronchial surface epithelium and in the bronchial mucous glands. Such cells in the lung like their counterparts elsewhere possess characteristic ultramicroscopic neurosecretory granules in their cytoplasm and the cells often exhibit argentaffinic staining properties during the first year of postnatal life. Afterwards the argentaffinity is lost though the cells usually retain argyrophilic properties. The Kulschitzky-like cells in the bronchial epithelium are the progenitor cells both of bronchial carcinoid tumours and the small cell lung cancers. Both tumours exhibit the same patterns of hormonal disturbances as well as showing neurosecretory-type granules in their cytoplasm on ultramicroscopy. It is also possible occasionally to show by light microscopy a transition of a bronchial carcinoid tumour to a truly malignant small celled lung cancer.

When considering any classification of lung cancer it must be realised that both bronchial and alveolar epithelial cells possess considerable metaplastic ability and both may on occasion give rise to squamous and mucus secreting cells and for this reason any lasting classification of lung cancer should be based on histogenesis and not solely on the predominantly light microscopic growth pattern.

DISCUSSION

LECTURER: Spencer   CHAIRMAN: Cumming

ROUSSOUW: Unfortunately there are only a few tumours, the
             bronchial adenomas, which develop without invading
             the bronchial wall. Sadly they account only for a
             minority of cases, then developing as "iceberg"
             tumours. I would like to ask you about those tiny
             tumours, measuring only 2 - 3 mm across which you
             showed us; were they discovered by chance during
             necropsies carried out for other purposes? And
             from what did the patient die?

SPENCER: May I answer the last question first? From what
             did the patients die. Patients died of something
             quite unrelated to the lung fibrosis, myocardial
             ischaemia or cor pulmonale, conditions totally
             unrelated to the lung lesion. How can one detect
             very small tumours in the bronchi, I think this was
             the question. The best clinical method of
             detection of early lung cancers, certainly those in
             the air passages in the bronchi, is probably
             biocytology. Radiology is not as successful in
             early diagnosis of lung cancer as cytology. By
             biocytology I mean not only sputum examination, but
             also bronchial brushings and examination of
             bronchial lavage fluids. Although I have no
             personal experience, bronchial brushing is a very
             valuable method, better than sputum examination,
             but do not be put off by one negative examination,
             go on examining the sputum, the more examinations
             of the sputum we do, the greater the chance of
             discovering lung cancer. So I'd say a minimum of 6
             - 7 sputum examinations, preferably bronchial
             brushing examinations. These I think are the best
             methods presently available. Does that answer your
             question.

SOLIMAN: I would like to make a comment from the clinical
             and radiological point of view in the early
             diagnosis of cancer. As Professor Spencer
             explained, the sequence of events in the
             development of lung cancer, especially carcinoma in
             situ, and especially the polypoidal form is a sort
             of nexus between pathology and radiology. With
             carcinoma in situ (especially the polypoidal form)
             may be associated a sort of air trapping, with a

non-return valve, some of the air being trapped in lung tissues and this might appear radiologically as a localised area of hypertranslucency. So if you find a localised area of hypertranslucency in a patient, provided that there is a background of predisposition, as for example, age and smoking history, pay attention that this patient may have an early cancer. Professor Spencer mentioned that damage of type I cells will lead to proliferation of type II cells and we have had the opportunity to study lungs from a heavy smoker and these papers have been published by Professor B. Corrin and myself in the fall 1978; we demonstrated specific changes in type II cells in patients attending the emergency department in Alexandria, whose chest had been opened due to other causes, for example, traumatic haemothorax. We were able to collect samples and study them under the electron microscope, and we saw a change in type II cells in the form of cytoplasmic clefts which had been thought to be cholesterol and described in other publications. These changes have been compared with the changes in samples of lung tissues collected from the relatively normal lung tissue distal to areas of lung cancer.

TAYLOR: Professor Spencer, I was glad to hear that you regard tumourlets as fundamentally hyperplastic lesions rather than neoplastic ones. You asked at the end of your paper what is the stimulus to hyperplasia of the lung endocrine cells which give raise to tumourlets, and you suggested that I should make a group research project. This project is actually underway and some of the results will be communicated on Friday morning. There is abundant evidence from animal work that these cells are sensitive to hypoxia; but whether or not tumourlets are the result of hypoxia is entirely another matter, and I hope I'll be able to enlarge on that on Friday morning. There is another point I'd like to make: many pathologists have difficulties in getting positive argyrophil reactions with pulmonary carcinoid tumours. I think the answer to this is: (a) use liquid fixation, and (b) do not distend the lung. I have a distinct impression which is supported by some animal work that when you distend the lung with fixing solution the staining of argyrophil lesions is much less satisfactory. If you do not distend the lung the stain is very satisfactory.

HEATH: I'd like to ask Professor Spencer this: there have
             been reports recently of raised levels of carcino-
             embryonic antigen in smokers. I was wondering if
             he would like to tell us what he makes of this,
             what interpretation does he place on that
             observation and can he relate raised levels of CEA
             to any of the ranges of histological changes that
             he showed in his paper.

SPENCER: Carcino-embryonic antigen was first associated with
             large biocancer and subsequently raised levels have
             been found in association with other forms of
             cancer. It seems a non-specific change and frankly
             I haven't had a great deal of experience with
             this, but what I have leads me to think it was not
             of very great value and I did not see carcino-
             embryonic antigen raised in a variety of tumours
             including lung cancer, therefore in my limited
             experience I don't think it is much help. Does it
             answer your question?

HEATH: I thought I could tempt you into being courageous
             enough to try and relate it to the early dysplastic
             changes in the bronchial cells that you showed, but
             I don't think I am going to be successful.

SPENCER: The staining of these cells I found very
             interesting. I haven't been doing much for the
             last few years on this, but I did find that it
             is much easier to show them under the age of one
             year. If one looked at the bronchiolar epithelium
             of neonates these cells were much more easily
             stained and were frequently argentophilic but they
             lose their argentophilic properties usually by six
             months, but remain argyrophilic since they were
             argentaffinic in early life. That presumably means
             they contain more pathway reducing agent at that
             stage. One should also remember that the fetus is
             living with a low oxygen tension, so anoxia may
             contribute to the reason why cells are more
             plentiful and more easily demonstrated in the
             neonate. I suspect their function has something to
             do with early life and the argyrophilic lesions
             which we see in the alveoli. They have little to
             do with normal function, they are pathological
             rather than physiological lesions.

KAY: Could you tell us whether Kultshitsky cells can be grown in culture. I imagine that it is difficult to separate them from the epithelium, but can carcinoid cells be maintained in culture and can you detect mediators.

SPENCER: I can't give you any help. I know of no work in which the cells have been cultured, but it might be difficult to separate them from the epithelium.

SPINA: In your presentation you mentioned scar tumours, do you think that these tumours might be related to a previous tuberculous infection? Did you find any such connection on an anatomic-pathological basis in the cases you observed?

SPENCER: Scar cancers may arise in relation to scars of all sorts of aetiologies, some of them have been tuberculous, but also infarcts, interstitial fibrosis, and so on. There is evidence from Australia, where Dr. Campbell showed that in ex-servicemen the incidence of cancer arising in the upper lobes in patients who had previously been diagnosed and treated for pulmonary tuberculosis was three times greater than in non-tubercuous patients. So I think that tuberculous scars increased the development of scar cancer, but not very much.

CHAIRMAN: May I ask if anybody has experience about the contribution of broncho-alveoli lavage to the diagnosis of cancer. Dr. Rossi.

ROSSI: We don't have so much experience in diagnosis since in our service we mainly deal with patients with interstitial lung disease but we occasionally see patients from other departments with lung cancer. In our experience brushing and transbronchial biopsy give at least as much information as bronchial lavage. Perhaps in the future we may have monoclonal antibodies against specific cell types and different types of lung cancer, and that could be a future application of lavage. I would like also to ask a question of Professor Spencer. Do you have any explanation why some interstitial lung diseases are associated with high incidence of cancer, like progressive systemic sclerosis and idiopathic pulmonary fibrosis and others not?

SPENCER: Dr. Rossi, the only suggestion I could make is that
                the more chronic the interstitial fibrosis the
                greater is the likelihood of development of lung
                cancer. On the whole in progressive systemic
                sclerosis of the lung, changes in the interstitial
                fibrosis seem to continue for a longer period of
                time than, for example, in some cases of rheumatoid
                disease and some of the cryptogenic interstitial
                fibrosing agents. Possibly it is a question of how
                long the interstitial fibrosis has persisted and
                the longer it persists, the greater the opportunity
                for lung disease to develop. So if it has only
                been present for a relatively short time the risk
                of disease is much less.

CHAIRMAN: Thank you very much Professor Spencer, for your
                presentation. The next paper is by Jean Bignon.

THE PLEURA: MESOTHELIAL CELLS AND MESOTHELIOMA

Jean Bignon, Jean-Francois Bernaudin, Marie-Claude
Jaurand and Marie-Claire Pinchon
Service de Pneumologie, Environnement, ERA CNRS N° 845
INSERM U 139, Centre Hospitalier Intercommunal
40, Ave de Verdun, 94010 Creteil Cedex

Whereas cell biology and biochemistry of the lung parenchyma is now well documented, little work has been carried out on the pleura. The biological reactivity of pleural cells and the physiology of the pleural space however are of paramount importance, in relation to three main conditions of the pleura: effusions, fibrosis and cancer. The highly significant association of these conditions with past exposure to airborn mineral fibres explains the interest of many workers in pleura, particularly in the field of experimental and human pathology and, more recently that of basic research. According to the work carried out in our laboratory and in others, the structure and function of the pleura will be examined in four parts: 1 - pleural cells in situ and in vitro; 2 - mesothelial barrier, in relation to fluid movement and particulate matters clearance; 3 - inflammatory lesions of the pleura; 4 - pleural carcinogenesis.

1. STRUCTURE AND FUNCTION OF PLEURAL CELLS

1. STRUCTURE OF PLEURAL CELLS

1.1. <u>Pleural cells in situ</u>. At least three types of cells can be identified in the pleural tissue: mesothelial cells (MC), submesothelial cells or fibroblasts (F) and interstitial free cells (IFC).

1.1.1. <u>Mesothelial cells</u>. With the scanning electron microscope (SEM), the surface of MC are covered with numerous digitations (Fig. 1). In transmission electron microscopy (TEM), these superficial digitations appear as typical microvilli, without cilia (exceptionally we observed one cilium in the parietal pleura

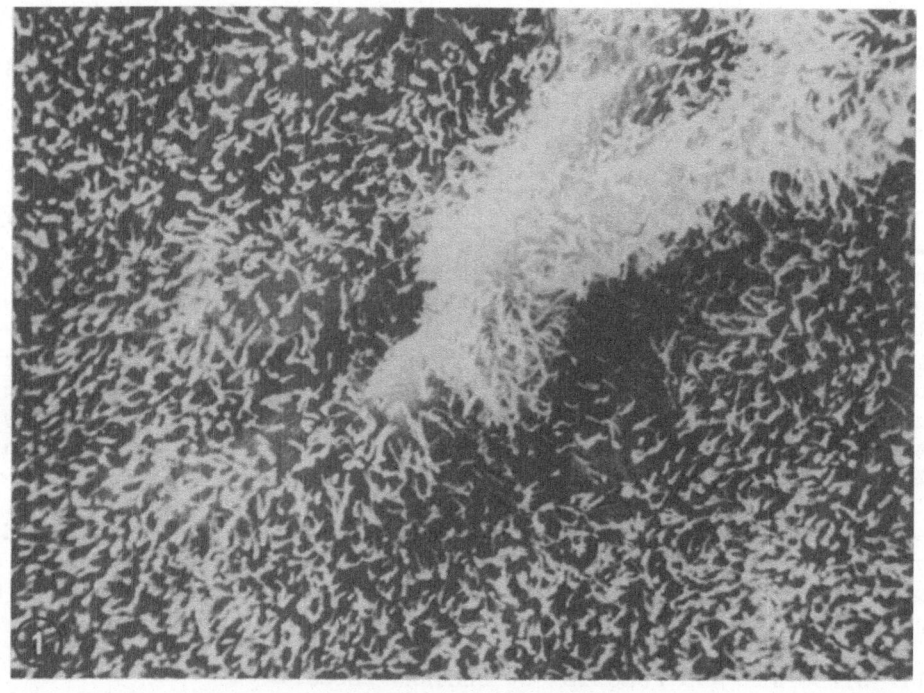

Fig. 1 : The surface aspect of the microvilli of the rat
         mesothelium (SEM).

of the rat); MC are separated from each other by an intercellular space usually with a tight junction near the apex, and often a few desmosomes in the nedial part of the space (Fig. 2). MC are very thin, having a thickness of 4 to 5 um in the nuclear area and 0.5 um in the peripheral cytoplasmic area. MC contain numerous plasmalemmal vesicles. 60 to 70 nm in diameter some intracytoplasmic, most of them opening toward the edges of the cells (Fig. 2). Large irregular vacoules (0.5 um in diameter) and coated vesicles are also seen in the cytoplasm (Fig. 2) (1). Plasmalemmal vesicles are more numerous in visceral than in parietal MC (2).

1.1.2. <u>Submesothelial cells.</u> In the area underlying the elastic lamina are found many interstitial cells which have a fibroblast-like aspect (F), some with long cytoplasmic extrusions that extend into contact with the basement membrane of the mesothelial cells. The ultrastructural pattern of this cell is no different from that of lung fibroblast (Fig. 3).

1.1.3. <u>Interstitial free cells.</u> In the matrix and associated with collagen and reticulin bundles are found numerous interstitial free cells (IFC), such as mast cells (which are frequent in the rat), polymorphonuclear leukocytes and monocytes or macrophages.

1.2. <u>Isolated pleural cells</u>

1.2.1. <u>From the pleural fluid.</u> Normally the pleural space is filled by a fluid film about 5 um thick, which in the rabbit contains about 1500 cells/mm^3 (3). Monocytic cells are the most common, representing about 70% (3, 4). The other cells are macrophages (7.5%), lymphocytes (10%), polymorphonuclear leukocytes (2%) and mesothelial cells (8.9%)., mostly degenerative. Fig. 4 shows a macrophage found in our laboratory at the surface of the parietal pleura in the rat. In an older study (5) concerning the fluid of the normal human pleura, there was 4500 cells/mm^3, corresponding to 54% of monocytes, 29.5% of degenerative mesothelial cells, 10% of lymphocytes and 3.5% of granulocytes.

This normal count is dramatically modified in <u>pleural effusions</u>, the relative proportions being related to the etiological factors of the disease. Free desquamated mesothelial cells are easily identified as rounded cells with central and oval nucleus; this shape is not totally specific and frequently it is difficult to differentiate MC from pleural macrophages.

1.2.2. <u>Cell culture from the pleura.</u> There are two sources from which one can culture mesothelial cells: pleural effusions and pleural tissue.

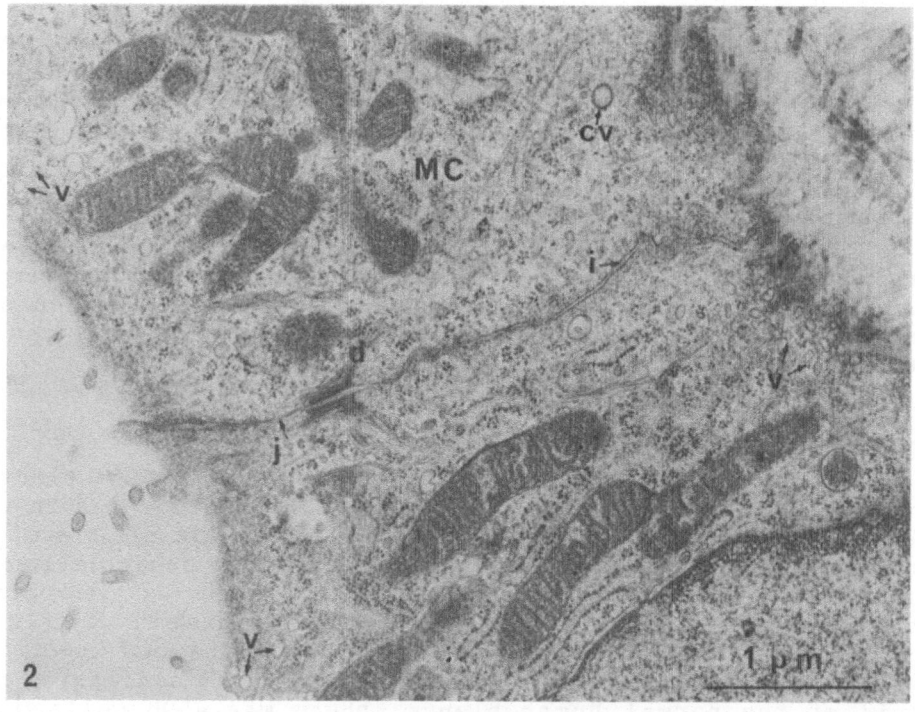

Fig. 2 : Rat parietal mesothelial cells (MC) : intercellular space
        (i), tight junction (j), desmosome (d), plasmalemmal
        vesicles (v) and coated vesicles (cv) (TEM).

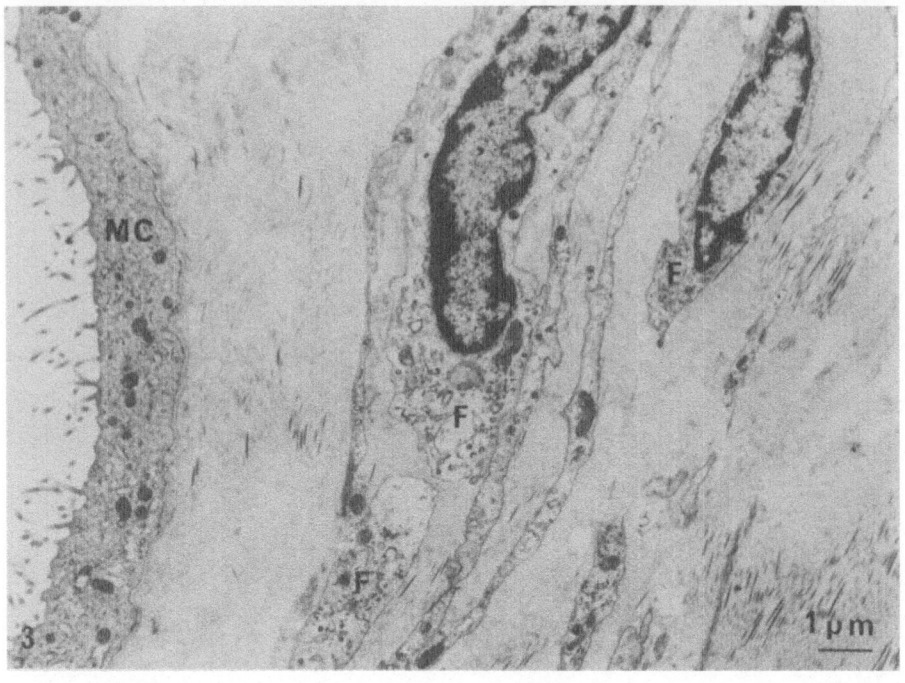

Fig. 3 : Fibroblasts (F) in submesothelial structures of rat
        parietal pleura (TEM).

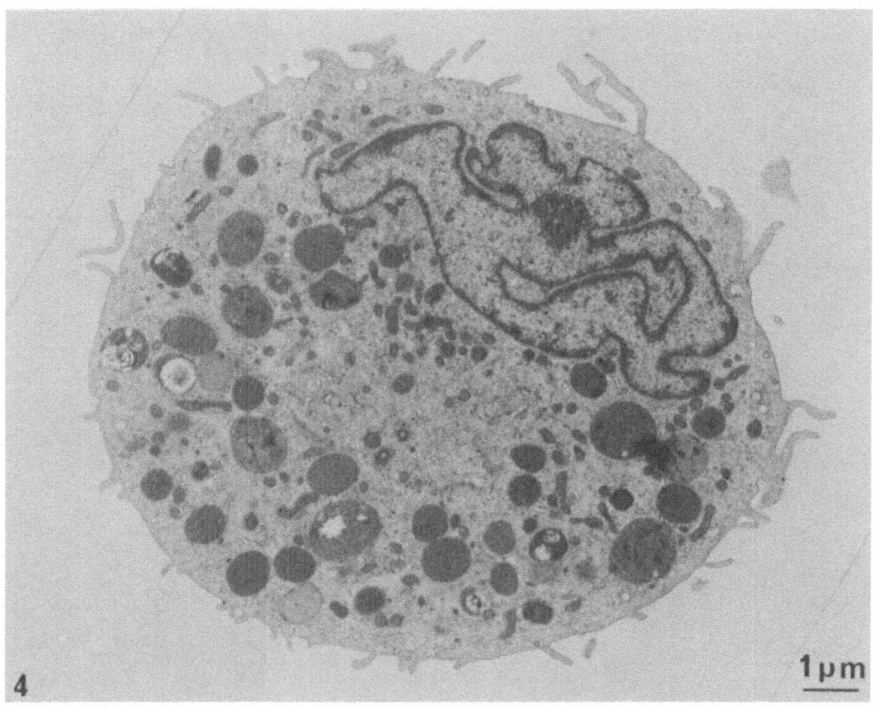

Fig. 4 : Rat pleural macrophage, similar to alveolar macrophages
        (TEM).

a. Culture of cells from pleural effusions. Different workers have succeeded in culturing cells present in human pleural effusions (6, 7). Some of these cells might be pleural macrophages, since it is easier to culture activated mononuclear phagocytes than mesothelial cells from experimentally induced pleural effusion (8).

b. Culture of cells obtained from pleural tissue. The second way for studying isolated pleural cells is to develop a culture system for long term maintenance of mesothelial cells (MC) and submesothelial cells (SMC). We succeeded in this type of culture, obtaining a continuous line of MC in 1978 (9); since then the method has been improved (10). For primary culture, the explant is obtained by scraping the parietal pleura and then cultured in NCTC 109 medium + foetal calf serum. After trypsinisation, it was possible to make subcultures with the same medium, up to 30 passages. Under the phase contrast light microscope, rat and human MC in culture appear as nonolayers of polyhedric cells in close juxtaposition (Fig. 5a). The electron microscopic study of rat MC sectioned perpendicular to the bottom of the plastic flask shows a pattern very similar to that of parietal mesothelial cells in situ (Fig. 6a and b). These cells have few microvilli, numerous mitochondria and ribosomes and an abundant endoplasmic reticulum (Fig. 7). MC in culture shows a growth curve reaching confluency in 4 to 5 days; at that time, in a flask of 25 cm^2, there was about 3×10^0 cells. The mean doubling time is about 30 hours. There exist interspecies variations in the size of MC, since we found that the surface area of human MC was 10 times more (6500 um^2) than that of rat MC (630 um^2).

More recently, we have developed a long term culture of submesothelial cells, obtained by scraping the submesothelial tissue of the parietal pleura. These cells are morphologically identical to fibroblasts (F) (Fig. 5b).

The selective culture of MC and SMC, which show specific patterns, gives a clue to the concept of mesothelium as an entity, as suggested by the histochemistry of resting MC (11).

2. FUNCTIONS OF PLEURAL CELLS

The experimental studies carried out in our laboratory and in others allow several functions to be attributed to mesothelial cells.

2.1. Endocytosis. The microvillous surface and the vesicular apparatus of MC suggest that these cells have important activities in relation to fluid macromolecules and small particle movement (2, 12). However, the role of microvilli and vesicles in fluid absorption and transfer remains theoretical (12). Thus, the

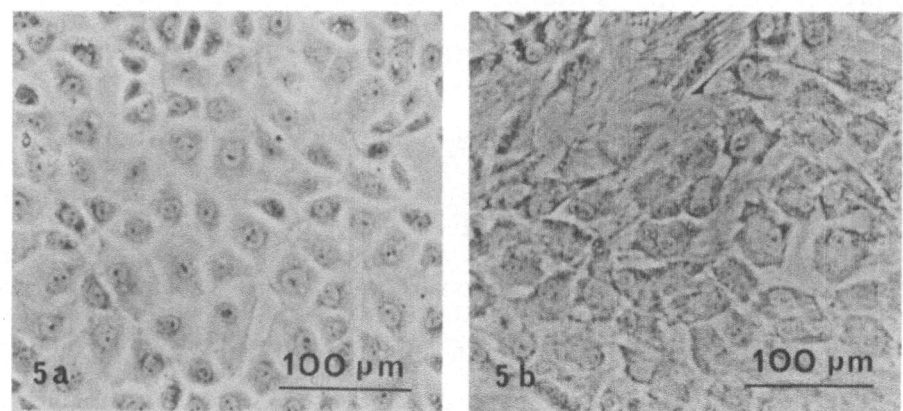

Fig. 5 : a. rat parietal mesothelial cells in culture;
          b. submesothelial fibroblasts in culture (light
          microscope).

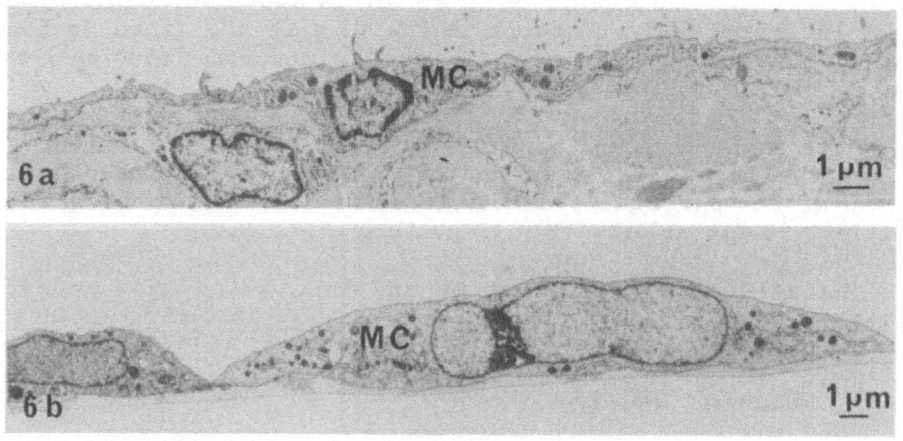

Fig. 6 : a. Rat parietal mesothelial cells (MC) in situ;
          b. Rat parietal mesothelial cells (MC) in culture,
          perpendicularly sectioned (TEM).

plasmalemmal vesicles, usually considered as pinocytotic, might also be considered as the resulto fo an endocytotic process for some nutritional or enzymatic functions.

2.2. **Phagocytosis.** Intraperitoneally injected particles have been shown to be taken up by mesothelial cells in situ (13, 14). In a recent experiment where strips of rat pleural tissue were incubated in vitro with asbestos fibres, we were also able to observe an active ingestion of fibres by MC (Fig. 8).

These experiments indicate that there is a gradient in the phagocytosing capability of MC, which is less in vivo than in vitro. These discrepancies are not clearly understood. It is possible that pleural cells (MC and SMC) can become phagocytic in culture, since, mouse gingival fibroblasts have been shown to be as phagocytic in culture as peritoneal macrophages, but with some specificity, fibroblasts ingesting more exogenous collagen debris and macrophages more latex particles (16). However rat MC and SMC in culture seemed to be equally phagocytotic for asbestos fibres and for latex particles (Fig. 9).

2.3. **Synthetic activities.** The dilated endoplasmic reticulum of MC suggests intense synthesic activity. So far, nothing is known about the molecules synthesized by the mesothelium, although it is well known that some primary pleural tumours are often associated with high concentration of hyaluronic acid (17). We have begun to identify some of the macromolecules synthesized by rat mesothelial cells, isolated cells in culture, as the ideal material for such biochemical studies. These cells were shown to be capable of synthesizing several connective tissue components, such as types I, III and IV (basement membrane) collagens, elastin, laminin and fibronectin (18). More recently, C. Lafuma, in Dr. Robert's laboratory, showed that rat MC and SMC were both capable of synthesizing acid hyaluronic.

2.4. **Cytodynamic behaviour.** The mesothelial cell population undergoes renewal at a slow rate, about 33 days turnover time for the rat mesothelium (19). This slow renewal of MC in vivo contrasts with their rapid growth in culture. It is possible that scraping the pleura to collect MC for culture is similar to a pleural injury. Indeed, after the injection of trypsin into the pleural cavity, Aronson et al (20) observed that the percentage of labelled cells increased dramatically : up to approximately 18% of cels were labelled instead of 1 out of 1800.

2.5. **Origin of mesothelial cells.** To assess technical feasibility this has been studied on damaged peritoneum in different species : the whole surface of the denuded basement membrane was simultaneously covered by a continuous layer of newly formed flattened cells looking like mesothelial cells (21, 22, 23, 24).

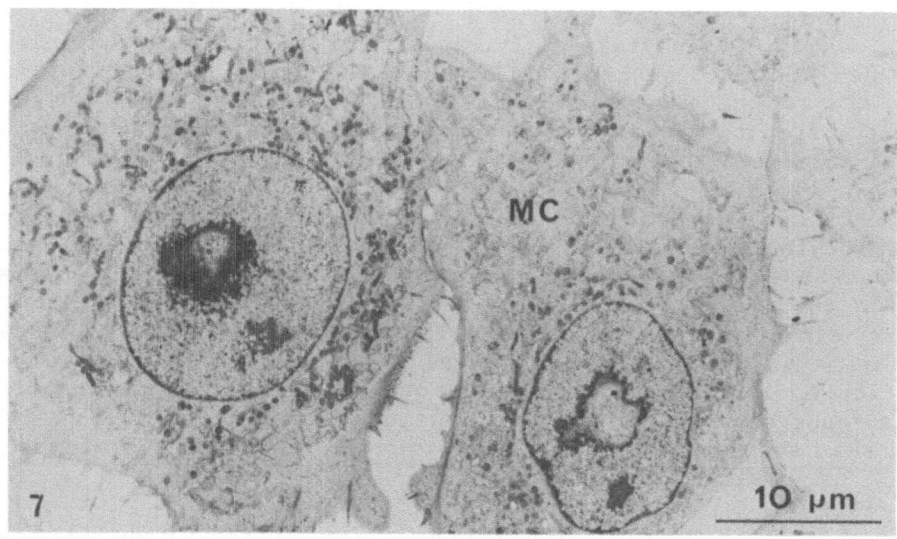

Fig. 7 : Rat parietal mesothelial cells (MC) in culture,
transversely sectionned (TEM).

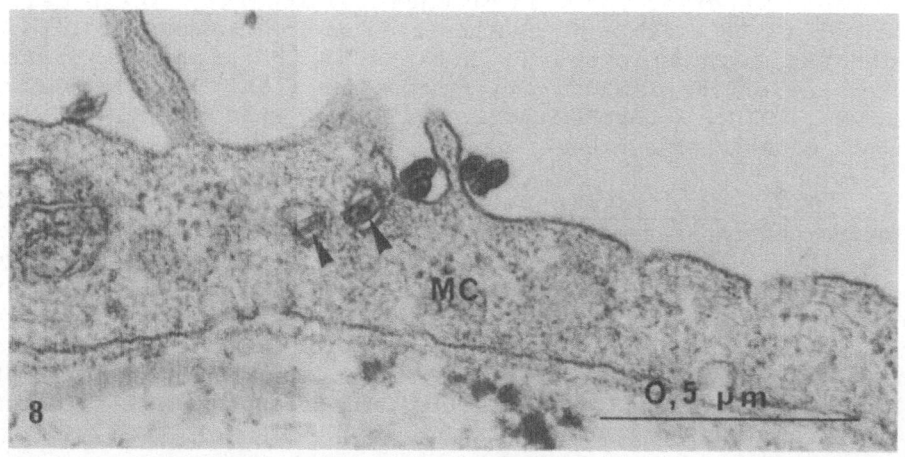

Fig. 8 : Ingestion of fibres by mesothelial cells in situ, in vitro
(TEM).

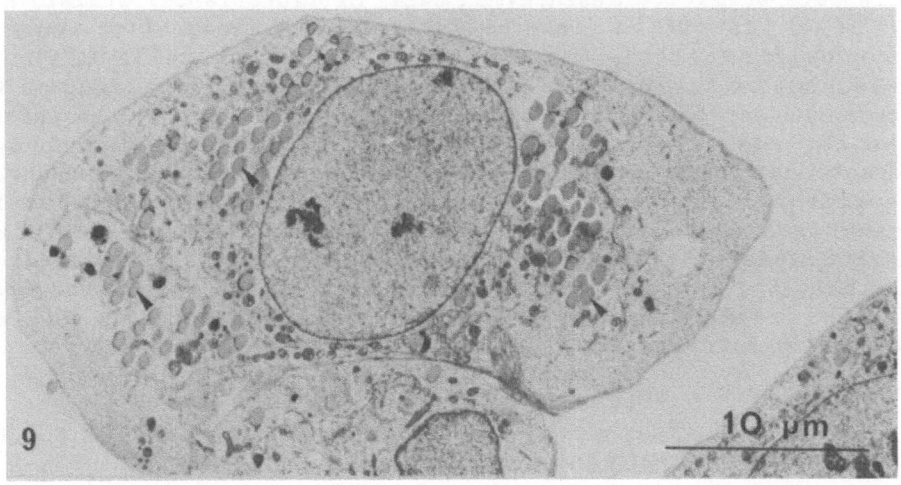

Fig. 9 : Ingestion of latex particles by mesothelial cells in
          culture (TEM).

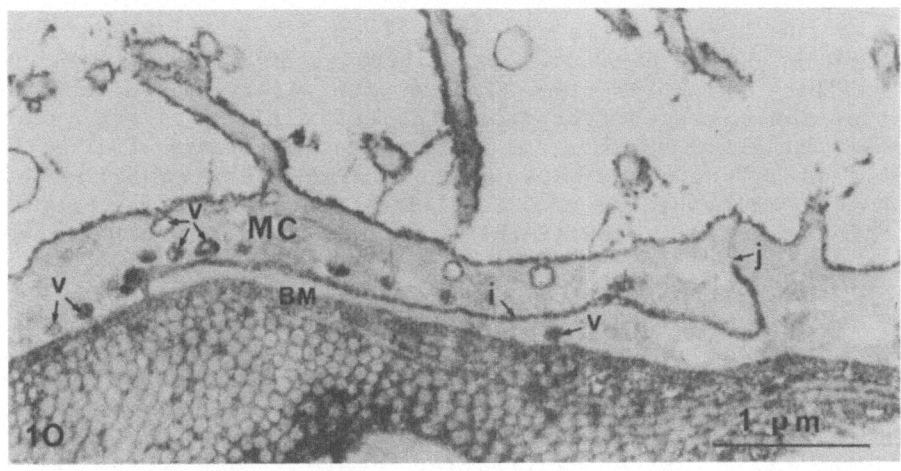

Fig. 10: Immunochemical localization of IgG in rat parietal pleura-
          intercellular space (i), tight junction (j), plasmalemmal
          vesicles (v), basement membrane (BM) (TEM).

The type of cells from which the new mesothelium arises is still under discussion. raftery (22) was convinced that these cells were primitive mesenchymal cells or fibroblasts; whilst, Ryan et al (24) identified these cells as macrophages, or at least as "mononuclear cells", in the sense that such cells might be precursors of several different kinds of cells, depending on the circumstances; however, the experiments carried out by the authors were not adequate to clarify this point. Indeed, macrophages present during the healing process can be replaced by mesenchymal subperitoneal connective tissue cells or by mesothelial cells coming from surrounding cells during the peritoneum regeneration. Nevertheless, the data concerning the phagocytotic function of mesothelial cells and submesothelial fibroblasts raised the basic question regarding the relationship of mesothelial and and submesothelial cells with the mononuclear phagocytic system.

II. STRUCTURE AND FUNCTION OF THE MESOTHELIAL BARRIER : FLUID CIRCULATION AND PARTICLES CLEARANCE

1. STRUCTURAL BASIS

1.1. The ultrastructure of the pleura has been extensively studied, by means of transmission and scanning electron microscopy (TEM and SEM) prticularly in rat, rabbit and man (1, 2, 25).

At the level of the visceral pleura, beneath a continuous basement membrane 50-60 nm thick, there is an elastic lamina 0.3-0.7 um thick, which is discontinuous with widely spaced fenestrae. Most of the collagen fibres and interstitial cells are located between the elastic lamina and the basement membranes of alveoli or capillaries, the space between the two basement membrane being reduced sometimes to 0.5 um. Subpleural lymphatics are rarely seen in this subpleural area of the lung. So far no stoma have been found.

At the level of the parietal pleura, beneath the mesothelial layer are found : a) a continuous basement membrane; b) collagen fibres; c) a discontinuous elastic lamina; d) thick and densely packed collagen fibres or bundles; e) a second discontinuous elastic lamina separating the submesothelial space from the muscular layer. Blood capillaries located close to the muscles have the structure of skeletal muscular capillaries. Submesothelial lymphatic channels are seen close to the mesothelial cell layer. Their lumen, frequently dilated were connected with the pleural cavity by the stomata described by Wang (26), which seemed to be closed by a circular expansion of the adjacent lymphatic endothelial cells, forming a valve. Red blood cells, fibrin clots and carbon particles have been seen in these stomata probably progressing through them from the pleural cavity to the lymphatic channels (1, 29). The diaphragmatic pleura has nearly the same structure. In the thinner fibrous central area of the diaphragm, the peritoneal and pleural mesothelial cells are separated only by a little connective tissue and by wide lymphatic channels (1).

1.2. In our laboratory, an immunochemical electron microscopic localization of two serum proteins (albumin, IgG) has been carried out in the rat pleura (27). Albumin and IgG were visualized in the pleural cavity. The pleural film, when well preserved by vascular fixation of the closed thorax, was found to be 5 um thick. The microvilli of MC appeared embedded within the serum proteins. The intercellular spaced between MC were labelled on all their length, except for the tight junctions, but some spaces possessing no tight junction were labelled over all their length, suggesting that they might be a possible transfer pathway for macromolecules across the mesothelium. By contrast, positive plasmalemmal vesicles were mostly observed at the edges of MC, suggesting more an absorptive endocytic phenomena than a pinocytotic transfer one. Basement membranes and submesothelial interstitial spaces were also labelled (Fig. 10). Parietal lymphatics were heavily labelled suggesting that they are a very likely reabsorption route of the pleural film components.

2. PHYSIOLOGICAL HEMODYNAMIC DEDUCTIONS

2.1. Movement of fluid and macromolecules across the pleura. The blood supply to the visceral pleura is only provided by the pulmonary capillaries in some species (rat), but in other species and particularly in man, the vascularization comes mainly from the bronchial arteries (28). By contrast, the patietal pleura is vascularized only by striated muscle capillaries. So, milky spots observed at the surface of the parietal pleura correspond in some species to special endings of intercostal arteries, where foreign bodies circulating in the blood can be trapped (29).

Thus, on the basis of physiological studies and of the Starling equation (30), it has been proposed that the pleural fluid should transudate from the parietal pleura where the systemic vascular pressure is high and be reabsorbed on the one hand by the visceral pleura where the hydrostatic pressure is low and on the other hand by the lymphatic channels through the Wang stomata. Indeed, the more numerous concentration of microvilli and pinocytic vesicles on the visceral pleura is possibly related to its greatrer fluid absorption (2). In hemodynamic pleural effusion, the fluid accumulates because of the increased post capillary pressure. In inflammatory exudates, the liquid comes from an increase of the mesothelial monolayer's permeability in relation to injury, which leads to the desquamation of dead cells.

      2.2. <u>Lymphatic clearance of particles</u>. The structural pattern and location of the Wang stomata (26) suggest that the lymphatic channels of the parietal wall and of the diaphragm work as suction pumps drawing fluid and particles from the pleura space to the lymph nodes of the parietal pleura and perhaps also of the abdominal space (1, 2); but at present there is no data documenting this route for biological or mineral particulate matter.

III. INFLAMMATORY RESPONSES OF THE PLEURA

1. HUMAN DATA

    Acute of subacute pleurisy is a common disease, which frequently consolidates by pleural fibrous thickening, whatever the causative factors. Two kinds of pleural fibrosis can be described: on the one hand, a diffuse fibrous thickening of the pleura, and on the other, a focal fibrosis with the characteristic appearance of pleural plaques, which can secondarily be calcified. Pleural diffuse thickening, often associated with obliteration of the pleura cavity, is usually observed as a scar after various inflammatory diseases (tuberculosis, microbial or viral infections), it seems that pleural plaques are mostly associated with past exposure to asbestos or to other mineral fibres (31, 32). There are still several unanswered questions concerning 1) the nature of the initial inflammatory reaction leading to pleural fibrosis 2) the mechanisms responsible for the formation of pleural plaques 3) relationship between pleural plaques and cancer, etc...

2. ANIMAL DATA

    A <u>great</u> <u>number</u> <u>of</u> <u>substances</u> have been used in order to create experimental pleural effusion : therebenthin, lugol, carboxymethylcellulose, formol, $AgNO_3$, dextran, dimethylsulfoxide, carrageenan, lactic acid, turpentine oil, etc... (33). These experiments carried out in different species (rat, rabbit) allowed us to explore the initial inflammatory reaction of the pleural cavity after different injuries, but none were designed to understand pleural fibrogenesis. The fibrogenic effects of <u>asbestos</u> <u>fibres</u> have been investigated in a number of studies in which dusts were administered into animals, either by the airways (inhalation or intratracheal instillation) (34, 35) or by intrapleural injection (36, 37, 38). Davis (36), after intrapleural injection of various dusts into the pleural cavity of mice, found that the production of collagen was related to the intensity of the granuloma, the final degree of pleural fibrosis being closely related to the initial cellularity of the lesions. Recent studies (34, 35, 37) support the contention that longer fibres are more fibrogenic than shorter ones, the long one inducing a marked granulomatous reaction. Moreover, it has been shown that, among asbestos, chrysotile is the most fibrogenic not only for lung parenchyma (34) but also for the pleura (38).

Two alternative hypothesis can be discussed to explain the formation of parietal and diaphragmatic pleural plaques. The first is that fibres, coming from the pleural cavity or from the parietal lymphatic channels by retrogression (39) may contribute to obliteration of the stomata, with subsequently cellular necrosis, fibrosis and calcification by focal calcium deposits (40). The same relation of stomata and plaques to the parietal pleura and to the diaphragm gives a clue to this suggestion. The other explanation is that fibres are transported to the parietal pleura by the systemic intercostal blood circulation down stream to the milky spots where they are entrapped : such embolization might lead to focal necrosis and fibrosis, as suggested by experimental data (29). Such milky spots have also been observed in man, at autopsy or by pleuroscopy, but, so far, measurement of fibres by TEM failed to reveal any asbestos accumulation inside the milky spots or pleural plaques sampled by thoracoscopy. Thus, the pathogenesis of pleural plaques is still obscure, as is the relationship between pleural fibrosis and early mesothelioma (41).

IV. MESOTHELIOMA

1. PRIMARY PLEURAL TUMOURS

Primary pleural tumours have been described as occurring in man, rarely as localized and benign, usually as diffuse and malignant. The term mesothelioma is applied to cell proliferation of mesoblastic origin which lines the coelomic cavities, including the pleura (17). Recently, Donna and Betta (42) proposed the term of mesodermoma to define neoplasms arising from undifferentiated and multpotential mesoderm which, in neoplastic disease, exhibits a wide range of differential activity and gives rise to tumours with prevaling myoblastic, angioblastic, lymphoblastic, chondroblastic, osteoblastic, fibroblastic or epithelial-like features. These two last patterns, the most frequent, correspond to the so called mesothelioma. Microscopically, the common feature of mesothelioma shows three main patterns : epithelial, fibroblastic and mixed, where both epithelial and sarcomatous elements are present and intermingled. This association of the two cell components has been described many years ago. In 1927, Maximow (43) culturing mesothelial tumours showed that mesothelial cells could develop into fibroblasts, demonstrating the interchangeable nature of the two type of cells. However, in our experimental work using normal tissue culture of mesothelial cells and of submesothelial fibroblasts we never observed a transition from one type to the other, even after 30 passages. But, it is possible that during malignant transformation, mesothelial cells can return to their mesoblastic origin and take on a fibroblastic appearance.

2. PLEURAL CARCINOGENESIS OF THE PLEURA

Since the initial observation of Wagner et al (44) in 1960, numerous epidemiological studies have shown that occupational exposure to asbestos dusts is associated with an enhanced incidence (7 to 10%) of pleural or peritoneal mesothelioma (45, 46) and that about 70 to 80 percent of registered mesothelioma cases are associated with past asbestos exposure (47, 48).

Mesotheliomas have been produced in animals by inhalation (49), by intrapleural implantation (50) or injection of mineral fibres (38, 51). All these data give primacy to asbestos dusts as the main aetiological factor for mesothelioma, although other types of fibres (zeolite, glass fibres) can also give mesothelioma in animals as well as in humans (52).

The present state of the art concerning cancerogenesis of the pleura raises the following questions:

a.   Is there any evidence that amphibole type fibres (crocidolite and amosite) are more associated with mesothelioma than chrysotile? This hypothesis was recently emphasized in the editorial of Lidell (54). Such a belief must be criticized because it is well known that chrysotile fibres which penetrate less easily into the deep lung are found in the parenchyma at a lower level than amphibole type fibres (41, 47). Moreover, Sebastien et al (47) found mostly short chrysotile microfibrils in the parietal pleura, which is the target tissue for mesothelioma, even when chrysotile fibres were found at a low concentration in lung parenchyma. Such an observation might have some relationship with the development of mesothelioma in man.

b.   How do mesothelial cells in culture react with fibres? Because of the controverys cited above, we have initiated a programme studying the behaviour of mesothelial cells, in culture, incubated with chrysotile fibres versus crocidolite fibres. So far, we have observed striking differences in the morphological changes of MC under the action of these two fibre types : the MC incubated with chrysotile showed a rapid and dramatic vacuolization, while those incubated with crocidolite (or quartz) showed nil or minimal changes. Although we do not yet know the exact biological significance of such vacuoles (which are not actually phagosomes) some understanding is provided by other experiments carried out on peritoneal macrophages : similar vacuolization was induced by concanavalin A (Con A) (55) and by the well known tumour promotor 12-o-tetradecanoyl phorbol 13 acetate (TPA) (56). Thus in our experiments, chrysotile fibres seemed to work on MC as chemicals (Con A, TPA) which activate macrophages and induce an inflammatory response. By contrast, crocidolite appeared much less reactive with MC in vitro.

Although these in vitro discripancies are apparent, we must recall that chrysotile and crocidolite gave always about the same proportion of tumours after an intrapleural injection in the rat (38, 51). The only two differences were that the initial pleural inflammation was more intense and the latency period before the appearance of the first mesothelioma shorter with chrysotile than with crocidolite (38).

As yet we have no biological explanation for these differencies in the inflammatory and carcinogenetic effects of chrysotile compared to crocidolite. However, this must be compared to the behaviour of acid leached fibres of chrysotile and crocidolite, in two in-vitro systems : haemolysis and release of enzymes by alveolar macrophages (57) : chrysotile was very reactive biologically, leading to the stimulation of alveolar macrophages, releasing only lysosomal enzymes, with no cytotoxic effect. However, crocidolite was cytotoxic and less activating for alveolar macrophages. After leaching asbestos cations by oxalic acid, the biological effects of fibres were reversed, chrysotile becoming cytotoxic (and non canerogenic) in rats (38) and crocidolite acting as an activating particle. It is possible that, although the starting events are strikingly different in vivo, some biodegradation of mineral fibres happens after some delay leading to a similar incidence of cancer during the whole life span of animals, the only difference being the latency period duration.

c.   Are other fibres able to produce mesothelioma? Among fibres, erionite-zeolite fibres appear actually probably more potent carcinogen for human and animal (52) than asbestos. Other fibres, such as clays (attapulgite) which are biologically active in vitro, must deserve a special caution in the near future because of their potential health hazards (58).

d.   Non fibrous carcinogens, such as radiations (59, 60) or chemicals (61, 62) can provoke pleural or peritoneal tumours in animals, but in these experiments, the carcinogens were given in high doses in close contact to mesothelial cells. In human, it is probable that 20 to 30 percent of mesothelioma cases are not related to past exposure to mineral fibres (48). But so far, the aetiological conditions are rarely found (63) and the causative factors remain unclear.

If we admit that asbestos fibres work as weak carcinogens but as strong promotors, activating mesothelial cells in a two stage carcinogenesis model (64), the association of a strong mutagenic initiation such as radiations or chemicals, with a powerful inflammatory promotor, such as chrysotile, should enhance the

tumour induction.   This has been recently shown in animals where
radiation were associated with asbestos fibres (60, 65).   In human,
the multiplicative synergistic effect of asbestos and tobacco smoke
is now well documented, concerning the risk of lung cancer (66) ;
regarding the risk of mesothelioma, there is at present no human
data, but such an hypothesis deserves further study.

REFERENCES

1. M.C. Pinchon, J.F. Bernaudin and J. Bignon, Biol. Cell., 37 : 269 (1980).
2. N.S. Wang, Amer. Rev. Resp. Dis. 110 : 623 (1974).
3. J.L. Stauffer, D.E. Potts and S.A. Sahn, Acta Cytol. 22 : 570 (1978).
4. G. Miserocchi and E. Agostini, J. Appl. Physiol. 30 ; 208 (1971).
5. S. Yamada, Z. Ges. Exptl. Med., 90 : 342 (1933).
6. C.W. Castor and B. Naylor, Lab. Invest. 20 : 437 (1969).
7. W. Domagala and L.G. Koss, Wirch. Archiv., 30 : 231 (1979).
8. J.Y. Chu and H.S. Lin, J. Reticuloendothel. Soc., 20 : 299 (1976).
9. J. Thiollet, M.C. Jaurand, H. Kaplan, J. Bignon and E. Hollande, Biomedicine 29 : 69 (1978).
10. M.C. Jaurand, J.F. Bernaudin, A. Renier, H. Kaplan and J. Bignon, In Vitro 17 : 98 (1981).
11. D. Whitaker, J.M. Papadimitriou and M.N.I. Walters, J. Pathol., 132 : 273 (1980).
12. J.R. Casley-Smith, J. Microsc., 90 : 251 (1969).
13. D.L. Odor, Am. J. Anat., 95 : 433 (1954).
14. H. Fukata, Acta Pathol. Japon, 15 : 309 (1963).
15. M.C. Jaurand, H. Kaplan, J. Thiollet, M.C. Pinchon, J.F. Bernaudin and J. Bignon, Am. J. Pathol., 94 : 529 (1979).
16. E.L.A. Svoboda and D.A. Deporter, J. Ultrastr. Res., 72 : 169 (1980).
17. H. Spencer, Pathology of the lung, vol 2, Pergamon Press, (1977), p. 918.
18. S. Rennard, M.C. Jaurand, J. Bignon, L. Saltzman, O. Kwanami, L. Stier, J. Davidson, V. Ferrans and R. Crystal, Amer. Rev. Resp. Dis., 123 : 224 (1981).
19. F.D. Bertalanffy, Int. Rev. Cytol., 17 : 213 (1964).
20. J.F. Aronson, L.W. Johns and G.G. Pietra, Lab. Invest., 34 : 529 (1976).
21. J.B. Bridges and H.W. Whitting, J. Pathol. Bacteriol., 87 : 123 (1964).
22. A.T. Raftery, J. anat., 115 : 375 (1973).
23. W.B. Watters and R.C. Buck, Lab. Invest., 26 : 604 (1972).
24. G.B. Ryan, J. Grobety and G. Majno, Am. J. Pathol., 71 : 93, (1973).
25. M. Legrand, R. Pariente, J. Andre, J. Chretien and G. Brouet, Presse Med., 79 : 2515 (1971).
26. N.D. Wang, Amer. Rev. Resp. Dis., 111 : 12 (1975).
27. M.C. Pinchon, J.F. Bernaudin, C. Sapin and G. Bignon, Biol. Cell., in the press.
28. A. Hirsh, J.F. Bernaudin, M. Nebut and P. Soler, Bull. Europ. Physiopathol. Resp., 12 : 387 (1976).
29. K. Kanazawa, F.J.C. Roe and J. Yamamoto, Int J. Cancer 23 : 858 (1979).

30. E.Agostini, The pulmonary circulatiuon and interstitial space.
     A.P. Fishman and H.H. Hechtids eds, University of Chicago
     Press (1969), pp. 65-77.
31. L. Meurman, Acta Pathol. Microbiol. Scand., 181 : 1 (1966).
32. E.N. Sargent, J. Gordonson, G. Jacobson, W. Birnbaum and M.
     Shaub, Am. J. Roetgenol, 131 : 579 (1978).
33. A. Hirsh, M. Nebut, P. Geslin, J.F. Bernaudin, P. Soler and
     J.J. Monsuez, Rev. Franc. Mal. Resp., 3 : 995 (1975).
34. J.M.G. Davis, S.T. Beckett, R.E. Bolton, P. Collings and A.P.
     Middleton, Br. J. Cancer 37 : 673 (1978).
35. G.W. Wright and M. Kuschner, Inhaled Particles IV, W.H. Walton
     ed., Pergamon Press, (1977), pp. 455-472.
36. J.M.G. Davis, Br. J. Exp. Pathol., 53 : 190 (1972).
37. M.F. Stanton and M. Layard, NBS Special Publication 506
     (1978), pp. 143-151.
38. G. Monchaux, J. Bignon, M.C. Jaurand, J. Lafuma, P. Sebastien,
     R. Masse, A. Hirsch and J. Goni, Carcinogenesis 2 : 229
     (1981).
39. E. Taskinen, K. Ahlman and M. Vilkeri, Chest 64 : 193 (1973).
40. J.M.G. Davis, Br. J. Exp. Pathol., 52 : 238 (1971).
41. P. Sebastien, X. Janson, G. Bonnaud, G. Riba, R. Masse and J.
     Bignon, Dusts on Disease, R. Lemen and J.M. Dement eds.
     Pathotox Publishers Inc (1979), pp. 65-85.
42. A. Donna and P.G. betta, Histopathology 5 : 31 (1981).
43. A. Maximow, Arch. Exp. Zellforsch, 4 : 1 (1927).
44. J.C. Wagner, C.A. Sleggs and P. Marchand, Br. J. Industr.
     Med., 17 : 260 (1960).
45. M.L. Newhouse and G. Berry, Br. J. Industr. Med., 33 : 147
     (1976).
46. I.J. Selikoff, E.C. Hammond and H. Seidman, Ann. N.Y. Acad.
     Sci., 330 : 91 (1979).
47. A.D. Mcdonald and J.C. Mcdonald, Cancer 46 : 1650 (1980).
48. J. Bignon, P. Sebastien, L. Di Menza, M. Nebut and H. Payan
     Rev. Franc. Mal. Resp., 7 : 223 (1979).
49. J.C. Wagner, G. Berry, J.W. Skidmore, V. Timbrell, Br. J.
     Cancer, 29 : 252 (1974).
50. M.F. Stanton, M. Layard, A. Tegeris, E. Miller, M. May and E.
     Kent, J. Natl. Cancer Inst., 58 : 587 (1977).
51. J.C. Wagner, G. berry and V. Timbrell, Br. J. Cancer 28 : 173
     (1973).
52. M. Artivinli and I. Baris, JNCI 63 : 17 (1979).
53. A.D. McDonald and J.C. McDonald, Environ. Res., 17 : 340
     (1978).
54. D. Liddell, Thorax, 36 : 241 (1981).
55. P.J. Edelson and Z.A. Cohn, J. Exp. Cell. Med., 140 : 1364
     (1974).
56. K. Brune, H. Kalin, R. Schmidt and E. Hecker, Adv. Inflam.
     Res., vol 1, g. Weissman ed., Raven Press (1979), pp. 467-
     475.
57. M.C. Jaurand, L. Magne, J. Bignon and J. Goni, Biological

effects of mineral fibres, vol 1, J.C. Wagner ed., IARC
Scientific Publication, 30 : 441 (1980).

58. J. Bignon, P. Sebastien, A. Gaudichet and M.C. Jaurand,
     Biological effects of mineral fibres, vol 1, J.C. Wagner
     ed., IARC Scientific Publication 30 : 163 (1980).

59. C.L. Sanders and T.A. Jackson, Health Physics, 22 : 755
     (1972).

60. C.L. Sanders, Radionuclide carcinogenesis. C.L. Sanders ed.,
     (1973), pp. 138-153.

61. A. Pelfrene and H. Garcia, Tumori 61 : 509 (1975).

62. K. Terao, Gann, 69 : 237 (1978).

63. J. Chretien, J. Delobel, J. Andre, J. Bignon, J. Modai and G.
     Brouet, J. Franc. Med. Chir. Thorac., 22 : 383 (1968).

64. D.C. Topping and P. Nettesheim, JNCI, 65 : 627 (1980).

65. J. Lafuma, M. Morin, J.L. Poncy, R. Masse, A. Hirsch, J.
     Bignon, and G. Monchaux, Biological effects of mineral
     fibres, vol 1, J.C. Wagner ed., IARC Scientific
     Publication, 30 : 311 (1980).

66. E.C. Hammond, I.J. Selikoff, H. Seidman, Ann. N.Y. Acad. Sci.,
     330 : 473 (1979).

DISCUSSION

LECTURER: Bignon   CHAIRMAN: Cumming

CHAIRMAN: The paper is open for discussion.

CORRIN: Bignon told us, correctly of course, that whatever
                 fibre is ejected into the pleural cavity, it will
                 result in mesothelioma and I am quoting the work
                 suggesting that by inhalation crocidolite is more
                 carcinogenic than chyrsotile.   I was very
                 interested in what I thought I heard him saying
                 about the analysis of type of fibres to be found in
                 the costal and diaphragamatic pleura which I
                 thought he said was chrysotile and I can't really
                 put these all together.   It would be easy to
                 understand that crocidolite when inhaled could
                 reach the pleura more easily than chrysotile which
                 would explain its greater carcinogenicity but it
                 seems to imply that chrysotile is the one which
                 drifts towards the pleura.

BIGNON: In the experiments carried out by Wagner chrysotile
                 and crocidolite by inhalation gave mesothelioma at
                 about the same frequency; in experimental animals
                 I am not sure that the classification I have given
                 is exact.   If you induce a mesothelioma by
                 intraparietal injection of type I fibres, you
                 observe the same amount of fibres of these two
                 types. The in vitro action of chrysotile and
                 crocidolite on the alveolar macrophage differs.
                 Chrysotile activates the macrophage releasing
                 lysozyme but without cytotoxicity.   Crocidolite
                 releases lysozyme with cytotoxicity. In animals
                 when you compare crocidolite and chrysotile one
                 sees the following: the animal injected intra
                 pleurally with chrysotile presents a very active
                 inflammatory reaction of the pleura and after about
                 18 months they develop a mesothelioma. By contrast
                 the inflammatory reaction with crocidolite is less
                 pronounced, but the time before the development of
                 mesothelioma is much longer.   It is possible that
                 fibres are modified in the biological system,
                 chrysotile being inactivated and crocidolite being
                 activated.   This is my interpretation.

TAYLOR: When you analysed the fibres in the pleura related
                 to inhalation studies either experimentally or

naturally did you find both chrysotile and crocidolite or did one or the other reach the pleura more quickly or in greater amounts than the other?

BIGNON: I have no data on this.

CORRIN: Jean, could I go to the beginning of your talk where you showed the microvillous surface of the pleural cells; I am interested by the practical problems of two microvillous surfaces rubbing over each other day after day. Could you tell us: how deep do you think is the layer of pleural fluid in comparison with the depth of microvilli and what are the microvilli doing? The depth of the pleural fluid is important because this is not the design of cells one would have chosen to have them sliding across each other.

BIGNON: It is very difficult to measure. We tried to measure it by freezing rats in liquid nitrogen, but when you cut the thorax by a saw there is a dissociation of pleura and lung so it is difficult to make measurement, but it could be about 1 mm. - it is difficult to say. There are some publications but the technology is difficult.

CORRIN: I wonder if I can ask about the cytology of normal cells in the pleural fluid. It's very difficult to obtain pleural fluid from the normal, so that one relies on pathological exudates and transudates. You showed at the beginning that most cells are not mesothelial cells but macrophages or monocytes. Many cytologists in England at least, refer to such mononuclear cells as mesothelial cells and my own electron microscopic studies suggest that this is erroneous and that they are really macrophages as you suggest. The question relates to some work published perhaps six years ago from Scandinavia where the workers scarified the mesothelium surface and then examined this by transmission electron microscopy and observed macrophages which differentiated into mesothelial cells and this was how a mesothelium was regenerated rather than by neighbouring mesothelial cells growing over to cover it. I wonder what your views are on that suggestion.

BIGNON: I am confused on the phagocytic properties of cells
               in vitro. I have read papers concerning the
               regeneration of cells and they are contradictory,
               some saying that the cell could be a macrophage.
               In the American Journal of Pathology it is
               suggested that we see a monocyte on a basement
               membrane of peritoneum. That is transforming
               mesothelial cells, but it is difficult solely from
               morphology to be sure that it is the same cell.
               Perhaps the regeneration comes from fibroblasts or
               mesoderm-like cells, but again it is difficult on
               the basis of morphology to say that it is the same
               series. The most probable origin of these cells in
               the wall of peritonium is from surrounding
               mesothelial cells. These cells maintain their
               morphology during 30 passages, and then after
               liquid nitrogen conservation we still have the same
               shape.

LAURENT: You added the mesothelial cells to the growing
               number of cells known to be capable of synthetising
               collagen. Do you have any evidence that it might
               be the mesothelial cells that synthesize pleural
               collagen?

BIGNON: We have no basis of comparison of the synthesis of
               collagen by mesothelial cells or sub-mesothelial
               fibroblast. Mesothelial cells can synthesize type
               I and type 4 but not type 3 and perhaps this could
               be a specific marker to distinguish the mesothelial
               cells from sub-mesothelial cells for instance in
               tumours. By contrast Madame Lafuma who checked the
               synthesis of collagen by mesothelial cells and
               sub-mesothelial cells found no difference between
               these cells and fibroblasts synthesizing collagen.

CORRIN: You showed some data suggesting that asbestos would
               be active as a carcinogen and other data that it
               could be acting as a promoter. Would you like to
               speculate as to why you feel it is not acting as a
               carcinogen but as a co-carcinogen?

BIGNON: Asbestos fibres and particularly chrysotile can
               absorb small molecules like carcinogens and
               transfer them into the cells. When we studied the
               interaction of these fibres with cells in vitro we
               discovered that they work mostly in activating the
               cells. I think it is important to differentiate in
               vitro the action of chrysotile from that of

crocidolite; this might explain the discrepancy between the epidemiological data and the animal data concerning the carcinogenic effects of chrysotile and crocidolite at the level of the pleura.

FABBRI: I would like to know whether for early detection of mesothelioma you prefer percutaneous biopsies or tissue samples taken in the course of an exploratory thoracotomy.

BIGNON: At present we are developing thoracoscopy. We mostly use thoracoscopy and take several biopsies; I think that it is better than doing open lung biopsy. It has been said that open biopsy can accelerate the evolution of mesothelioma.

CHAIRMAN: Now it seems it is time to have a cup of coffee. We start again at 11.30.

ACTIVATED MONONUCLEAR CELLS

W. Jones Williams

Welsh National School of Medicine
Pathology Department
Llandough Hospital
Penarth, S. Glam. CF6 1XX

The bone marrow derived circulating monocytes, together with the activated (effector) tissue cell-the macrophage, constitutes the Mononuclear Phagocytic System, (van Furth et al, 1972). Recognition of the MPS has considerably increased our knowledge of the processes of chronic inflammation and understanding of the body's defence mechanisms.

Recent experimental work (van der Rhee, et al, 1979) confirms previous results and proves that macrophages at the site of chronic inflammation are derived from bone marrow monocytes and their precursors.

The stimulus to activation includes all the innumerable causes and mechanisms of chronic inflammation (Allison, 1978). Activation may result from phagocytosis of a vast variety of substances ranging from products of haemoglobin breakdown to infective agents, inert and toxic minerals. Monocytes may also be activated under the direct stimulus of an antigen or immune complex antigen/antibody/complement. The latter is dependant on the production of antibodies from antigen stimulated B lymphocytes with the participation of lymphokines from T lymophocytes.

Coupled with the fact that macrophages may occasionally be secretory cells (vide infra) they are not always identifiable by the morphological finding of recognisable phagocytosed material. A number of criteria, including "in vivo" functional markers, are therefore required (Table 1), though not all may be demonstrable in a given cell at a given time (van Furth, 1980a).

Table 1. Outline for the identification of mononuclear phagocytes[*]

Structure	features in light, phase-contrast, and electron microscopy.
Cytoplasmic and lysosomal enzymes	non-specific esterase, lysosyme, peroxidase
Ectoenzymes	5' -nucleotidase, leucine aminopeptidase, alkaline phosphodiesterase I
Membrane characteristics	Fc and C receptors specific antigens
Function	Immune phagocytosis, pinocytosis
Cell proliferation	^{3}H-thymidine incorporation, DNA content of nucleus, survival or multiplication in culture
[*] R. van Furth	Mononuclear Phagocytes. Martinus Nijhoff, The Hague, 1980.

<u>Morphology</u> The typical macrophage (Fig. 1) contains a single, often lobed, nucleus, with abundant cytoplasm and numerous processes. The cytoplasm shows variable organelles, increasing in number and complexity with the degree of cell activity. Microtubules, Golgi complexes and various membrane bound vesicles, including primary and secondary lysosomes and residual bodies are frequent. Mitochondria are numerous and both smooth and rough endoplasmic reticulum are present. The cell membrane is characteristically thrown into folds and processes which, together with both pinocytotic and phagocytic vesicles, indicate the essential mobile, phagocytic nature of the cells.

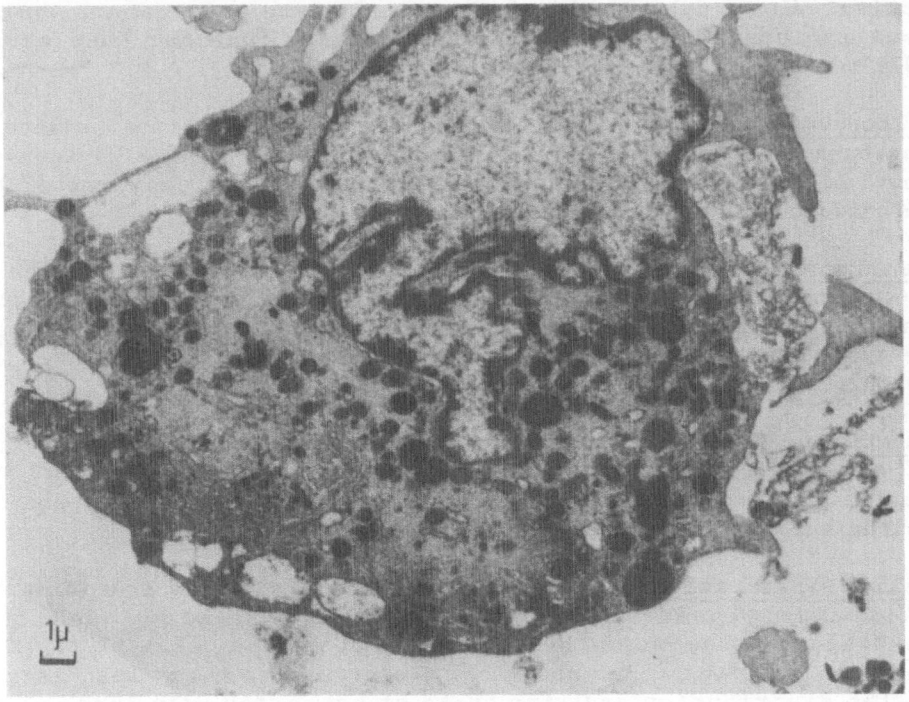

Fig. 1.   Pulmonary Macrophage showing numerous phagosomes and cell
           processes.

Cytoplasmic and lysosomal enzymes constitute the most important
hallmark of the MPS.   Non specific esterases are diffusely
distributed throughout the cytoplasm, distinguishable from the
scanty granular form found in thymic (T) lymphocytes.   Membrane
bound lysosomal enzymes, e.g., acid phosphatases, are of particular
value in identification.

        Lysozyme, detectable by immune fluorescent antibody
techniques, is of importance – particularly since in a variety of
inflammatory diseases its production leads to raised serum levels
(vide infra).

        The demonstration of peroxidases is of assistance in
distinguishing resident (inactive) tissue macrophages from the
activated monocyte.   Granular staining is a feature of the

activated macrophage, monocyte and monocyte precursours, but is usually absent in the resident macrophage (van der Rhee et al, 1979).

Ectoenzymes, such as 5 nucleotidase, are plasma membrane markers of macrophages. This activity is high in resident tissue macrophages and low in activated monocytes, and is thus the converse of the peroxidase activity (van der Rhee et al, 1979).

Membrane Characteristics   Immunological receptors for the Fc component of IgG and C3 are useful markers and present in the majority of mature MPS cells. Suprisingly (?) these markers are nearly always absent in alveolar macrophages (van Furth et al, 1980b).

Function   The demonstration of immune phagocytosis, e.g., uptake of opsonised bacteria or antibody coated (IgG/IgM) coated red cells, is also of value. Like endocytosis this property is not unique but is a prominent feature of MPS.

Cell Proliferation   The incorporation of tritiated thymidine has been widely used as a marker of cell division, and thus distinguishes between the immature and mature cells of the MPS. Combined with the other described properties much has been learned of the distribution and life cycle of monocytes (van Furth et al, 1980b). It has been shown that about 15% of monocytes leaving the circulation become alvelar macrophages which is interesting to compare with a figure of 56% which become Kupffer cells in the liver.

Surprisingly not so much is known of the fate of macrophages but it is likely that they end up in the "graveyard" of lymph nodes and those in the lung are also "excreted" into the bronchi.

ACTIVATED MONOCYTES/MACROPHAGES IN THE LUNG

Monocytes are normally found in the interstitium of the lung in the activated form. Following a suitable stimulus they enter the alveolar space and are recognisable as macrophages. Within alveoli, macrophages have to be distinguished from the flattened TypeI, epithelial cell and the Type II, surfactant secreting, granular pneumocyte. The Type II cell is distinguished from the macrophage by the presence of microvilli, excess of lamellar bodies and paucity of lysosomes. Though both Type I and Type II cells are capable of limited pinocytosis and phagocytosis this is not their primary role. The alveolar macrophage, perhaps due to its continuing exposure, is characterised by a very conspicuous content of a variety of dense bodies. These may be of endogenous or exogenous origin and the result of phagocytosis or end products of antiphagocytosis (Table 2).

Table 2.   Identifiable Material within Pulmonary Macrophages

Endogenous	Exogenous
Products of Hb breakdown Haemosiderin Lipofuscin	Mineral Particles Fibrous, Non Fibrous
Lipids	Lipids
Lamellar Bodies	Organic Compounds
Elastic	Micro-Organisms
Calcium	Other Organisms

Endogenous

Haemosiderin and non-iron containing lipofuscin, breakdown products of haemoglobin are among the commonest. The "heart failure" cell of chronic venous congestion is the classical example. In mitral stenosis and idiopathic (? immune complex) pulmonary haemosiderosis, the iron deposits produce micronodular radiographic opacities and are associated with pulmonary fibrosis.

Lipids of various forms are also common. "Lipid pneumonia" is a common sequel of bronchial and bronchiolar obstruction when both droplet and crystalline fats, including cholesterol, are present. Lipid, myelin lamellar bodies (surfactant) originating from the Type II cells are often phagocytosed by macrophages.

Elastic tissue, usually encrusted with calcium and iron, elicits a strong macrophage and often foreign body giant cell reaction. It is seen in a variety of destructive lung diseases and is especially prominent in the rare veno-occlusive disease.

Calcium is frequently present in lung macrophages and is often associated with iron and complexed muco-lipoproteins. Calcispherules in the lung interstitium are common in a variety of fibrotic diseases and probably originate from accumulation of "residual bodies", end products of indigestible phagocytosed material. Lung calculi in mitral stenosis are well recognised and frequently take the form of the alveolar spaces. In the rare disease microlithiasis alveolaris pulmonum the whole lung is destroyed and becomes a solid mass of calcified tissue. In the end stage of microlithiasis, macrophages are however inconspicuous. Schaumann bodies, common in sarcoidosis and some other granulomatous diseases, are thought to develop within epithelioid cells (vide infra) and may again be end stage products of incomplete intracellular digestion (Jones Williams, 1960).

Exogenous

There are numerous examples of inorganic dusts capable of
activating pulmonary macrophages (Raymond Parkes, 1974). Some,
like coal, are relatively inert, and other, e.g., silica are highly
toxic. The extent and degree of resulting lung diseases are
related in particular to the fibrogenic activity of the dusts.
This may be a direct action and/or mediated by the release of
a variety of macrophage produced enzymes.

Coal/carbon particles are ubiquitous and present to some
degree in any chance histological lung section examined. Coal is
practically inert and causes lung damage more by bulk than by any
toxic or fibrotic action. The three main types of coal workers'
pneumoconiosis are (a) simple, (b) progressive massive fibrosis
(PMF) and (c) Caplan lesions. The simple variety consists of focal
collections of free and macrophage containing dust. PMF is a
massive collection of dust and though the cause is still not
completely agreed is likely to be a reaction to dust complicated by
tuberculosis. The Caplan lesion is coal dust in association with
rheumatoid disease. The macrophages ae activated or stimulated by
a combination of rheumatoid (immune complex) factors and denatured
collagen. In the active stage they "incidentally" phagocytose coal
dust. Eventually the coal dust becomes deposited in fibrous tissue
in a ring form as a marker of previous macrophage activity (Gough
et al, 1955).

Silica, as distinct from coal, is a potent activator of
monocyte recruitment into the lung. Silica is toxic to macrophages
which, though attracted to the site, are rapidly killed (Allison et
al, 1966). The persistence of the undegradable particle acts as a
constant stimulus leading to the hallmark of silicosis, the dense
whorled fibrous nodules.

The fibrous silicates, such as asbestos, are also potent
stimulators of macrophage activity. Asbestos fibres, dependant on
size, are phagocytosed into the cytoplasm, or if too long become
surrounded by macrophages with frequent giant cell formation. The
fibres rapidly become covered with iron/protein/calcium complexes
produce the characteristic beaded asbestos body. As distinct from
silica, the resulting fibrosis is diffuse and recognisable by the
persistence of asbestos fibres and bodies.

Beryllium metal and its salts differ from the above mineral
dusts in that an immune, hypersensitivity reaction plays an
essential role in pathogenesis. The resulting granulomas consist
of monocyte derived epithelioid cells and will be discussed (vide
infra) in relation to sarcoidosis.

Exogenous lipid accumulation in pulmonary macrophages is

exemplified by the 'lipid pneumonia' following inhalation of a variety of oily substances, including nasal sprays and even excessive intake of liquid paraffin. Macrophages, though laden with fats, are incapable of digesting them, die and are continually replaced. In some instances the oils are surrounded by macrophage giant cells producing an 'oleo-granuloma'.

Inhalation of organic (non mineral) dusts result in extrinsic allergic alveolitis. They are all characterised by a complex immunologically mediated monocyte, macrophage, epithelioid cell response. The prototype, Farmers' Lung Disease (Seal et al, 1968) results from inhalation of mouldy hay containing a variety of fungal antigens, predominantly spores of Micropolyspora faeni.

Micro-organisms, as stimulators of macrophage activity, are classically exemplified by Mycobacteria tuberculosis. Cocci and bacilli usually evoke a leucocyte response, though persistence of the organism associated with continuing tissue damage, may evoke a macrophage reaction. More complex organisms, including parasites such as actinomycetes, histoplasmosis, coccidiomyosis, etc., produce a macrophage response. It is likely that the macrophage producing organisms share the common features of undegradability and complex lipolpolysaccharide structure.

The recent development of broncho-alveolar lavage (BAL) is providing useful information on the cellular content of the bronchi in a variety of diseases. There are, however, a number of drawbacks:- (a) Many of the cells are effete and may be "on the way out"; (b) Sampling obviously does not include cells from the lung interstitium; (c) Sampling errors are frequent; (d) Cells are altered at the time of sampling and indeed may be "activated". In view of the above BAL is complementary to, but does not supercede, tissue examination.

Activated monocytes, epithelioid cells and granuloma formation

We have seen that many known agents produce epithelioid cell granulomas, including M.Tuberculosis, M.Faeni and beryllium. Sarcoidosis is a disease characterised by similar granulomas but of unknown cause. Let us examine the nature, source and possible role of epithelioid cells.

Epithelioid cells are mononuclear cells with pale vacuolated cytoplasm and complex interdigitating cell processes. On light microscopy and histochemistry they are morphologically indistinguishable in sarcoidosis, Kveim test granulomas, tuberculosis, chronic beryllium disease and extrinsic allergic alveolitis. (Williams, et al, 1969). On electron microscopy we confirmed the similarity and showed that the most prominent features are numerous membrane bound muco-glycoprotein containing

vesicles and associated Golgi complexes. (Fig. 2). Phagocytosis
and lysosomal bodies are normally inconspicuous. (James and Jones
Williams, 1974; Muller-Hermelink et al, 1980). However,
epithelioid cells are obviously capable of phagocytosis so that in
tuberculosis the above description applies to non bacterial
containing cells, which are surprisingly common. Similarly in
Kveim test granulomas the cell and tissue debris included in the
crude injected preparation evokes epithelioid cells rich in
lysosomes and phagosomes.

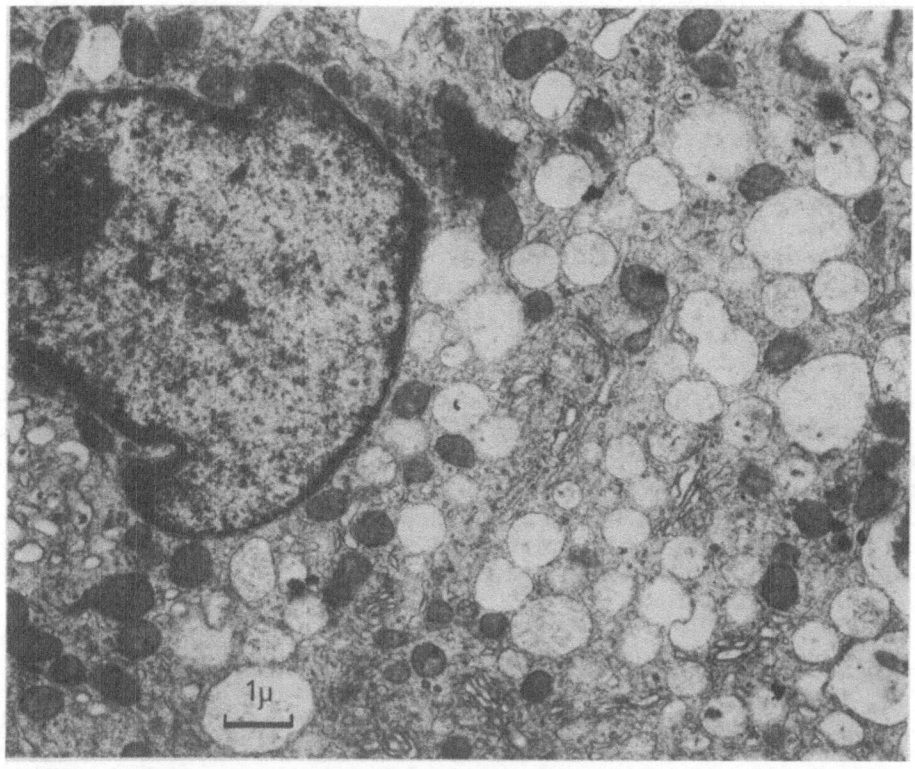

Fig. 2.   Epithelioid cell with numerous vesicles and active Golgi
          complexes.

Based on experimental evidence (Papadimitrou and Spencer,
1971; van Rhee et al, 1979) it is considered that epithelioid cells
are derived from macrophages of bone marrow monocyte origin. The
morphology of the experimentally produced epithelioid cells,
however, differ from those in human granulomas. They show many
features of an active macrophage and are rich in lysosomes,
residual bodies and evidence of phagocytosis and seldom show the
characteristic vesicles.

It is now agreed that epithelioid cells may be of two types, phagocytic or secretory. Intermediate forms exist and their function is dependant on the nature of the stimulus and the general immunological state of the host.

Macrophages, and monocyte derived cells have been shown both in experimental and human studies to secrete a variety of substances, (Davies et al, 1977). These include, T and B lymphocyte stimulators, prostaglandins, cytotoxins, lysozyme, collagenase, elastase and acid hydrolases. In addition it has been shown that epithelioid cells in sarcoidosis are the source of the raised serum angiotensin converting enzyme, SACE (Silverstein et al, 1980). Increased levels of serum lysozyme (Selroos et al, 1980) and collagenase have also been found in sarcoidosis. Though raised levels of these enzymes are not diagnostic of the disease they are of considerable value in following the course and response to treatment. We have shown that serum in sarcoidosis contains T lymphocyte depressant factors (Davies et al, 1980) and suggested that these also originate in epithelioid cells.

The other important cell in the above granulomas is the lymphocyte which is in close contact and admixed with epithelioid cells. Jones Williams (1977) has previously discussed the functional interdependance of the two cell types and postulated that secretory products of both may play an important role in perpetuating granulomas in sarcoidosis even after disappearance of the original stimulus. Both T and B lymphocytes are involved in the process.

Let us suppose that the causative agent of sarcoidosis is antigenic, either extrinsic or immune complex. (Fig. 3). The antigen stimulates committed T lymphocytes to produce lymphokines. These are known monocyte/macrophage activators, increase general cell activity, adhesive properties and hence aggregation, phagocytosis, and decrease migration. All these actions may be of importance in forming the closely packed focal granuloma.

The causative "antigen" will also stimulate B lymphocytes to produce immunoglobulins and macrophage conversion to epithelioid cells. Some epithelioid cell products have a "lymphokine"-like action and will secondarily stimulate T and B lymphocytes. The neutral proteases, phosphatases and lysozyme react on tissue protein which, combined with the increased locally produced immunoglobulins and complement, result in the formation of immune complexes.

In this way, products of the cellular response to the original stimulus may well play a part in the persistence of the granuloma.

This brief review of a vast subject outlines the importance of the activated monocyte and macrophage in chronic inflammation. As

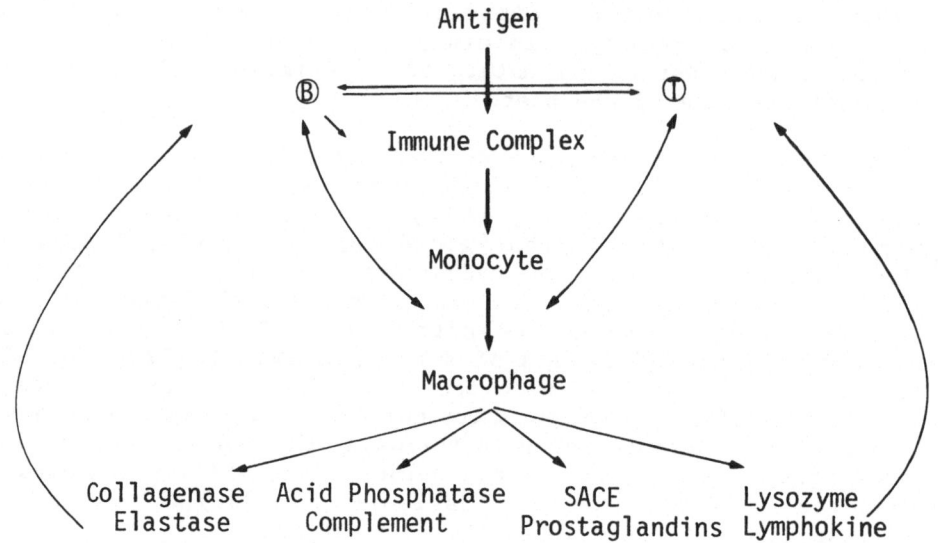

Fig. 3.   Cellular interaction (Perpetuum mobile) in granuloma
          formation.

so often in pathology, basic defence mechanisms may, in themselves,
lead to further tissue damage and perpetuation of disease.

REFERENCES

Allison, A.C., Harington, J.S., and Birbeck M., 1966, An
    examination of the cytotoxic effect of silica on macrophages,
    J. Expt. Med., 124:141.
Allison, A.C., 1978, Macophage activation and non-specific
    immunity, in: "Internal. Rev. Expt. Pathol"., vol.8., G.W.
    Richter and M.A. Epstein, eds., Academic Press, New York.
Davies, P., Bonney, R.J., Dahlgren, M.E., Pelus, L., Kuehl, F.A.,
    and Humes, J.L., 1977, Recent studies on the secretory
    activity of mononuclear phagocytes with special reference to
    prostaglandins, in: "Perspective in inflammation", D.A.
    Willicombe, J.P. Giroud and G.P. Velo, eds., MTP., London,
    p.179.
Davies, B.H., Jones, K., Evans, P., Jones Williams, W., and
    Williams, J.D., 1980, "Thymic lymphocyte responses in
    sarcoidosis: effect of sera and levamisole", in: "Proc. 8th.
    Internat. Conf. Sarcoidosis and other granulomatous diseases",
    W. Jones Williams and B.H. Davies, eds., Alpha and Omega
    Publishing Ltd., Cardif, Wales, p.477.
van Furth, R., Cohn, Z.A., Hirsch, J.G., Humphrey, J.H., Spector,
    W.G., and Langevoort, H.L., 1972, "The mononuclear phagocyte
    system. A new classification of macrophages, monocytes and
    their precursor cells", Bull. World Health, 46:845.

van Furth, R., 1980a, Cells of the mononuclear phagocyte system. Nomenclature in terms of sites and conditions, in: "Mononuclear Phagocyte," Part I., R. van Furth, ed., Martinus Nijhoff, The Hague, p.279.

Gough, J., Rivers, D., and Seal, R.M.E., 1955, Pathological studies of modified pneumoconiosis in coal miners with rheumatoid arthritis, (Caplan's syndrome), Thorax, 10:9.

James, E.M.V., and Jones Williams, W., 1974, Fine structure and enzyme histochemistry of epithelioid cells in sarcoidosis, Thorax, 29:115.

Jones Williams, W., 1960, The nature and origin of Schaumann bodies, J. Path. Bact., 79:193.

Jones Williams, W., 1977, Sarcoidosis, 1977, Beit. z. Pathol., 160:325.

Muller-Hermelink, H.K., Kamiyama, R., Kaiserling, E., and Lennert, K., 1980, Lymph node findings in sarcoidosis. Light microscopial, ultrastructural and enzyme histochemical results in: "Proc. Internat. Conf. Sarcoidosis and othe granulomatous diseases", W. Jones Williams and B.H. Davies, eds., Alpha and Omega Publishing Ltd., Cardiff, Wales, p.23.

Raymond Parkes, W., 1974, Occupational lung disorder, Buterworths, London.

Papadimitrou, J.M., and Spector, W.G., 1971, The origin properties and fate of epithelioid cells, J. Path., 105:187.

van der Rhee, H.J., van der Burgh de Winter, C.P.M., and Daems, W.Th., 1979. The differentiation of monocytes into macrophages, epithelioid cells and multinucleated giant cells in subcutaneous granulomas I, Fine Structure. Cell Tissue, Res., 197:355.

Seal, R.M.E., Hapke, E.J., Thomas, G.O., Meek, J.C., and Hayes, M., 1968. The pathology of the acute and chronic stages of farmer's lung, Thorax, 23:469.

Selroos, O., Tiiten, H., Gronhagen-Riska, G., Fyhrqvist, F., and Klockars, M., 1980, Angiotensin-converting enzyme and lysozyme in sarcoidosis, in: "Proc. 8th. Internat. Conf. Sarcoidosis and other granulomatous diseases", W. Jones Williams and B.H. Davies, eds., Alpha and Omega Publishing Ltd., Cardiff, Wales, p.309.

Silverstein, E., Friedland, J., and Pertschuk, L.P., 1980, Sarcoidosis pathogenesis: Mechanisms of angiotensin-converting enzyme elevation: Epithelioid cell localisation and induction in macrophages and monocytes in culture, in: "Proc. 8th. Internat. Conf. Sarcoidosis and other granulomatous diseases", W. Jones Williams and B.H. Davies, eds., Alpha and Omega Publishing Ltd., Cardiff, Wales, p.246.

Williams, D., Jones Williams, W., and Williams, J., 1969, Enzyme histochemistry of epithelioid cells in sarcoidosis and sarcoid-like granulomas, J. Path. 97:705.

DISCUSSION

LECTURER: Williams   CHAIRMAN: Bonsignore

ROUSSOUW: Dr. Williams, may I ask you what is the normal
                 evolution of monocytes into macrophages, into
                 epithelial cells and then giant cells. What is the
                 possibility of macrophages becoming giant cells
                 without epithelial cells intervening.

WILLIAMS: The experimental work has been done very carefully
                 and I have been able to look at some of the
                 material, particularly from Professor Spencer. He
                 was quite sure that in his experimental conditions
                 with a variety of different antigens and antibody
                 mixtures, that he was producing in the mouse
                 peritoneum activated macrophages which then became
                 epithelioid cells. I looked at this material when
                 he was first doing it because I wasn't very happy
                 about this interpretation, since what he was
                 calling epithelioid cells I would have called
                 macrophages. The fact that they fused to form
                 giant cells has been shown many times by a number
                 of different people, Sattin and Wise were also
                 doing it at the same time and there is no doubt
                 that the circulating mononuclear cell activated to
                 become a macrophage, is fused to become a giant
                 cell and not an epithelial cell.

ROUSSOUW: I saw that you mentioned lipotoxin under your
                 heading of immune globulin breakdown products. Is
                 that correct?

WILLIAMS: Lipotoxin is a product of tissue breakdown and is a
                 whole mixture of things, but one of the types of
                 lipotoxin does come from non-iron-containing
                 immunoglobulin. Lipotoxin is a fascinating
                 substance that I studied in some detail years ago
                 and then realised that it was a dead end job,
                 becuase it is composed of dead ends of all types of
                 different materials, so there are twenty
                 histochemical ways of demonstrating lipotoxicity,
                 it varies all the time depending on where the water
                 has come from.

SOLIMAN: In sarcoidosis, which is liable to change
                 pathologically to tuberculosis, how could this

pathological change affect the immune response, especially regarding the T cells and B cells.

WILLIAMS: Some people feel that all sarcoidosis is a type of tuberculosis. There is no doubt that there are cases, and I have some in my own collection, of patients with proven tuberculosis, which has been cured and have come back five years later with classical non-tuberculin reactive sarcoidosis. The fact that you get a positive tuberculin reaction does not completely exclude sarcoidosis. Most of these transfer cases are from sarcoid and in the end they get classical tuberculosis with a positive reaction. One explanation is that whatever the "causative immune complex" in sarcoidosis, it has covered every T lymphocyte at every part of its membrane, so that there is no room for any other antigen to get on to it to produce the positive skin test.

TAYLOR: Two questions, if I may. First of all you said that you used carbon black test as a marker for the enqiry test in the skin; if you do that how can you be sure that you haven't got a foreign body granuloma to the carbon rather than a positive carbon reaction?

WILLIAMS: By the fact that I have injected carbon many times into the skin and it produces a very indolent lesion with one or two macrophages with carbon in them and they won't produce a proper sarcoid-type granuloma which is the feature of a positive test.

TAYLOR: I think it is useful to know about. I think Kveim takes the view that it will produce a good-going granuloma as a reaction, and I am interested to hear you saying that it doesn't. Secondly berylliosis; we have a patient at the moment who has a very strong history of exposure to beryllium and he has granulomas only in the liver, because there are many causes of liver granulomas – the chest x-ray is clear. Have you ever seen a case of proven berylliosis in which the granulomas appear in the liver before they appear in the lungs?

WILLIAMS: Berylliosis is by definition a multisystem disease and though I worked with Professor Harlett Hardy years ago in the U.S., who was known as the beryllium queen, she used to say beryllium is not a hypersensitivity disease despite all the work being done by her American colleagues in other centres. I do not have a case of beryllium disease with granuloma only in the liver, so what I might suggest to you is that you might get tissue analysed for you or we can do the "ace" test (Angiotension Converting Enzyme) which will be negative in beryllium disease. This is another useful distinction with the "ace" test, most cases of beryllium disease haven't had an "ace" test and Nancy has done about 40 now and the "ace" is not raised. What I could do for you, if you would like to send me some fresh lymphocytes, is the beryllium-lymphocyte transformation test and that could be additional proof; then you should look at the urine to see if there is beryllium in the urine.

JANSEN: You mentioned the existence of immune complex deposits in the Kaplan nodules in the lung. Did you look for different immunoglobulin classes bound to the tissue in these immunofluorescence studies? Were they mostly IgM or IgG and secondly do you have any idea about the antigen enclosed in the complexes? Is it the same antigen as we see in rheumatoid arthritis? Is it denaturated IgG and, if so, how do you explain that we don't find any immunofluorescence on lung tissue in other lung involvement in rheumatoid disease?

WILLIAMS: This is not my own work, I was illustrating with someone elses material. They were demonstrating IgG.

CORRIN: I do not have a question, I merely wish particularly to criticise your terminology. Like all people interested in sarcoid, you use "ace" as an abbreviation of one of the two trivial names of an enzyme which is correctly termed dipeptidase. I am sure we will hear later in the week that this enzyme in addition to converting angiotensin by breaking away two terminal aminoacids, is also a very important kininase. Has the epithelioid cells something to do with the regulation of blood pressure? I think that it is much more likely that if epithelioid cell is secreting this type of peptidase it is because of its kinin splitting properties and you should not use trivial names but the correct biochemical name which is quite long but terminates with dipeptidase.

WILLIAMS: Thank you Brian, for that comment. I was using
             this term because it is now in the literature and
             is usually referred to as "ace". However I quite
             agree with you, it is a kininase.

KUHN: One of the questions which you have raised is how
             or perhaps I should say why the macrophage once in
             the air spaces gets back to the interstitium or
             into the lymph nodes. There have been a number of
             studies recently, perhaps the most elegant was that
             of Surropian with iron particles, that suggests
             that inert dust can be transported across the
             epithelium by phagocytic activity and hence could
             get into the interstitium and the lymphatics
             relatively easily as free particles and make their
             way into the interstitial macrophages. So my
             question is: is there any real evidence that once
             in the alveolar space a macrophage has any way of
             getting back into the interstitium or whether they
             are, once there, simply passively cleared up the
             tracheo-bronchial tree?

WILLIAMS: I don't have direct evidence relative to that. I
             imagine, as you do, that the majority of them will
             find their way in the interstitium via the bronchi.
             I have no actual evidence as to how they get back.

TUCCON: May I enlarge upon the question put by Dr. Kuhn? I
             have a pupil with the name of Y. Hanson and she has
             treated rabbits with iron-containing nickel dust
             and examined the macrophage in the lungs; then she
             has also fixed and examined the lymph nodes from
             the axilla of the rabbits. She had few cases where
             she found macrophages in the lymph nodes.

SIMSON: Is the sarcoid reaction that we see in response to
             some tumours an indication that the patient is
             different or that the tumour is different or that
             there is something happening to the immune response
             to that particular tumour?

WILLIAMS: A local sarcoid reaction in the tumour?

SIMSON: The sarcoid reaction that one sees in the area of
             influence in the presence of a peripheral tumour.

WILLIAMS: It is a common finding to get sarcoid reaction at
                the site of neoplastic tumours and I have looked at
                many of these and I don't know the explanation.
                The best examples have been in cases of carcinoma
                of the stomach. The usual answer given is that it
                has been caused by breakdown products of tumour
                degeneration, tumour necrosis, a mixture of broken
                down natural tissues, altered collagen and so on.
                So one imagines that it is some immunological
                immune complex triggering mechanism and I have no
                direct evidence of this. Professor Corrin is about
                to say something.

CORRIN: To return to the point raised by Dr. Kuhn whether
                alveolar macrophages have taken up dust at their
                return into the interstitium, one of the earliest
                altered structure study on the lung was that of
                Policard in Paris. He described this phenomenon -
                dust-laden cells silted up in the alveoli and there
                was an erosion of the epithelium underneath the
                alveolar epithelium so that dust-laden macrophages
                were in contact with the interstitium; then he
                described the epithelium re-growing over this as an
                endothelium does over a muralthrombus. That was 30
                years ago and this has never been confirmed. I
                wonder if any of our French colleagues has any
                confirmation of this.

BIGNON: It is difficult to obtain evidence in disease and
                on experimental animals with beryllium. How can
                one explain the difference between animal and mans
                response towards beryllium migration. There are
                some papers in the literature showing that no
                granulomatous disease results when cats inhale
                beryllium. How do you explain this difference
                between the results in animal and in man?

WILLIAMS: It is always the problem with experimental models.
                I have tried to produce beryllium disease myself by
                tracheal installation of a variety of beryllium
                cells into rats by direct implantation and I
                followed these for 18 months at monthly intervals.
                I also did a group of sensitised rats and then
                repeated the intratracheal injections and I
                produced no permanent granulomas at all. Before I
                did the work I had been working with Professor
                Skippers in Salt Lake City and I looked at his
                experimental beryllium lungs. He did produce some
                granulomas, so it could well be that it is a

species difference. It has also been tried in primates with more success, but I don't really know the explanation except to say it may be species difference.

ROSSI: Just to say that we have preliminary results in producing berylliosis in guinea-pigs. The work has been done by Dr. Morritz in Dr. Crystal's laboratory and he is able to produce what we think are granuloma either injecting beryllium salts in the trachea or under the skin and he produced granuloma in the lung, we don't yet know why.

WILLIAMS: I used guinea-pigs as a control model for my lymphocyte transformation experiments. I was able to sensitise the guinea-pigs with some difficulty and was able to show tht I had produced lymphocyte transformation in vitro. Obviously one needs a sensitised animal but it is the same problem again, that you can produce the local lesion but you need a "general systemic hypersensitivity".

CHAIRMAN: Thank you very much. Now have a good lunch and we start again at 4.30.

BRONCHIAL MUCUS

Peter Mitchell-Heggs

Consultant Physician Epsom
Hon. Sen. Lecturer Charing Cross Hospital
Medical School

Bronchial secretion or sputum?

Sputum, the material which is usually collected for examination of respiratory mucus, is a mixture of bronchial mucus from goblet cells, bronchial gland secretions and various transudates from the bronchial mucosa. In addition, there are bacteria, lysozymes, alveolar macrophages and lymphocytes together with cellular debris. Above the larynx sputum gathers its envelope of salivary and possibly nasal mucins. Sputum is therefore a mixture of mixtures.

The material which makes bronchial secretion different from other secretions is chiefly mucus and there are features of respiratory mucus which are useful and used in the respiratory tract.

Normal bronchial mucus is produced at rates which have been described as between 0.1 and 0.75 ml/kg body weight/24 hours - about 10 to 50 ml a day in the average adult human. However, animal studies suggest mammals produce nearer 100 ml of bronchial mucus per day. An even more recent calculation estimated the 24 hour volume of bronchial mucus as 355 ml. Whatever the volume of bronchial mucus secreted, it is certain that normal subjects do not cough up sputum. Bronchial secretion in normal subjects is moved to the larynx and swallowed.

Function of bronchial mucus

In the normal subjects bronchi there is a continuous barrier which protects the respiratory tract from noxious stimuli. This

barrier moves continually upwards towards the nasopharynx, and the
sticky and viscoelastic properties of the mucus, carry inhaled
particles towards the nasopharynx. Thus bronchial mucus acts as a
first line defence mechanism for the lung, and, there are
particular features of respiratory mucus which make it ideal for
its function as a barrier moved by ciliary action.

In disease, both the quality and quantity of the mucus
produced is altered. There is, moreover, a loss of efficiency of
the mucociliary escalator. Particles stay trapped in the lung,
bacteria and viruses are not cleared and may proliferate. The
number and quality of the secondary defence mechanisms - the
antibodies and phagocytes which are normally entwined in the
strands of mucus - are diminished. As the natural barrier is
removed, infection, irritation and inflammation of the respiratory
tract occur, further diminishing respiratory function and
efficiency.

Site of production of bronchial mucus

Mucus in the normal respiratory tract is formed in the small
submucosal glands and in the goblet cells of the bronchial mucosa
of the trachea and major airways. In the chronic bronchitic
subject there is a proliferation of mucus secreting tissue in
response to chronic irritation from tobacco smoke. Hyper-secretion
of mucus occurs in two ways; the bronchial submucosal glands
hypertrophy and grow down to the muscle layers of the airways, the
increase in volume decreasing the diameter of the lumen of the
airways, an increase in volume of secretion further decreasing the
size of the airway lumen. Furthermore, the site of mucous glands
alters. In subjects with chronic bronchitis mucus secreting
glands both submucosal glands and goblet cells, are present in
more distal airways. In mucus hypertrophy and diseases related to
it, not only is more mucus produced but more airways are narrowed
by the increased mass of mucus-secreting cells.

In disease the quality of mucus is altered and the number and
beating efficiency of the cilia is decreased, there is too much of
the wrong sort of mucus in the wrong places - the mucus 'moving
escalator' becomes a smothering blanket.

The control of mucus production

The rate of production of mucus is under parasympathetic
nervous control, and stimulation of the vagus leads to mucus
secretion in a number of mammalian species. Parasympathomimetic
agents increase the secretory rate of mucus and serous cells of the
human bronchial glands while atropine causes a reduction in
secretory rate. It appears that serous secretion, a mixture of
lower molecular weight mucins and other proteins is continuous

whereas mucous gland secretion of 'thicker' mucus patches is intermittent and under neurological control. Goblet cell secretion has not clearly been shown to be under any nervous control. There may be local nervous reflexes or hormonal factors related to the function of these glands; nerve fibres have been demonstrated in close association with these cells but there is no definite evidence of agonist activity or blockade. Prostaglandin inhibition by salicylate diminishes uptake of mucus precursors and thus the rate of their agglutination into mucins and their secretion as glycoproteins. Local stimuli appear to increase the volume of serous secretion but not other mucosal gland secretions.

Collection of bronchial mucus for analysis

The physical and chemical properties of bronchial mucus change rapidly after removal from the respiratory tract, experience has shown that for reproducible results fresh mucus samples must be examined.

Normal mucus is almost impossible to obtain. The nearest extrapolations from the normal are subjects with early mucus hyper-secretion or normal subjects induced to cough and produce sputum by inhalation of acetylcholine, prostaglanding $F2\alpha$ and histamine or even citric acid.

Samples for chemical (but not physical) analysis may be obtained by tracheobronchial lavage from bronchoscopy or tracheostomy.

Sputum produced immediately on waking should not be used for analysis. It has usually remained in the airways all night, is dehydrated and often physically partially autolysed. Morning samples are usually examined but it should be remembered that there is a diurnal variation in mucus content of glycoprotein-associated sulphate and neuraminic acid: materials responsible for the tensile strength of the tertiary structure of the glycoprotein molecule.

Mucus structure

Mucus consists of strands of glycoprotein molecules. A central protein core has many side chains like the brush segments on a laboratory bottle brush. The protein core is twisted and twirled by the process of its secretion and also by the physical effects of differing 'acid' and 'alkali' radicals as end pieces on the 'brush bristles'. These entangle and hold the various strands of glycoprotein as a springy ball of wool. The interstices are filled with electsolytes from serum exudate and transudate. Variation in pH of these pockets of electrolytes also determine the siting of the glycoprotein parts, the springiness of the molecule – its tertiary structure.

Chemical composition of bronchial mucus

In 'normal' mucus there is little plasma component as exudate.
In disease especially where there is infection, the plasma exudate
component rises. Aminosugars, sugars and sulphate linkages are
present in form of specific mucin only in plasma or mucus and their
site of origin may thus be determined.

Bronchial mucus contains little free water, whilst the bound
water accounts for 95% of mucus by weight. The remaining 5% of
macromolecular material is 2-3% glycoprotein, proteins (including
immunoproteins) 0.1-0.5% and fats and lipid-based materials 0.3-
0.5%. It is this macromolecular component which is responsible for
the physical properties of the material.

Physical properties of bronchial mucus

Bronchial secretion is a viscoelastic material with properties
of both a solid and a liquid. As a solid material, its shape may
be deformed by an applied force and the energy which is stored
determines the elasticity. As a liquid, bronchial mucus flows at a
rate proportional to any applied force. Viscosity is a function of
this flow rate; increased shearing or applied force causes a fall
in viscosity of bronchial mucus which is therefore a non-Newtonian
or pseudoplastic material.

Some materials such as car oil have a constant viscosity at
any applied shear rate - these materials are Newtonian. Other
materials change viscosity with shearing and are described as non-
Newtonian. An example of a material which increases in viscosity
on shearing (dilatant materials) include 'whipping cream';
materials which decrease in viscosity on shearing (pseudoplastic)
include thixotropic paint and many biological materials e.g.
synovial fluid, blood and respiratory mucus.

In order to measure the viscosity of bronchial mucus (or, as
this varies with shear rate in non-Newtonian materials, we should
correctly call it apparent viscosity; (η')) it is necessary to test
the material at a number of shear rates, assess the apparent
viscosity at these levels and outline the profile of the behaviour
of apparent viscosity over this range. Many studies have been
carried out, testing apparent viscosity after homogenisation of
bronchial mucus (which removes any Pathophysiological correlation)
after freezing, pooling, acidifying and other physicochemical
procedures which alter the physical nature of mucus; these studies
unfortunately have, in general, only passing interest, no
physiological relevance and have been at times dangerously
interpreted clinically.

More recently sophisticated cone and plate viscometers and rheogoniometers such as the Ferranti-Shirley and Weissenberg apparatus have been used to evaluate the flow characteristics of bronchial mucus.

In Professor Lynne Reid's departmental laboratories at Brompton I carried out a number of investigations of respiratory mucus using freshly obtained material examined on a Weissenberg rheogoniometer. The results of this work form the basis for much of this presentation.

The viscosity of bronchial mucus decreases with applied shear; at very low shear rates probably related to the shearing applied during secretion or irritation of mucus 'raft' formation at the site of secretion, apparent viscosity alters little with shearing. After a remarkably constant (around 0.1sec^{-1}) shear rate the apparent viscosity decreases more rapidly and smoothly until at very high shear rates (over 1000 sec^{-1}) the material is almost homogenised and becomes a Newtonian liguid. The site of change of apparent viscosity is not a feature of all mucus, is not related to the measuring apparatus and is probably related to a specific feature of the alteration of the tertiary structure of mucus with shearing.

At low levels of shear purulent secretion has a higher apparent viscosity than mucoid sputum. This applies to sputum from all conditions including asthma (the physical properties of mucus plugs will be discussed below). Shearing at very low shear rates, allowing the material to stand for a time and then reshearing, produces an increase in apparent viscosity of mucus over this and higher shear rates.

There is no level of apparent viscosity any disease group or at any stage of a disease process which is diagnostic. There is great variation between the apparent viscosity of sputum samples from patients with the same disease. It appears however that a particular individual will have mucoid sputum which will from day to day remain of a similar viscosity profile, will increase in viscosity profile with purulence and return to the same level after clinical recovery. The only relationship between pulmonary function and apparent viscosity is at higher shear rates where an inverse relationship has been found in chronic bronchitic subjects with mucoid sputum. FEV_1 and $FEV_1/FVCx100$ were inversely related to apparent viscosity at this high shear rate.

I found that elasticity of sputum varies with the shear rate. The elasticity profile of bronchial secretion has two zones over a range of low shear rates, after which it is destroyed. In the

first zone (0.01-0.025 Hz) elasticity increases slowly, over the
junctional region (0.025-0.397 Hz) it changes little, and then
increases sharply over the second zone (0.397-0.791 Hz).

There is no significant difference in mean elasticity at any
frequency between purulent and mucoid samples or between diseases.
The difference in the elasticity profile compared with the
viscosity profile suggests that elasticity reflects the basic
physiochemical properties of sputum rather than variation in the
relative proportions of the plasma and bronchial components which
vary widely with the presence of pus.

At low shear rates few studies have been undertaken of other
physical properties of mucus; such studies are needed to understand
the relation between structure and function of mucus and its
variation with disease.

There is a good relationship between apparent viscosity levels
at low shear rates and pourability of sputum. The relative ease of
flow of sputum in a container may be graded, an index of
'pourability' established and an index of degree of clinical change
obtained. Pourability is loosely related to FEV_1.

Other physical properties of respiratory mucus are easily
observed and clinically significant but have never been
investigated. This is usually because there is no way to
investigate the property, or workers have not yet used available
methods of study fully. I refer to the surface tension,
spinnability, stickiness and tailiness of sputum, for example.
Unfortunately viscosity, initially used in the lay sense from
'viscous or sticky' and later as the formal scientific measurement,
has been used as a specific physical property which will describe
all features of sputum. The ease or difficulty with which various
agents alter expectoration and volume of mucus in a number of
subjects, has led to great confusion over the significance of
'sputum viscosity', and the place in therapy of agents which
allegedly diminish this variable. It is to be hoped that further
investigation detailing the physical properties of the native
material, bronchial mucus, and how pharmaceutical agents may affect
these will be carried out in the light of which physical properties
have physiological, and therefore clinical, relevance.

'Asthmatic' airway plugging

The plugs which form in the smaller airways of subjects with
lung diseases, especially in the smaller airways and especially in
subjects with atopic asthma, are of considerable physiological and
clinical interest. Their aetiology is unknown. It may be that
they form as a result of the disorganised ciliary beating common in
the small airways in these conditions. Bouts of alternately

shearing and then leaving a mucus sample increases the apparent
viscosity of the material until it becomes impossible for the cilia
to generate the force to move the material – continuing irritation
of the airways leads to continuing production of mucus and plugging
of the small airways.

Conclusion

Bronchial mucus is mixture of high molecular weight glycoproteins. The tertiary structure is determined by the physical
properties and chemical structure not only of the glycoproteins,
but also of the smaller protein molecules and electrolyte ions
trapped in the matrix of the mucus. The tertiary structure will
determine the reaction of the mucus to shearing forces such as
ciliary action and coughing. It is the efficiency or inefficiency
of interaction between mucus and the ciliary escalator which
determines the efficiency of protection and clearance of the
respiratory tract.

Mucus is a fascinating and important material whose full
properties await scientific evaluation.

Selected reading:

Lopez-Vidriero, M.T., and Reid, L. (1978) British Medical Bulletin
        34 (1) 63.
Mitchell-Heggs, P. (1977) Advances in Exp. Med. and Biology (89)
        203.
Puchelle, E., Girard, F., Polu, J.M., Aug, F. and Sadoul, P. (1979)
        Sem. Hop. Paris, 55, 273.

DISCUSSION

LECTURER: Mitchell-Heggs   CHAIRMAN: Bonsignore

CHAIRMAN: The paper is open for discussion. May I ask you
                    to comment briefly on the therapeutical
                    implications of this paper, particularly on the
                    role of mucolytic drugs.

MITCHELL-HEGGS: It is a very difficult field confused by
                    commercial considerations. In such a situation
                    perhaps there is no definitive material which
                    one should use or recommend. However there are
                    patients who might be helped, I have no idea
                    why, by a number of materials, perhaps they
                    would be helped by the use of simple agents such
                    as steam inhalations or warm drinks which
                    stimulate the vagus and produce apparently a
                    looser type of mucus. The occasional patients
                    who improve on mucolytic agents may not have any
                    alteration in their respiratory function but may
                    feel symptomatically better – there may be
                    patients whose respiratory function is improved
                    but don't feel symptomatically any different.

CUMMING: May I make a comment firstly on the quantity of
                    normal human bronchial secretion and then go on
                    to discuss the quantity in disease and in the
                    end make some comments about mucolytic drugs.
                    It is possible to calculate in the human what is
                    the likely normal daily production by knowing
                    the rate at which the muco-ciliary escalator
                    moves in the human trachea and that every
                    estimate gives us values of about 25 mm. per
                    minute. The diameter of the trachea is readily
                    found by a variety of techniques and so is the
                    thickness of the mucus blanket in the trachea.
                    Knowing the perimeter and the rate at which the
                    mucus flows over it, simple mathematics give us
                    the value of between 8 and 10 ml. per day.
                    That's an estimate and it may be variable but we
                    are talking about 10 not 100 and certainly not
                    150 ml. per day. The second question relates to
                    disease; is this amount per diem increased?
                    This is a difficult question, because the
                    clearance rate along the trachea in chronic
                    bronchitis is difficult to determine, but there

is no good evidence that there is a remarkable
increase in mucus secretion in disease, there is
certainly something wrong in disease. Can it be
due to the characteristics of the mucus? The
characteristics of mucus, both normal and
abnormal, which may be complicated by
infections,are due to macromolecular
aggregation. The protein nucleus of mucus has
its side chains of glucosamine, galactosamine
and so on with terminal groups and they make
cross linkages by sulphydryl groups. The nature
of the macromolecular aggregation and its size
determines the physical behaviour of the sputum
and what we know about that is simply expressed
– nothing. It is certainly true that mucolysis
would involve the degradation of a macromolecule
and make it more easily dealt with by the
mucociliary escalator. So firstly in order to
define what we mean by mucolysis we must define
what we mean by a macromolecule and since we
cannot do this, mucolysis has no real meaning.
I personally do not believe any such drug
exists.

MITCHELL-HEGGS: May I first take you up on one feature of that?
The first thing is that although I fully
appreciate your extrapolation in relation to the
volumes of mucus, I think that there is a
variation between sexes, with age and probably
with different times of the day. The second
point is: I absolutely agree with you but would
take it even further. Even if the macromolecule
broken up into two or three smaller molecules,
since we do not know how cilia work we cannot
know if the change would be beneficial. it is a
sort of fine common sense idea that if we make
it nice and runny it's going to be all right,
but actually it may be less tractable.

HEATH: You pointed out that there were two layers to
mucus. A thin one in which the cilia were free
to move easily and a thick one which would trap
material, and you also pointed out that in the
periphery there were just a few rafts and later
on they joined together to form a more
continuous escalator and this means that out in
the periphery it is more probable that particles
would fall on free cilia than those covered by
rafts and then you would have a problem; how

would you remove particles that have fallen into the free cilia? What do you think would happen? Do you think you are going to send macrophages hunting for them like a dog looking for a rabbit in a cornfield?

MITCHELL-HEGGS: I don't know. That's obviously a weak link in what is a neat little story. You may well be right, I think there are probably a number of other features. It may be that things drift around until they hit a patch of cilia and then are chopped up with it. I don't really know. Maybe Peter Jeffery has a better idea in relation to this, as he is the person whose ciliary activities the film and my discussion are really stolen from. Peter, have you an answer to Donald's question.

JEFFERY: Nothing based on evidence, because we know that there are large numbers of macrophages which are still active as they come up the airways and it is quite possible that these macrophages may pick up the particles which have fallen between the cilia. Certainly on electron microscopy you can see macrophages which are in the right region to pick up such particles. It might not follow that these particles should necessarily fall through the swimming pool down to its bottom, but might actually lie on the top of the swimming pool and any raft of mucus that floated by would pick it up and so transport it. I think there is some evidence for that from the study of Silverberg and colleagues.

DENISON: Over what range of shear do you feel that the cilia in the respiratory tract act physiologically on the mucus and what would be the range of elasticity which would be relevant to mucociliary transport.

JEFFERY: Extrapolations are made depending on the beat of the cilia, the link of the cilia, the number of the cilia and the distance between them. Roughly I think the figures that I showed you as differences between viscosity levels are probably related either to the ease or difficulty of grasp of the cilia in relation to how quickly and easily the material is secreted and moves up in the various streams. We are

talking about very low rates of applied shear.
It is probable that on that scale cilia work I
would expect somewhere between about 10 and 500
Hz, varying on the position of the cilium that
one is dealing with. But again like the volume
of mucus there is a great dispute about this.

CHAIRMAN: I think the discussion can be continued after
the second paper, so I invite Dr. Pavia to give
his paper on mucociliary transport.

LUNG MUCOCILIARY TRANSPORT IN MAN

Demetri Pavia, Philip P. Sutton, John E. Angew, Maria T.
Lopez-Vidriero, Stephen P. Newman and Stewart W. Clarke
Department of Thoracic Medicine
The Royal Free Hospital
London NW3 2QG, England

INTRODUCTION

The human lung has several defence mechanisms against all
types of inhaled deposited material. The three main clearance
mechanisms are (a) mucociliary transport, (b) cough and (c)
alveolar clearance. In this article we review the various methods
currently available for the objective measurement of the efficiency
of the mucociliary clearance mechanism in vivo in man. We also
summarize the effect of (a) physiological factors, (b) disease
states, (c) therapeutic factors and (d) environmental pollutants on
human lung mucociliary transport.

MUCOCILIARY TRANSPORT

The mucociliary escalator is composed of ciliated epithelium
extending from the larynx to the terminal bronchioles (16th airway
generation of Weibel's model of the lung). The cilia are bathed in
a watery layer of fluid - the periciliary layer. On top of the
cilia lies a viscid layer of mucus, which may be continuous or
consist of discrete islands. In addition to the ciliated columnar
cells, the tracheobronchial epithelium contains mucous cells which
secrete epithelial glycoprotein, the ratio between the two being 5
to 1, with a decrease in the number of mucous cells from the
trachea to the terminal bronchioles. On the surface of each
ciliated cell, there are approximately 200 cilia, each with a
diameter of approximately 0.3 μm. The length of the cilia
decreases from 6-7 μm in the trachea to 3.6 μm in the 7th
generation bronchi[1] and possibly more as one approaches the
terminal bronchioles. The cilia beat at a frequency of 700-1000
times min^{-1}, with an active stroke and a recovery phase; a quarter

of each cycle is taken up by the cilia beating "forward", thus pushing the secretions towards the mouth, and three-quarters in bending "backwards" ready for the next stroke. The cilia beat in a coordinated fashion, one after the other, so producing metachronal waves travelling in the opposite direction to the effective stroke of the cilia, i.e., to the mucous flow. Cilia propel lung secretions against gravity and are able to carry weights up to 10 g cm^{-2} without affecting their performance. Cilia have been found to beat in excised ciliated epithelium for several hours after death. The source of energy for cilia is believed to be adenosine triphosphate (ATP), which is distributed along the length of the cilium with one molecule of ATP per 700 Å.

Lung secretions originate from several sources:[2] (a) The submucosal glands (of approximately 4 ml in total volume) which are found in the cartilagenous airways. The submucosal glands are simple tubulo-acinar structures and contain mucous and serous tubules. The collecting duct cells regulate the concentration of ions and water of the mucus secreted by the glands. Myoepithelial cells are found close to the mucous and serous cells and may serve to extrude secretions. (b) Secretory cells of the surface epithelium - mucous cells, Clara cells and probably ciliated cells. The mucous cells have a total volume of 1/40th of that of the glands. The Clara cells are found in the terminal bronchioles, and on irritation they transform into mucus-secreting cells. (c) The type II alveolar cells produce a lipoprotein substance (surfactant). The total volume of secretions cleared per day in healthy man is not clearly known due to the difficulties in collection. However, estimates range from 10 ml, based on data from tracheotomized accident victims,[3] to 100 ml from animal experiments extrapolated to man. It is well known that patients with chronic bronchitis can produce 30-100 ml day^{-1} of sputum during exacerbations. Tracheobronchial secretions contain water (94%), electrolytes (1%) and macromolecular components (4%), the latter including mucus glycoprotein (2-3%), proteins (0.1-0.5%) and lipids (0.3-0.5%). Little is known about the differences, if any, in the biochemistry of the periciliary layer in which the cilia beat and of the viscous layer which they propel.

The flow rate of mucus in the trachea has been reported to be in the range of 5-20 mm min^{-1} and progressively diminishes towards the lung periphery. The absence of a significant build-up of secretions in the proximal airways during clearance has been attributed to reabsorption of secretions. This is supported by the fact that secretions in the peripheral airways are believed to be more watery than those in the central airways.[4]

The majority of inhaled particles and debris deposited on the ciliated airways will be swept cephalad and in health removed from the lungs by the mucociliary escalator within a matter of several hours and certainly within one day.[5] The same, however, is not true in the case of lung disease, where tracheobronchial clearance via the mucociliary escalator can take much longer than in health. Further, there is increasing evidence from animal experiments[6] that some of the deposited particles on the ciliated airways will be engulfed by macrophages and subsequently end up beneath the epithelium. This is particularly so where ciliated columnar areas have become cuboidal or flattened, with loss of ciliated cells and infiltration with lymphocytes.

MEASUREMENT OF MUCOCILIARY TRANSPORT IN MAN

There are essentially three basic methods, with several variations of each method, available for measuring mucociliary transport in man.[7]

1. Bronchographic Method

This technique involves the introduction into the lungs of a radio-opaque substance, which may be either a powder (e.g., tantalum) insufflated through a catheter placed at the carina[8] or a broncho-graphic contrast substance injected intratracheally[9]. X-rays allow visualization of representative airways from the trachea to bronchioles. Serial X-rays at, say, 0, 4, 8, 16, 24 and 48 h following the introduction of the radio-opaque substance give an indication of the amount remaining (based on scoring by an observer). This technique is capable of measuring whole lung mucociliary transport as well as local mucus velocity.

This technique, however, has been applied only to a limited number of investigations, primarily because (a) it is invasive, (b) it is semi-quantitative and (c) it involves a considerable absorbed radiation dose to the lungs from sequential X-rays.

2. Local-Mucus Velocity Methods

(i) Cinebronchofibrescopic technique. Sackner et al.[10] described the cinebronchofibrescopic technique to measure the rate of movement of small (0.13 mg) teflon discs blown through the inner channel of the fibreoptic bronchoscope onto the bronchial mucosa. The cephalad motion of these discs is filmed through the bronchoscope, and effectively the discs serve as markers of the mucous flow in the trachea. A measure of tracheal mucus velocity (TMV) is achieved by measuring the time taken for, say, 10-20 discs to move a given distance.

(ii) Radioisotopic bronchoscopic technique. Chopra et al.[11] described this technique in which a graduated fibreoptic bronchoscope is used to ascertain the length of the trachea from the carina to the larynx. A small volume (40 µl) of albumin microspheres (5-7 μm diameter and numbering approximately 50,000) tagged with 99Tc[m] is deposited near the carina by a catheter inserted through the fibreoptic bronchoscope, and the bronchoscope is then withdrawn. Sequential pictures with a gamma camera at, say, 1 minute intervals over a period of 30-60 minutes (the approximate time taken for the particles to traverse the trachea) yields information regarding the average speed of tracheal mucociliary transport.

(iii) <u>X-ray bronchoscopic technique</u>. This technique is a modification of the cinebronchofibrescopic in as much as the teflon discs in this instance are mixed with the radio-opaque substance bismuth trioxide and insufflated through the vocal cords via the inner channel of a fibreoptic bronchoscope[12], which is then withdrawn. The cephalad motion of these particles in the trachea is recorded with a fluoroscopic unit provided with an image intensifier, television monitor and videotape.

(iv) <u>Radioaerosol bolus technique</u>. Yeates et al.[13] were able to deposit a bolus of aqueous aerosol containing 99Tc[m]-labelled albumin microspheres (0.5 μm median diameter) in the trachea and the first few airway generations by inhaling near total lung capacity at a high flow rate. Thereafter the movement of the bolus was recorded by a gamma camera, and an estimate of TMV was obtained by measuring the distance moved by a bolus in a given time.

Fig. 1 illustrates the values of the tracheal mucus velocities which have been reported using these four techniques for healthy subjects. It is of important interest to note that as one moves from the most invasive technique (i.e., cinebronchofibrescopic) through (ii) and (iii) to the non-invasive (i.e., radioaerosol boli) there is an approximately fourfold decrease in the reported mean TMV. The respiratory epithelium is fragile and has been shown to be easily injured by gauze, cotton swabs, aspiration and instrumentation.

3. Radioaerosol Method

Albert and Arnett[14] were the first to use inhaled radioaerosols for assessing objectively lung mucociliary transport in man. This method involves the inhalation of an aerosol unleachably tagged with a gamma-emitting radioisotope, which can be monitored by external counters located over the chest. The initial lung burden (deposition) can be ascertained, as can the subsequent clearance resulting in a decrease of radioactivity within the lungs. After correction for the physical decay of the radioisotope and background radiation, the typical resulting curve of radiation present in the lungs against time (Fig. 2) may be conventionally divided into two phases, an early fast phase and a later slow phase. The fast phase is due to the clearance of deposited

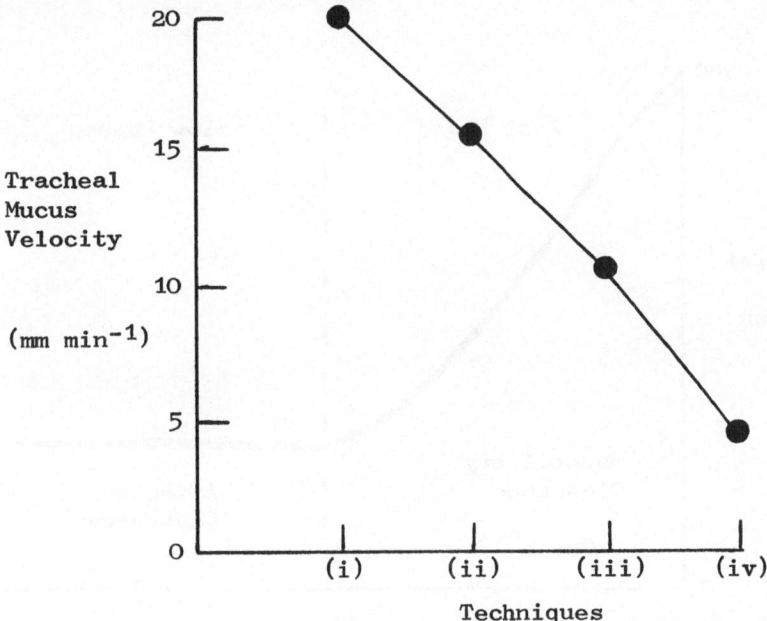

Fig. 1.   Reported values of tracheal mucus velocities for the
          following four techniques: (i) cinebronchofibrescopic,
          (ii) radioisotopic bronchoscopic, (iii) roentgenographic
          bronchoscopic and (iv) radioaerosol boli.

particles via the mucociliary escalator, augmented by cough on
occasion.   The slow phase represents alveolar clearance and is the
radioactivity remaining at 24 h and longer.   This is assumed to be
due to deposition on non-ciliated terminal airway units.   If the
estimated alveolar deposition is subtracted from the fast phase
mucociliary clearance, this results in a measure of
tracheobronchial clearance (Fig. 3).

The rate of clearance of the deposited particles from the
lungs is directly related to their initial site of deposition,
which itself is dependent on (a) the physical properties of the
aerosol, (b) the mode of inhalation of the aerosol and (c) the
patency of the airways.

The important properties of aerosols that affect their site of
deposition in the lungs are (i) particle size - the bigger the
particle size the more likely it is to deposit by impaction near
the mouth; (ii) density - greater density increases the aerodynamic
diameter of the particle, causing it to behave like a bigger unit
density particle; (iii) hygroscopicity - absorption of water will
result in a bigger particle; and (iv) electric charge - the effect
of which is not yet well defined.

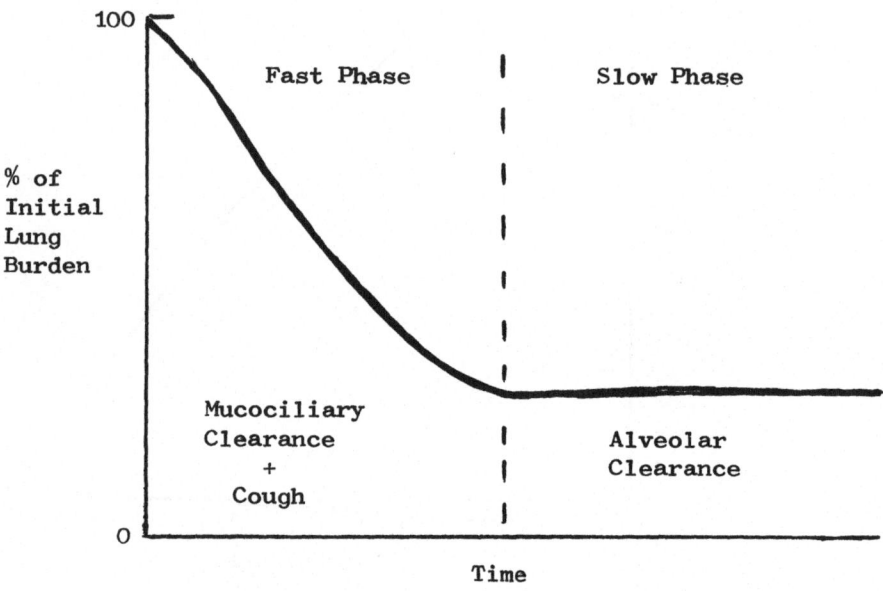

Fig. 2. Schematic diagram of whole lung clearance of deposited
particles.

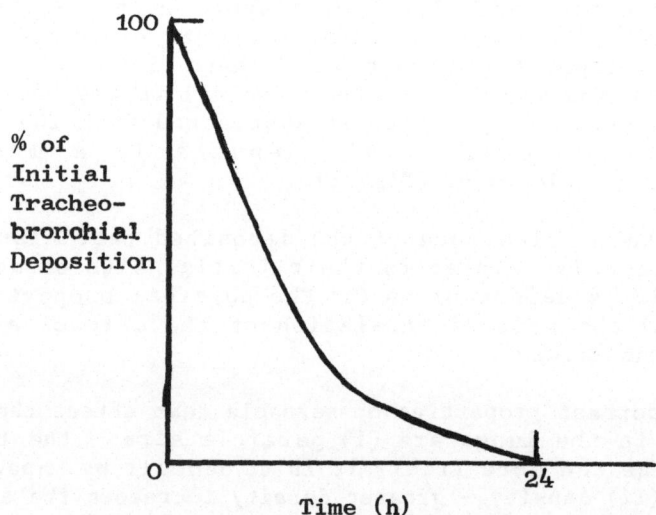

Fig. 3. Schematic diagram of tracheobronchial clearance.

The manner of inhaling an aerosol affects its deposition within the lungs.[15] (i) A large volume breath ensures more particle deposition in the peripheral airways, (ii) a high flow breath results in a more central deposition as a result of increased inertial impaction, (iii) breath-holding after inspiring an aerosol permits particles to settle out at the furthest point of entry and (iv) the lung volume at the commencement of an aerosol inhalation also influences the site of deposition.

Airflow obstruction (i.e., a reduction in airway patency) leads to more central deposition due to impaction.[16]

The rate of removal of deposited particles on the ciliated airways is directly related to the length of the transit pathway. In general, particles deposited in the peripheral airways take proportionally longer to be cleared than particles deposited proximally. Therefore, when using this method for measuring mucociliary transport in man, it is important that account be taken of the initial topographical distribution of the tracer aerosol in the analysis of tracheobronchial clearance. This can be achieved in three ways: by monitoring the initial deposition pattern with a rectilinear gamma scanner or a gamma camera or, alternatively, by measuring the amount of aerosol remaining in the non-ciliated airways at 24 h or longer. Probably the best way of analyzing the clearance data is with a combination of measurements of 24 h retention and distribution in the lungs using a gamma camera linked to a computer.

The most popular types of aerosols that have been used to study mucociliary transport are polystyrene, iron oxide, Teflon, albumin microspheres and resin (Amberlite IRP 67).

Some of the radionuclides which are used for the study of mucociliary transport in man are ^{18}F (with a physical half-life, T1/2, of 112 min), $^{99}Tc^m$ (T1/2 = 6 h), ^{198}Au (T1/2 = 2.7 d) and ^{131}I (T1/2 = 8.04 d).

The type of monitor used for measuring the radionuclides depends on the dose of radioactivity used. For low lung burdens (e.g., less than 1 µCi of $^{99}Tc^m$) a whole body scintillation counter in a well shielded room can be used. For medium doses of lung activity (e.g., less than 30 µCi of $^{99}Tc^m$) a suitably collimated single central anterior chest counter suffices to measure clearance. It is better, however, to use twin axially opposed scintillation counters placed antero-posteriorly or laterally or a series of detectors placed round the chest wall in order to minimize variations in geometrical counting efficiencies as a result of varying distribution of deposited radioaerosol within the lungs with time. If a relatively high dose is acceptable (about 100 µCi of $^{99}Tc^m$ or greater), the gamma camera (with a large field

of view preferably) is the instrument of choice because it also
yields information on the distribution of aerosol within the lungs
at the same time, thus permitting estimates of clearance from
particular zones of interest of the lungs.

Several review articles[17-19] give detailed information
regarding the effect of various factors on lung mucociliary
transport; below we summarize briefly the effect of some of the
more important of these factors.

EFFECT OF PHYSIOLOGICAL FACTORS ON LUNG MUCOCILIARY TRANSPORT

No difference in lung mucociliary transport has been found to
exist between the two sexes. Posture also appears to have no
effect on the clearance of lung secretions in health, although this
is not true in disease. Healthy subjects were studied in the 25°
head down position, and their tracheal mucous velocities were found
to be no different to those when they were in the upright position.
Controlled studies carried out in our laboratory on healthy
volunteers demonstrated that lung mucociliary transport measured
during daytime did not differ from that during night-time when the
volunteers were awake, thus eliminating any possible effect of
circadian rhythm on mucociliary transport.

Exercise (e.g., half an hour briskly pedalling on a bicycle
ergometer) has been shown to result in an increase in mucociliary
clearance. However, the effect of eucapnic hyperventilation is
approximately half of that observed during exercise. This increase
is attributed to either (i) mechanical effect of increased lung
movement or (ii) an effect on the autonomic nervous system
producing stimulation of airway glands and/or stimulation of
ciliary beat.

It has been reported by at least two independent groups of
workers using different techniques that in older subjects
mucociliary transport is slower than that of a younger group of
subjects. Controlled studies carried out by us on healthy subjects
and on patients with asthma have demonstrated increased retention
of lung secretions during sleep. This finding may be of relevance
in patients with chronic bronchitis or asthma who have early
morning productive cough and wheeze.

EFFECT OF DISEASE ON LUNG MUCOCILIARY TRANSPORT

Patients who have emphysema with alpha$_1$-antitrypsin deficiency
but no chronic bronchitis have been found to have an intact lung
mucociliary transport. Patients with a restrictive type
occupational lung disease (asbestosis) were also found to have a
normal mucociliary clearance. We have studied patients with
Sjogren's syndrome, which is characterized by deficient secretion

of the lachrymal, salivery and other glands, chiefly confined to postmenopausal women and often associated with connective tissue disease; they were also found to have a normal clearance. Patients with active sero-positive rheumatoid arthritis have been reported to have an increased incidence of respiratory infection and increased airflow obstruction. In a controlled study of 11 patients with rheumatoid disease, we found no evidence of impairment of their lung mucociliary clearance.

Scanty evidence exists on two patients with bronchogenic carcinoma who were reported to have normal tracheal and lung mucociliary clearance. There is some information on the clearance of particulate matter from the tracheobronchial tree of patients with chronic pulmonary tuberculosis. It appears that mucociliary clearance was present, but areas were found of alleged decreased ciliary activity resulting in local accumulation of the tracer particles.

Influenza A virus can destroy ciliated cells in the respiratory tract, and patients infected with this virus and studied one week after the onset of symptoms showed a definite retardation in clearance which was, however, reversible, with recovery approximately 2-3 months later. Infection with Mycoplasma pneumoniae also damages the ciliated tracheal epithelium. In patients with serologically verified infection by M. pneumoniae, a significant reduction in clearance was noted 10-15 days after the onset of disease, with a tendency to impairment in clearance up to one year later.

Studies from several centres claim that clearance of lung secretions in patients with chronic bronchitis is normal. However, cough can play a major role in the clearance of secretions and may compensate for an otherwise impaired mucociliary clearance mechanism. Carefully controlled studies have shown that lung mucociliary transport is decreased in chronic bronchitis. There appears to be no relationship between impairment in lung clearance and pulmonary function indices. There is evidence from autopsies carried out on asthmatics that there is extensive sloughing of ciliated cells from the bronchial epithelium and of mucus plugging; such histological changes might give rise to or result from an impaired mucociliary clearance in patients with bronchial asthma. Marked reductions in lung mucociliary transport and tracheal mucous velocities have been reported in symptomatic and asymptomatic asthmatics. In patients with cystic fibrosis, the evidence from several studies regarding involvement of lung mucociliary transport is equivocal.

A careful study of patients with bronchiectasis has demonstrated abnormal mucociliary clearances which were not dissimilar to those found by the same workers in patients with

chronic bronchitis without bronchiectasis. There is also some
evidence from Polynesian bronchiectatics with abnormal lung
mucociliary clearances attributing this impairment to
ultrastructural abnormalities of the cilia.

Patients with Kartagener's syndrome (situs inversus with
bronchiectasis and sinusitis) have been found to have grossly
abnormal lung mucociliary clearances. This impairment has been
attributed to ultrastructural abnormalities (missing dynein arms)
of the cilia. However Kartagener's syndrome is believed to be a
special case of a more general "immotile-cilia syndrome", relating
defects in the ultrastructure of the cilia, male sterility and
chronic infections of the lower and upper respiratory tract. This
proposed syndrome lacks the features of situs inversus. Some
evidence exists suggesting that absent mucociliary transport
predisposes patients with "immotile-cilia syndrome" to develop
obstructive lung disease. Young's syndrome comprises infertile
male patients with primary obstructive azoospermia (a type of
infertility) and chest involvement but no dextrocardia. We carried
out a controlled study of patients with Young's syndrome and found
that they had reduced lung mucociliary transport. The degree of
impairment was similar to that seen in patients with chronic
bronchitis who were twice the age of the patients with Young's
syndrome and who smoked eight times as much and had pulmonary
function indices half those of the patients with Young's syndrome.

EFFECT OF THERAPEUTIC FACTORS ON LUNG MUCOCILIARY TRANSPORT

The efficacy of mucolytic/expectorant drugs in man has been
assessed (i) subjectively, (ii) by their effect on pulmonary
function indices and (iii) by the changes in the chemical and
viscoelastic properties of sputum. Objective in vivo techniques
for measuring clearance of lung secretions in man has provided a
fourth, direct technique for evaluating the efficacy of these
drugs. Bromhexine has been shown statistically to enhance
mucociliary transport. Three independent laboratories have been
unable to find any effect of S-carboxymethylcysteine on mucociliary
transport. The evidence for guaiphenesin, glyceryl guaicolate,
from two laboratories is equivocal. Water aerosol administration
has been shown to result in a small but significant acceleration in
mucociliary transport. Aerosolized half-normal saline increased
tracheal mucociliary transport rates in only 30% of normal subjects
studied and produced no change in patients with cystic fibrosis.
Hypertonic saline (7.1%) aerosol has been reported by our group to
result in marked increases in clearance of lung secretions in
patients with chronic obstructive bronchitis, predominantly as a
result of more productive coughing. By contrast, aerosolized
sodium 2-mercaptomethane sulphanate administered to chronic
bronchitic patients resulted in an enhancement of mucociliary
transport which failed to reach statistical significance.

Circulating catecholamines might have a role to play in the therapy of chronic obstructive lung disease by enhancing the transport of mucus in the airways. The effect of several adrenergic agents (isoprenaline, epinephrine, fenoterol, salbutamol, terbutaline) administered to man has been investigated. In general, the numerous studies on this topic have shown that beta adrenergic drugs administered orally or parentally result in a stimulatory effect on lung mucociliary transport. However, when these drugs are administered in aerosol form from pressurized canisters, in general they have either a very small effect or no effect on clearance. It appears that the small doses (less than 10% of the metered dose) delivered directly to the lungs, although adequate for bronchodilatation, are not high enough to stimulate clearance. Acute (single) doses of both selective (atenolol, 100 mg) and non-selective (propanolol, 160 mg) beta blockade on healthy subjects had no effect on lung mucociliary transport. Any long term effects of beta blockers on the clearance of lung secretions have yet to be determined. The long term effect (one week's administration) of a sustained release aminophylline (Phyllocontin, 450 mg b.d.) on lung mucociliary clearance in patients with obstructive lung disease was an enhancement of clearance.

Although the bronchodilator properties of the anticholinergic drugs have been known for many years, their clinical use has been restricted due to their side effects in patients already suffering from retention of lung secretions, particularly the drying of secretions and inhibition of the ciliary beat. The anticholinergic drugs hyoscine and atropine have been shown to retard lung mucociliary transport. Conversely, the cholinergic drug bethanecol gives rise to an enhancement of tracheobronchial clearance. Ipratropium bromide, a synthetic anticholinergic agent developed specifically to relieve airways obstruction by bronchodilatation but without the undesirable effects of other anticholinergic drugs, is available in aerosol form (20 µg per puff) from a pressurized canister. Its acute and long term effects have been studied in healthy subjects and in patients with chronic bronchitis, and it has been found not to affect clearance of secretions adversely.

One possible mechanism of the pharmacological action of corticosteroids in asthma may be enhancement of lung mucociliary clearance. Studies in our laboratories have shown that a two week course of oral corticosteroids (prednisolone) 15 mg daily followed by another two weeks at 30 mg daily resulted in an improvement in mucociliary transport.

Experiments in animals have suggested that the release of chemical mediators may play a role in mucociliary dysfunction in allergic asthma. In asymptomatic asthmatic patients treated with sodium cromoglycate, a decrease in tracheal mucous velocity after bronchial provocation with ragweed antigen extract was prevented.

It has been suggested that pulmonary oxygen toxicity in healthy volunteers may result in acute tracheobronchitis. In a study on 10 healthy subjects breathing 90-95% O_2 for 6 h, endoscopic examination at the end of that time revealed evidence of tracheitis in all the subjects, and tracheal mucous velocity was depressed 3 h after beginning to breathe O_2.

Postoperative chest complications may arise as a result of damage to the ciliated epithelium during anaesthesia and hypoventilation, resulting in impaired clearance and undue retention of secretions. Premedication with atropine-like drugs has been shown to result in impaired mucociliary clearance; further, during sleep clearance is also reduced. Analgesics, humidification of anaesthetic gases and the insertion of intratracheal tubes might also affect clearance adversely. For example, the respiratory epithelium is fragile and easily injured by gauze, cotton swabs, aspiration and instrumentation. Some anaesthetic gases are also known to be cilio-toxic. The combined effect of all these factors acting adversely on lung clearance may be ameliorated by the administration of chest physiotherapy postoperatively to enhance the clearance of lung secretions, particularly since voluntary coughing might well be suppressed because of associated pain (e.g., in abdominal operations).

Studies of the effect of chest physiotherapy have demonstrated enhancement of clearance of lung secretions, although there is conflicting evidence from various centres as to the role of productive coughing.

EFFECT OF ENVIRONMENTAL POLLUTANTS ON LUNG MUCOCILIARY TRANSPORT

Sulphur dioxide, a major component of urban pollution, has been shown to affect human nasal clearance adversely. Breathing 5 ppm of SO_2 for 3 h has been reported to have no effect on the lung mucociliary transport of healthy non-smoking subjects.

Another ambient pollutant is submicron sulphuric acid aerosol, which is often found in urban areas with an atmospheric concentration in the range of 10-20 μg m^{-3}. No consistent effect of this pollutant at concentrations of 100, 300 and 1000 μg m^{-3} breathed via a nasal mask for 1 h by healthy subjects was found on their lung mucociliary transport.

There is some evidence to suggest that fluorocarbon propellants, contrary to their reputation as being inert, might affect the heart and lungs. Fluorocarbons are used in pressurized canisters to administer bronchodilator and corticosteroid drugs to patients' lungs. Their effect, therefore, on lung mucociliary clearance has attracted attention. In fact, several workers have shown that there are no adverse acute or long term effects on lung mucociliary clearance from administration of such propellants to the lungs.

A commercially available hair spray directed towards the hair of healthy subjects for 20 s has been reported to result in a reduction of tracheal mucous velocity. The tracheal mucous velocity of subjects exposed to the freon propellants only, under the same exposure conditions, was not adversely affected.

Cigarette smoke has been shown to stop ciliary action temporarily, possibly resulting in a prolonged residence of inhaled carcinogens in the lungs. This plus the numerous histological changes noted in the respiratory tract epithelia of smokers compared to non-smokers has prompted a number of investigators to look at the effects of cigarette smoking on lung mucociliary clearance. Studies on the long term effects of tobacco smoking have shown, on the whole, a slowing of tracheobronchial mucociliary clearance. The effect of giving up smoking has been investigated in asymptomatic smokers. There is evidence to suggest that their impaired clearance returns to normal 3 months after giving up the habit. The acute effect of cigarette smoking on lung mucociliary transport is more controversial. Some workers have reported no effect, others have noted an enhancement in clearance, and we ourselves have observed a reversible slowing effect on lung mucociliary clearance.

REFERENCES

1. S.M. Serafini and E.D. Michaelson, Length and distribution of cilia in human and canine airways, Bull. Eur. Physiopath. Respir. 13:551 (1977).
2. S.W. Clarke and D. Pavia, Lung mucus production and mucociliary clearance: methods of assessment, Br. J. Clin. Pharmacol. 9:537 (1980).
3. N.G. Torelmalm, The daily amount of tracheobronchial secretions in man, Acta Otolaryng. 158 (suppl):43 (1960).
4. T. Asmundsson and K.H. Kilburn, Mucociliary clearance rates at various levels in dog lungs, Am. Rev. Respir. Dis. 102:388 (1970).
5. P. Camner and K. Philipson, Human alveolar deposition of 4 μm Teflon particles, Arch. Environ. Health 36:181 (1978).
6. C. Stirling and G. Patrick, The localisation of particles retained in the trachea of the rat, J. Pathol. 131:309 (1980).
7. D. Pavia, J.R.M. Bateman, N.F. Sheahan, J.E. Agnew, S.P. Newman and S.W. Clarke, Techniques for measuring lung mucociliary clearance, Eur. J. Respir. Dis. (in press).

8. G. Gamsu, R.M. Weintraub and J.A. Nadel, Clearance of tantalum from airways of different caliber in man evaluated by a roentgenographic method, Am. Rev. Respir. Dis. 107:214 (1973).

9. K.E. Seyfarth, Accelerated elimination of bronchographic contrast media by means of mucolytic preparation, Praxis Pneumol. 18:803 (1964).

10. M.A. Sackner, M.J. Rosen and A. Wanner, Estimation of tracheal mucus velocity by bronchofiberscopy, J. Appl. Physiol. 34:495 (1973).

11. S.K. Chopra, G.V. Taplin, D. Elam, S.A. Carson and D. Golde, Measurement of tracheal mucociliary transport velocity in humans - smokers versus non-smokes (preliminary findings), Am. Rev. Respir. Dis. 119 (suppl):205 (1979).

12. R.M. Goodman, B.M. Yergin, J.F. Landa, M.H. Golinvaux and M.A. Sackner, Relationship of smoking history and pulmonary function tests to tracheal mucous velocity in non-smokers, yourn smokers, ex-smokers and patients with chronic bronchitis, Am. Rev. Respir. Dis. 117:205 (1978).

13. D.B. Yeates, N. Aspin, H. Levison, M.T. Jones and A.C. Bryan, Mucociliary tracheal transport rates in man, J. Appl. Physiol. 39:487 (1975).

14. R.E. Albert and L.C. Arnett, Clearance of radioactivity dust from the lung, Arch. Environ. Health 12:99 (1955).

15. D. Pavia, M.L. Thomson, S.W. Clarke and H.S. Shannon, Effect of lung function and mode of inhalation on penetration of aerosol into the human lung, Thorax 32:194 (1977).

16. M.L. Thomson and M.D. Short, Mucociliary function in health, chronic obstructive airway disease, and asbestosis, J. Appl. Physiol. 26:535 (1969).

17. D. Pavia, J.R.M. Bateman and S.W. Clarke, Deposition and clearance of inhaled particles, Bull. Eur. Physiopath. Respir. 16:335 (1980).

18. A. Wanner, Clinical aspects of mucociliary transport, Am. Rev. Respir. Dis. 116:73 (1977).

19. P. Camner, Clearance of particles from the human tracheobronchial tree, Clin. Sci. 59:79 (1980).

DISCUSSION

LECTURER: Pavia   CHAIRMAN: Bonsignore

CUMMING: Thank you Demetri, I enjoyed your presentation.
                  I have two questions, both of them
                  extraordinarily simple. Why is it that in the
                  bronchitic the particles of the size you use
                  lodge in the major airways and do not go
                  peripherally?

PAVIA: The particles in patients with chronic
                  bronchitis tend to deposit more centrally than
                  in healthy persons for the following simple
                  reason: the airways are narrowed because of the
                  increased mucus and therefore for the same
                  inhalation flow rate there is an increase in
                  linear velocity resulting in turbulence and
                  deposition by impaction more centrally than
                  would otherwise have been.

CUMMING: May I ask a supplementary question? In a normal
                  subject with twice the tidal volume would he
                  have a similar central deposition.

PAVIA: I agree with you Professor Cumming and indeed
                  this is what we are doing in the United States.
                  When you wish to compare healthy subjects with
                  patients, we get healthy people to inhale
                  rapidly to simulate the conditions in the
                  bronchitic lung in order to achieve deposition
                  patterns which are comparable to the bronchitic
                  for comparison purposes. This is a very valid
                  point.

CUMMING: My second question: when you are giving the
                  patients therapeutic oxygen is it humidified?

PAVIA: I must add that the study on oxygen therapy was
                  not carried out by us, it was carried out in the
                  United States and indeed it was humidified to
                  between 96 and 100%.

CUMMING: At room temperature?

PAVIA: At room temperature.

AFZELIUS: I couldn't understand how patients with Sjogrens syndrome and possibly scleroderma did not have reduced mucociliary clearance when they probably have less sputum secretion, they have dryness of the sputum.

PAVIA: That's a very good point and indeed that was puzzling Professor Turner-Warwick whose patient were kindly sent to us to be assessed. These patients with Sjogrens syndrome had dry secretion in the eyes, in the mouth, in the nose, in the vagina and therefore they were wondering whether there was a dry secretion in the lungs which might adversely affect lung mucociliary transport, which is a host defence mechanism. Much to our surprise and Professor Turner-Warwick's, we were able to show that the transportation of secretions in these patients' lungs was not different to that of a healthy matched control group of subjects.

KAY: In your last slide you listed a number of agonists-antagonists which had an effect on mucociliary clearance, but you didn't include in your list leucotriene D or a slow reacting substance as it was originally called, this agent has a profound effect on mucociliary clearance and this can be reversed using the specific SS antagonists FPO 55712. Do you agreed that this is an important observation which should be included in this context?

PAVIA: I am not familiar with the particular study that you refer to. It was probably done in animals as opposed to humans.

KAY: It has been done recently in humans.

PAVIA: From Dr. Wonder's laboratory maybe.

KAY: Dr. Wonder's laboratory.

PAVIA: I am sorry, I am not familiar with this.

KAY: I think it is an important observation because these compounds, as you know, are extremely potent and to have an effect like this on mucociliary clearance in such small concentrations is quite intriguing.

SOLOMON: What are the effects of mechanical ventilation? You mentioned that oxygen therapy will affect mucociliary clearance. It is oxygen per se or is the mechanical aspect of oxygen inhalation?

PAVIA: We have not investigated this particular study on cells and I am only quoting other people's work with mechanical ventilation. The only study that I know of is that of Dr. Newhouse who asked subjects to hyperventilate voluntarily and this appeared to stimulate clearance. He thought it was due to stimulation of autonomic nervous system either stimulating ciliary beat frequency or the secretion of mucus into the airways. As far as the oxygen therapy goes there is the work which was carried out in Chicago and also in Miami where they asked patients to inhale oxygen for six hours from a mask. At the end of the six hours they all had a high tracheitis, but within three hours of administering oxygen, they were able to observe retardation in the flow of mucus in the trachea.

COHEN: Do you believe there is a threshhold dose for radiation carcinogenesis?

PAVIA: I am sorry I cannot give an answer.

GIUNTINI: You mentioned that in chronic bronchitis the deposition of the aereosol may be central and so that could explain the relative rapidity of clearance. This may also occur in asthma especially during clinically active asthma where there is a very significant restriction of central airways. So you have to define the state of asthma before going on measuring the clearance with this approach.

PAVIA: I agree with you, I didn't wish to get involved with the deposition problem too much but in our studies with asthmatic patients even the patients who were in remission showed a statistically significant more central deposition than the healthy matched control group under the same inhalation conditions. So what you say is true and has been observed.

GIUNTINI: Just to proceed one further step. We have observed that it's much more frequent in asthma than in chronic bronchitis to have a central deposition of these inhaled aereosols, which I believe have about the same characteristics that you have showed. So I wonder whether it is necessary actually to produce a picture of the original deposition in order to proceed with the analysis that you have so widely applied.

PAVIA: In all the studies that I have shown you we have evidence of the initial deposition pattern which was obtained either using a gamma camera immediately after inhalation or using a rectilinear gamma scanner. For example, in the Bisolvon study we have observed only a small effect on clearance. Having taken Bisolvon the patients we observed had deposited particles more distally in the lung, therefore you would expect the clearance would be slower, but it was not slower. So I agree with your point that no clearance study should be carried out without adequate information regarding the initial topographical distribution of the tested particles within the lungs.

SPENCER: My question is to the last two speakers but particularly the first speaker; has he any information on the quality of bronchial mucus and its ability to sustain bacterial growth. I ask this because one of the most important things in chronic bronchitis is that the mucus seems to be capable of sustaining continual bacterial growth. Does he have any facts relating to the quality of the mucus and its ability to sustain bacterial growth?

MITCHELL-HEGGS: May I answer Professor Spencer's question? One of the things which is very interesting is that white cells produce neuraminadases which destroy part of the glycoprotein links within mucus. Certainly it appears that mucoid sputum allows bacteria to grow better than purulent sputum, but why this should be I am not sure; it should be in fact the other way round, but it isn't. It may be that there are in fact more destructive enzymes contained within purulent than in the mucoid secretion.

CHAIRMAN: Thank you. The next paper will be presented by
                    Dr. Bienenstock on Bronchus associated with
                    lymphoid tissue.

BRONCHUS–ASSOCIATED LYMPHOID TISSUE

John Bienenstock

Department of Pathology
McMaster University
Hamilton, Ontario
Canada

It is well known and appreciated that the lungs contain organized lymphoid tissue. A few years ago while investigating the effect of the introduction of immune complexes into rabbit lungs, we observed that there were present in the bronchial walls lymphoid aggregates which resembled those in the intestine.[1-3] Because the Peyer's patches had been loosely included in the term gut-associated lymphoid tissue (GALT) we termed this bronchus-associated lymphoid tissue or BALT. On becoming much more interested in the nature of this tissue, and upon doing a literature survey, we found that this lymphoid tissue had been well described some 100 years previously but for obscure reasons in more recent years reference to it was infrequent. Klein,[4] more than 100 years ago, observed such aggregates in the bronchial tract and noticed their similarity to Peyer's patches. Similar lymphoid aggregates have been observed in primitive mammals by Miller[5] and have now been identified in humans, rabbits, guinea pigs, dogs, rats, mice, swine and chickens.[1-15]

Much is now known about the morphology of BALT in animals. Little to nothing has been written in recent years on BALT in man, although its presence has been described for about 100 years.[1-5] Recently, Meuwissen and colleagues (personal communication) have described a group of nine children who suffered from severe recurrent respiratory infections and in whom no immune deficiency could be identified. On biopsy of the right middle lobe, characteristic findings of lymphoid nodules in the small bronchi were seen protruding into the lumen. We have recently been shown another such case by Professor Corrin referred by Dr. Henry. In this patient there was also evidence of a lymphoepithelium.

Because of the abundance of this type of lymphoid tissue in the rabbit most work has been performed in this species. The BALT occurs along the whole length of the bronchial mucosa and appears to be concentrated around bifurcations.[1] These maybe positions where antigens impinge in relatively greater amounts than at other sites due to the turbulence of air flow.[3] The BALT consists of lymphoid follicles which, at some point, make contact with the epithelium above. The epithelium is rather specialized and termed a lymphoepithelium which is characteristic also of the GALT and bursa of Fabricius. There are no goblet cells in the epithelium. The cells constituting this lymphoepithelium are attenuated, non-ciliated and possess microvilli.[1,9,16] These cells have numerous crevices and microfolds and have been termed M cells.[17,18] The extent of the lymphoepithelium may well be influenced by immunological stimulation[18-20] and it seems possible that products of lymphocytes and macrophages may directly influence M cell growth.

The lymphoepithelium is heavily infiltrated with lymphocytes. No characteristic differences between these and other lymphocytes have been observed but in the rabbit and mouse,[21,22] the epithelial lymphocyte population of the gut has peculiar granulations and may represent an unusual type of T cell.[22] In the rabbit, we have described[16] granulated basophil-like cells in the BALT lymphoepithelium as well as in the normal bronchial epithelium. These cells were present more commonly in animals in which respiratory tract infection occurred. They resembled blood basophils which at least in the rabbit appeared to bear a T cell marker.[23] Other authors have shown that basophil-like (basophiloid) cells are present in the bronchial lavage fluid of dogs, monkeys and man.[24,25] Since the epithelial T cell in the intestine is thought to be related to the mast cell found in the lamina propria[21,22,26] and since these mucosal mast cells have morphologic characteristics which distinguish them from connective tissue mast cells[26] and are said to be "atypical" both in lung and gut, it is possible that the granulated T cell, mucosal mast cell and basophil-like cell in the bronchial epithelium are related in some way. Guy Grand et al[22] have observed that the Peyer's patches give rise to a cell which migrates via the mesenteric lymph node and the thoracic duct to eventualy populate the intestinal epithelium. We have observed the traffic of mesenteric node T blast cells into the bronchial epithelium (McDermott and Bienenstock, Unpublished).

BALT follicles possess post-capillary venules with a high endothelium particularly in the parafollicular areas.[7,9] These areas are associated with the traffic of small recirculating lymphocytes from blood into the lymphoid tissue.

The BALT is organized into lymphoid aggregates. The follicles are generally composed of small lymphocytes but large lymphoblasts are often found in the domes. In vivo tritiated thymidine labelling has shown that there may be a movement of cells from the follicles toward the luminal surface and out across the lumphoepithelium into the lumen.[2]

In the rabbit, 18% of BALT cells carried a T-marker which was comparable to the content of Peyer's patch lymphocytes whereas some 50% appeared to be B-cells as identified by surface immunoglobulin staining.[18] The predominant immunoglobulin class found in these lymphocytes was IgA. Although antibody containing cells are not found in Peyer's patches, occasional plasma cells are seen in BALT, both in the follicle and in the dome.[1,7,9]

From studies performed in chickens which were bursectomized or thymectomized at hatching, few effects were seen in the morphology of BALT.[29] Following neonatal thymectomy in the rabbit we were similarly unable to observe any effect on the BALT. When bursectomy was coupled to sublethal X-irradiation, which produces agammaglobulinemic birds, germinal centers in all lymphoid tissues were significantly depleted. Only when bursectomy was coupled to thymectomy without irradiation, did we observe that germinal centers in the BALT were totally disorganized but still present in the caecal tonsil.[29] This latter effect was accompanied by a total absence of serum and secretory IgA[30] suggesting that the BALT contained cells destined for IgA production.

BALT contains macrophages, and some dendritic macrophages have been observed. When antigen is instilled into the bronchial lumen there is selective antigen uptake by the lymphoepithelium.[20] In both rats and rabbits soluble antigen is absorbed by pinocytosis[8,20] as has been shown for M cells in the GALT[17,31,32] and bursa. The M cells also seem able to selectively transport large particulate matter such as BCG organisms.[20] It has recently been suggested that alveolar macrophages capable of antigen presentation in immune responses, may migrate across the epithelium into BALT follicles.[35] Brundelet[6] showed some years before that the introduction of particulate matter and certain dyes led to uptake by alveolar macrophages and tracking up along presumed lymphatics in the longitudinal axis of the bronchus to these bronchial lymphoid follicles. This would suggest that much of BALT would have afferent lymphatics but this has never been formally investigated.

In most animals which have been examined BALT development correlates with the degree of antigenic stimulation.[10-13,15] However, it is poorly developed in germ-free animals[3] and is even present when transplanted fetal lungs develop in the subcutaneous space in a presumed antigen-free miliew.[36] In children, this type of lymphoid aggregate has not been found at birth but appears at approximately one week of age and continues to develop throughout infancy and childhood.[15] We can assume that BALT development is dependent upon antigenic stimulation but that this is not essential for its appearance. Similar observations have been made with regard to GALT.[3,37]

Attempts to explore in more detail the nature of BALT lymphocytes led to repopulation experiments. Craig and Cebra[38] had shown that lymphocytes derived from GALT when transferred into lethally irradiatied rabbits repopulated the spleen and gut mucosa six days later with predominantly IgA-containing cells. We repeated these experiments[39] and looked at the bronchial tract. We showed that the bronchial tract was repopulated similarly with IgA although the numbers of such cells were markedly fewer. We then investigated the population capacity of BALT cells and showed that these had an almost similar potential to populate the gut with IgA-containing cells as did those derived from GALT. We suggested that there was sufficient similarity between these two types of lymphoid tissues to consider the presence of a common mucosal immune system.[1-3,26] More recently, we have shown that blast cells from the mesenteric node selectively localized in the lung[40] and other mucosal sites such as the female cervix, where this localization is regulated by the hormonal state of the animal.[41] Similar observations have been made for the breast by Lamm and Phillips-Quagliata coworkers.[42-44] Experiments with blast cells from the bronchial lymph node have shown that those cells subsequently found in mucosal tissues again contain predominantly IgA.[40] We are currently investigating the possibility that there is a similar traffic for T-cells between mucosal tissues as is snow accepted for B-cells, particularly of the IgA class.

The implications of this sort of study suggest that upon priming either in the BALT or the draining lymph node, cells destined to make IgA will return to the bronchial mucosa and other mucosal sites. Should they meet the antigen to which they are primed at these sites, there is likely to be proliferation and selective retention. Because the gut provides the larges source of mucosally-derived cells in the body, and since the localization of these blasts in mucosal tissues is dependent on the number injected in the experimental state (Mirski, McDermott, Befus, Bienenstock, Submitted). It would appear that seeding of the lung might best be achieved either by local immunization with a replicating agent or by immunization of the gut. Indeed, satisfactory resistance to adenovirus infection has been obtained by feeding of the virus in enteric-coated capsules.[45,46] Although much work has been done on the best approaches to immunize the gut,[47,48] these types of experiments are only just beginning to be performed[49] in the lung. By analogy with the gut, it may be that oral immunization followed by lung challenge may provide the most efficacious immunity.

Acknowledgement

The work described in this paper was supported by the Medical Research Council of Canada.

References

1. J. Bienenstock, N. Hohnston, D.Y.W. Perey, Bronchial Lymphoid Tissue. I. Morphologic characteristics, Lab. Invest. 28:686-692 (1973)

2. J. Bienenstock, N. Johnston, D.Y.E. Perey, Bronchial lymphoid tissue. I. Functional characteristics, Lab. Invest. 28:693-698 (1973)

3. J. Bienenstock, R.L. Clancy, D.Y.E. Perey, Bronchus-Associated Lymphoid Tissue (BALT): Its Relationship to Mucosal Immunity, in "Immunologic and Infectious Reactions in the Lung", C.H. Kirkpatrick and H.Y. Reynolds, ed., Marcel Dekker, Inc., New York (1976)

4. E.Klein, The anatomy of the lymphatic system. II. The lung", London, Smith, Elder, and Co., (1875)

5. W.S. Miller, "The Lung", C.C. Thomas, Springfield, III. (1973)

6. P.N. Brundelet, Experimental study of the dust-clearance mechanism of the lung, Acta Path. Microbiol. Scand., 175 (Suppl.) (1975)

7. D.W. Chaberlain, C. Nopjaroonsri and G.T. Simon, Ultrastructure of the pulmonary lymphoid tissue, Am. Rev. Resp. Dis. 108:621 (1973)

8. M. Fournier, F. Vai, J.P. Derenne and R. Pariente, Bronchial lymphoepithelial nodules in the rat: morphologic features and uptake and transport of exogenous protein, Am. Rev. Resp. Dis. 116:685-694 (1977)

9. P. Racz, K. Tenner-Racz, Q.N. Myrvik and L.K. Fainter, Functional architecture of bronchial associated lymphoid tissue and lymphoepithelium in pulmonary cell-mediated reactions in the rabbit, J. Reticuloendothel. Soc. 22:59-83 (1977)

10. R.L. Gregson, M.J. Davey and D.E. Prentice, Post natal development of bronchus associated lymphoid tissue (BALT) in the rat, "Rattus norvegicus", Lab Animals 13:231-238 (1979)

11. R.L. Gregson, M.J. Davey and D.E. Prentice, The response of rat bronchus-associated lymphoid tissue to local antigenic challenge, Brit. J. Exp. Path. 60:471-482 (1979)

12. K.W.F. Jericho, P.K.C. Sustwick, R.T. Hodges and J.B. Dixon, Intrapulmonary lymnphoid tissue of pigs exposed to aersols of carbon particulates, of salmonella oranienburg, of Mycoplasma granularum, and to an oral inoculum of larvae of Metastrongylus apri, J. Comp. Pathol. 81:13-21 (1971)

13. K.W.F. Jericho, J.B. Derbyshire and J.E.T. Jones, Intrapulmonary lymphoid tissue of pigs exposed to aerosols of haemolytic streptococcus group L and porcine adenovirus, J.

Comp. Pathol. 81:1-11 (1971)

14. C.C Macklin, Pulmonary sumps, dust accumulations, alveolar
     fluid and lymph vessels, Acta Anat. 23:1-33 (1955)

15. J.L. Emery and F. Dinsdale, The postnatal development of lymph
     nbodes in infants' lungs, J. Clin. Path. 26:539-545 (1973)

16. J. Bienenstock and N. Johnston, A morphologic study of rabbit
     bronchial lymphoid aggregates and lymphoepithelium, Lab.
     Invest. 35:343-348 (1976)

17. R.L. Owen and A.L. Jones, Epithelial cell specialization
     within human Peyers's patches: an ultrastructural study of
     intestinal lymphoid follicles, Gastroenterology 66:189-203
     (1974)

18. B.H. Waksman and H. Ozer, Specialized amplification elements
     in the immune system. The role of nodular lymphoid organs in
     the mucous membranes., Prog. Allergy 21:1-113 (1976)

19. P.Racz, K. Tenner-Racz, Q.N. Myrvik, B.T. Shannon and S.H.
     Love, Sinus reactions in the hilar lymph node complex of
     rabbits undergoing a pulmonary cell-mediated immune response:
     sinus macrophage clumping reaction sinus lymphocytosis and
     immature sinus histocytosis, J. Reticuloendothel. Soc.
     24:499-525 (1978)

20. K.Tenner-Racz, P. Racz, Q.N. Myrvik, J.R. Ockers and R.
     Geister, Uptake and transport of horseradish peroxidase by
     lympoepithelium of the bronchial associated lymphoid tissue in
     normal and bacillus Calmette-Guerin-immunized and challenged
     rabbits., Lab. Invest. 41:106-115 (1979)

21. O.Rudzik and J. Bienenstock, Isolation and characteristics of
     gut mucosal lymphocytes, Lab. Invest. 30:260-266 (1974)

22. D. Guy-Grand, C. Griscelli and P. Vassali, The mouse gut T
     lymphocyte, a novel type of T cell. Nature, origin, and
     traffic in mice in normal and graft-versus-host conditions.,
     J. Exp. Med. 148:1661-1677 (1978)

23. R.P. Day, D.P. Singal and J. Bienenstock, Presence of thymic
     antigen on rabbit basophils, K. Ummunol. 114:1333-1336 (1975)

24. R. Patterson, Y. Tomita, S.H.Oh, I.M. Susko and J.J.
     Pruzansky, Respiratory mast cells and basophiloid cells Clin.
     Exp. Immunol. 16:223-234 (1974)

25. R. Patterson, I.M. Susko and K.E. Harris, The invivo
     transferring of antigen induced airway reactions by bronchial
     lumen cells, J. Clin. Invest. 62:519 (1978)

26. J. Bienenstock, The physiology of the local immune response
     and the gastrointestinal tract, in: "Progress in Immunology
     II" Bol. 4, L. Brent and J. Holborow, ed., North-Holland
     Publishing Company (1974) pp. 197-207

27. L. Enerback, Mast cells in gastrointestinal mucosa. I. Effects
     of fixation, Acta Path. Microbiol. Scand. 66:289-302 (1966)

28. O. Rudzik, R.L. Clancy, D.Y.E. Perey, J. Bienenstock and D.P.
     Singal, The distribution of a rabbit thymic antigen and
     membrane immunoglobulins in lymphoid tissue, with special
     reference to mucosal lymphocytes, J.Immunol. 114:1-4 (1975)

29. J. Bienenstock, O. Rudzik, R.L. Clancy and D.Y.E. Perey, Bronchial lymphoid tissue, in: "The Immunoglobulin A System", J. Mestecky and A.R. Lawton, ed., Plenum Publ. Corp., New York (1974) pp. 47-56 Advances in Experimental Medicine and Biology vol. 45

30. D.Y.E. Perey and J. Bienenstock, Effects of bursectomy and thymectomy on ontogeny of fowl IgA, IgG and IgM, J. Immunol. 111:633-637 (1973)

31. R.L. Owen, Sequential uptake of horseradish peroxidase by lymphoid follicle epithelium of Peyer's patches in the normal unobstructed mouse intestine: an ultrastructural study, Gastroenterology 72:440-451 (1977)

32. D.E. Bockman and W. Stevens, Gut-associated lymphoepithelial tissue: bidirectional transport of tracer by specialized epithelial cells associated with lymphoid follicles, J. Reticuloendoth. Soc. 21:245-254 (1977)

33. D.E. Bockman and M.D. Cooper, Pinocytosis by epithelium associated with lymphoid follicles in the bursa of Fabricius, appendix and Peyer's patches. An electron microscopic study. Am. J. Anat. 136:455-477 (1973)

34. T. Schaffner, J. Muller, M.W. Hess, H. Cottier, B. Sordat and C. Ropke, The bursa of Fabricius: a central organ providing for contact between the lymphoid system and intestinal content, Cell. Immunol. 13:304 (1974)

35. M.F. Lipscomb, G.B. Toews, C.R. Lyons and J.W.Uhr, Antigen presentation by guinea pig alveolar macrophage, J. Immunol. 126:286 (1981)

36. R.W. Milne, J. Bienenstock and D.Y.E. Perey, The influence of antigenic stimulation on the ontogeny of lymphoid aggregates and innunoglobulin-containing cells in mouse bronchial and intestinal mucosa, J. Retic. Soc. 17:361-369 (1975)

37. M. Pollard and N. Sharon, Responses of the Peyer's patches in germ-free mice to antigenic stimulation, Infect. Immun. 2:96-100 (1970)

38. S.W. Craig and J.J. Cebra, Peyer's patches: an enriched source of precursors for IgA-producing immunocytes in the rabbit, J. Exp. Med. 134:188-200 (1971)

39. O. Rudzik, R.L. Clancy, D.Y.E. Perey, R.P. Day and J. Bienenstock, Repopulation with IgA-containing cells of bronchial and intestinal lamina propria after the transfer of homologous Peyer's patch and bronchial lymphocytes, J. Immunol. 114:1599-1604 (1975)

40. M.R. McDermott and J. Bienenstock, Evidence for a common mucosal immunologic system. I. Migration of B Immunoblasts into intestinal, respiratory and genital tissues, J. Immunol. 122:1892-1898 (1979)

41. M.R. McDermott, D.A Clark and J. Bienenstock, Evidence for a common mucosal immunologic system. II. Influence of the estrous cycle on B immunoblast migrtion into genital and intestinal tissues, J. Immunol. 124:2536-2539 (1980)

42. M.R. Roux, M. McWilliams, J.M. Phillips-Quagliata, P. Weisz-Carrington and M.E. Lamm, Origin of IgA-secreting plasma cells in the mammary gland, J. Exp. Med. 146:1311-1322 (1977)

43. P. Weisz-Carrington, M.E. Roux, M. McWilliams, J.M. Phillips-Quagliata and M.E. Lamm, Hormonal induction of the secretory immune system in the mammary gland, Proc. Nat. Acad. Sci. USA 75z;2928-2932 (1978)

44. P. Weisz-Carrington, M.E. Roux, M. McWilliams, J.M. Phillips-Quagliata and M.E. Lamm, Organ and isotype distribution of plasma cells producing specific antibody after oral immunization: evidence for a generalized secretory immune system, J. Immunol. 123:1705-1708 (1979)

45. W.P. Edmonson, R.H. Purcell and B.F. Gundelfinger, Immunization by selective infection with type 4 adenovirus grown in human diploid tissue culture. II. Specific protective effect against epidemic disease, J. Am. Med. Asso. 195:453 (1966)

46. T.J. Smith, E.L. Buescher, F.H. Top, W.A. Altemeier and J.M. McCown, Experimental respiratory infection with type 4 adenovirus vaccines in volunteers: clinical and immunological responses, J. Infect. Dis. 122:239-248 (1970)

47. N.F. Pierce and J.L. Gowans, Cellular kinetics of the intestinal immune response to cholera toxoid in rats, J. Exp. Med. 142:1550-1563 (1975)

48. A.J. Husband and J.L. Gowand, The origin and antigen-dependent distribution of IgA-containing cells in the intestine, J. Exp. Med. 148:1146-1160 (1978)

49. D.D. Joel, Ad.D. Chanana and P. Chandra, Immune responses in pulmonary lymph of sheep after intrabronchial administration of heterologous erythrocytes, Am. Rev. Resp. Dis. 122:925 (1980)

DISCUSSION

LECTURER: Bienenstock   CHAIRMAN: Bonsignore

CHAIRMAN: This paper is open for discussion.

DENISON: John, in that slide you showed of a lymph node
                 beneath the cartilage and just pushing its way
                 through to the surface, I suppose that the cells
                 have to get to the surface to do a useful job, or
                 material has to come to them. Now, two separate
                 things: how free are those cells to move around
                 within the node? Are they rushing around? How
                 quickly will they reproduce themselves?

BIENENSTOCK: Both the patches and the tissue turn over extremely
                 rapidly, much faster than in a peripheral node.
                 Your concept of them rushing around is alien to an
                 immunologist, they are just crawling, not rushing
                 around. From studies using pulse labelling we have
                 been able to show that one can see these cells out
                 in the lumen, but it's a very hard business to
                 guess as to how many there are. There is no
                 question that this lymphoid tissue is in large part
                 driven by antigens; but that's no surprise. The
                 birth of fibrosis is now even thought to be driven
                 to some extent by antigen as well; it's just that
                 we feel that this tissue in small amounts is
                 present in a relatively normal lung and just
                 extends under direct stimulation.

JEFFERY: John, earlier on there was a remark regarding
                 ciliary insufficiency at the bronchial bifurcation.
                 It's our experience in the rat where we are looking
                 at the distribution of lymphoid tissue, that when
                 it forms a lymphoepithelium it does often have
                 bronchial epithelium and in that event there is a
                 deficiency of ciliated cells. Would this be your
                 finding also in a larger number of species?

BIENENSTOCK: There is almost invariably a cilium over the few
                 cells that form the true lymphoepithelium. The
                 difficulty is that the follicle which might be
                 constituted by several hundreds to several
                 thousands of cells is potentially only covered by a
                 smaller number of five or six lymphoepithelium
                 cells; it's very difficult sometimes to cut it.

There is a lady called Benita Plesh who works in Holland, who has been studying the ontogeny of this type of lymphoid tissue in the rat. She would agree with that, it is there in part at birth and gradually develops. Grigson at Chelsea Polytechnic has also done some work in this area and again he does not see the cilia at the start. The same thing as the lymphoepithelium is found in the appendix in the rabbit, where careful studies have been done and again this lymphoepithelium appears in those situations to precede the actual invasion of lymphocytes.

KAY: John, you have told us that there are intraepithelial cells which look like mast cells and you have told us that these are very intimately associated with T cells. Although you haven't said it, I wouldn't be surprised if they are of T cell origin. Would you like to speculate or would any lung pathologists like to comment on what is the association between these intraepithelial mast cells in the lung and bronchial asthma?

BIENENSTOCK: Well, I'll have a crack, because when I first saw these cells in the cat, I assumed that they have a lot to do with bronchial asthma, because antigen would have to meet them assuming that they have any commitment to antigen in the intraepithelial cellular space. When I looked at the morphological nature of these things in the rabbit bronchus I really looked very hard but we never saw those types of lymphocytes. We have seen occasionally cells looking like mast cells, quite a few cells that look like the classic description of globular lymphocytes which were thought by some to be related to mast cells, and we have seen basophils and of course lymphocytes. Whether they are intimately related to firing and then are influencing the local nervous system is a big question and I wonder if anybody else has some comment on that.

CAPELLI: You carried out most of your studies with animals and I would like to know whether we can assume that these bronchial lymph nodes are involved in sarcoidosis. If so, what sort of role would they play as a lymphocyte-supplying reservoir to the air passages that we usually sample in broncho-alveolar lavage?

BIENENSTOCK: To take the last question first, some of those
                cells enter the lumen of the respiratory tract;
                certainly in the rabbit and in the rat we have been
                able to trace them there. The exact extent to which
                this occurs and therefore how many of the cells
                found in bronchial lavage come from this route is
                not clear. Daniel has found that the majority of
                the cells in the rat lumen have recently divided
                and this would fit with some of our work as well.
                The nature of this type of lymphoid tissue was
                first investigated in about 1860, the first studies
                being carried out with tuberculosis in the guinea
                pig and what I am saying is an extrapolation from
                what I read and from the drawings I showed you. If
                this is the same tissue, then these are areas where
                there is accumulation in those early days of
                tuberculosis. I can't claim except for what I have
                shown you, that I know anything about man in
                relation to this type of lymphoid tissue and I
                think that it remains yet to be ascertained in man
                for the initial findings that I have described to
                you. The exact form which it takes is clearly not
                the same as it is in, for example, the rabbit. I
                think the question will only be resolved by a very
                careful examination of the human respiratory tract
                looking for this type of tissue. So far it has
                been found in all species where it has been sought.
                The early German literature suggested that it is
                present in man, but I think it is at the moment
                disputable.

CUMMING: John, I would like to ask you two quetions. The
                first one relates to the slide in which you showed
                data, on the ordinate did I read correctly that the
                number of cells you counted were one per thousand
                fields, so you have seen 20 cells in one thousand
                fields?

BIENENSTOCK: What we do is to lay out the bronchi on a
                longitudinal axis cut so the bronchus is laid right
                across the side. This data is from a mesenteric
                node, a blood cell which has a tendency to go back
                not into the lung but into the gut. If we use
                bronchial lymph node cells we see much greater
                numbers.

CUMMING: Can you give me some feel for the number of those
                cells, they are quite small.

BIENENSTOCK: Well, the numbers are quite small but, you see, according to Gigant who has done with Vassalli a lot of work on the intestinal cells, it takes quite some time for these cells to get into the epithelium. This is what we are seeing at 24 hours; from her work it would appear that it's between 48 and 96 hours that these cells appear and if you use tritiated thymidine labelled cells, you can't chase them out before four days and be certain that you are seeing the correct population because of the utilisation of label. So I am giving these small numbers, which I can be sure about but I think the real number is much greater.

CUMMING: Thank you. The second question relates to trying to find a solution to a problem that I currently have, which suggests that some curious phenomenon is going on somewhere between the subsegmental bronchus and the terminal bronchioles and I wonder therefore whether you are able to tell me yet, whether you have any idea about the distribution of mucosal lymphoid tissue in this area in the human.

BIENENSTOCK: Absolutely no. I have no word in the human.

ROSSI: Regarding the distribution of lymphocytes in broncho-alveoli lavage in normal subjects, we found that the majority of spontaneous secreting immunoglobin in BAL is IgA which in some ways was surprising; then we made another study trying to look at lung biopsies and we found, in agreement with your study, that bronchial tissue is secreting more IgA than IgG. In terms of sarcoidosis we have only preliminary results but in the bronchi there is the same proportion between T and B cells and there is no major difference in terms of immunoglobulin secretion.

KUHN: Just a comment about the distribution of lymphoid tissue in man. Quite commonly we see it in young subjects at the level of the respiratory bronchioles and terminal bronchioles. You don't ordinarily see any lymphoid tissue in adults in cartilage-containing bronchi. However, in some old studies by Emery on sudden infant death syndrome it was quite clearly described, and illustrated in cartilage-containing bronchi. So I think that in children at the time at which lymphoid tissue is at its peak of development, it is present in human bronchi.

ROSSI:
Thank you. I didn't mention Emery and Dienstal who published that report because they didn't actually show specifically a lymphoid epithelium although I interpret their studies in this way, I think they would have to be done again to see if they have a lymphoid epithelium.

CORRIN:
Relevant to the last point, I have recently visited Professor Emery with a view to reviewing those lymphoreticular aggregates and in the course of the examination it became obvious we were looking at what is covered by bronchoepithelium as well. I think he is seeking both types of lymphoid tissue in human infant bronchus. The question I wanted to pursue is the differences between epithelial and connective tissue mast cells. I have two questions, perhaps the first I could direct to Barry Kay. John Bienenstock has suggested that the epithelial mast cell is of T lymphocyte parentage; what is known of the derivation of connective tissue mast cell? Is it different from that and then John, could we have some more details from you about the structural, functional and tinctorial differences you said exist between epithelial and connective tissue mast cells and then hopefully what do the granules contain in the epithelial mast cells?

KAY:
I know very little about those particular cells in the bronchus. We are just setting up to do some studies with Peter Jeffery along this line. What has been said is that they are atypical from several points of view. They often have bilobed nuclei, the granules were described by Anniback in Sweden as being different from the granules of other cells. The tinctorial and fixation properties are rather bizarre in that it is not possible to detect mucosal mast cells in the lamina propria after fixation in formalin and since the vast majority of tissues have been studied in formalin that really presents a major problem. We have just carried out a very careful study in the human intestine in which we can drop the count of mast cells in the lamina propria twentyfold when we compare formalin to other fixation. We are just in the process of trying to see if that holds true for

the lamina propria. What we have pointed out is in the lamina propria, I can't speak for the epithelium in man. There are other histochemical properties, it is a question of degree of sulphation, in fact cells have less sulphation than do cells in the peritoneal cavity and therefore stain differently with blue at different pH's, which can be strengthened by addition of magnesium ions. There are large numbers of such differences. Regarding the derivation of those cells, Akitomura in Japan has the best information, and rather surprisingly this goes against the dogma that the connective tissue mast cell is derived locally. His very careful study was able to show that the connective tissue mast cell appears to be derived rather surprisingly from the bone marrow and it is distinctly possible that some of these cells in the intestine and the lung are also derived from bone marrow, but that some of them may be heavily influenced by T cells if not related themselves to T cells.

ROSSI: You spoke about immunisation with antiadenovirus vaccine. Do you have any information if this immunisation is mediated by B cell monocytes or T cells and if the traffic between the intestine and the lung is mediated by any of those cells?

BIENENSTOCK: The best information is that every virus that we are aware of in the respiratory tract has been directly associated with the presence and quantity of specific IgA antibody; this doesn't mean to say that IgA antibody mediates that particular defence mechanism, and it could well be that T cells are equally important if not more important; as for monocytes and macrophages, there is virtually no information in this area.

PULMONARY EMPHYSEMA

D.C.S. Hutchison

Chest Unit
King's College Hospital Medical School
London, SE5 8RX

The cardinal feature of pulmonary emphysema is progressive destruction of the lung substance. Patients with this disease suffer from increasingly severe shortness of breath on exertion which may result in loss of working capacity, early retirement or premature death. The condition may present at any period of adult life from the age of 30 onwards.

The disease has been known for a century or more, but it is only in the last 15 years that a detailed understanding of its pathogenesis has been possible. Great impetus was given to research in this field by the discovery of alpha$_1$antitrypsin (AT) deficiency by Laurell and Erikson[1]. AT is one of the main serum inhibitors of proteolytic enzymes and the association of this hereditary deficiency with pulmonary emphysema was quickly recognised[2]. At about this time Gross et al[3] discovered that emphysema could be produced experimentally in rats by intratracheal administration of the proteolytic plant enzyme, papain, but it is now thought that lysosomal elastase is responsible for human emphysema.

These discoveries have led to the idea that a balance is normally maintained between proteolytic enzymes and their inhibitors. It is now recognised that these enzymes are released from circulating neutrophil leukocytes and from alveolar macrophages as part of the normal functioning of the organism. The enzymes are concerned with the removal of cell debris and coagulation products which accumulate in the pulmonary capillaries and they also participate in the equilibrium between structural protein breakdown and repair. In AT deficiency, it appears that the balance is tilted sharply in favour of enzymatic degradation of

239

pulmonary elastic tissue; in cases of emphysema with normal serum levels of AT, the pathological and physiological features are very similar to those of AT deficiency and we must therefore conclude that this much commoner form of the disease is brought about by excessive proteolytic activity.

ALPHA₁ANTITRYPSIN DEFICIENCY

AT deficiency, although not a common disorder, has provoked wide interest as a 'model' for pulmonary emphysema. AT can exist in a number of different biochemical forms and over 30 have now been described. The majority are rare and for practical purposes we need only consider 3 variants, M, S and Z from which 3 homozygous types (MM, SS and ZZ) and 3 heterozygous types can be derived, each variant contributing a characteristic quantity of AT to the serum (Table 1). The majority of the population are homozygous for type M. Homozygotes of type Z have only 10-20% of the normal serum AT concentration and have a strong tendency to develop severe emphysema. Subjects of type SZ also have a low serum AT concentration, and have a theoretical risk of developing emphysema. (Homozygotes of a rare phenotype 'Z null' who have no circulating AT, may present with emphysema in childhood).

The precise risk for type Z homozygotes is difficult to assess because subjects of this phenotype are usually identified by attendance at a chest clinic with established lung disease. The rarity of the condition in the UK means that the only practical method of detecting 'unselected' individuals is by family studies.

About 5 years ago, the British Thoracic Association started a multicentre survey to study the disorder in more detail and in particular to define the influence of smoking, sex and other factors. The Pi phenotype was assessed by starch gel electrophoresis and a completed clinical questionnaire and the results of lung function tests were obtained in AT deficient

Table 1

Serum AT concentration[4] and prevalence[5] in various phenotypes

Phenotype	AT conc (gl⁻¹) Mean	SD	% contribution (MM = 100)	Prevalence in UK population %
MM	2.86	0.73	100	86
MS	2.15	0.47	75	9
MZ	1.64	0.44	57	3
SS	1.49	0.23	52	0.25
SZ	1.06	0.34	37	0.2
ZZ	0.45	0.08	16	0.029

Table 2

Age and FEV_1 in type ZZ index cases

	Males		Females	
	Smoked	Never Smoked	Smoked	Never Smoked
n	57	5	24	7
Mean age (years)	48	51	47	58
FEV_1 (% predicted) ± SD	27.6 ± 15.3	48.4 ± 31.9	33.4 ± 15.3	55.9 ± 24.2
p	NS		<0.05	

subjects.   Patients identified through attendence at a chest clinic were classified as the 'index cases' and those identified through family studies (who were almost all full sibs of the index cases) were classified as the 'non-index cases'. Some of the initial data from cases of phenotypes ZZ and SZ are presented here. This data has already been published in abstract[6]. Lung function tests have been obtained in 93 index cases of type ZZ (Table 2). The FEV_1 (as % of predicted) was greater in those who had never smoked, though the difference was only significant at the 5% level in females. The members of the non-smoking group were older than those who had smoked with a difference of 11 years in the females. The age of onset of Grade 2 dyspnoea was also related to the smoking history; symptoms occurred later among patients who had never smoked, though again the difference was only significant in females. The numbers in the non-smoking groups were relatively small and there was a wide scatter of results in all the groups of patients.

Patients with the SZ phenotype were much fewer in number than those of type ZZ, to date only 11 index cases being included in the survey,   although type SZ is more prevalent in the general population (Table 1). The onset of Grade 2 dyspnoea in 7 male index type SZ smokers was not significantly different from that of emphysematous patients of non-deficient phenotypes (all male smokers or ex-smokers).   Type ZZ patients however developed dyspnoea significantly earlier than either of the above two groups, suggesting that the SZ phenotype is much less important than the ZZ phenotype in predisposing towards emphysema.

Emphysema associated with type ZZ AT deficiency differs from the much commoner non-deficient form in a number of respects:

1. The symptoms are of earlier onset.

2. Females are more commonly affected, whereas among non-deficient cases the male to female ratio is about 9:1.

3. Subjects who have never smoked may develop emphysema, whereas the non-deficient form is virtually confined to smokers.

4. The lower zones of the lungs are almost always the most seriously affected; in contrast, upper zone disease is commoner in non-deficient cases.

EMPHYSEMA WITH NORMAL AT PHENOTYPE

The role of cigarette smoking in the pathogenesis of the common non-deficient form of emphysema is well established and this disease is rare among those who have never smoked. In working populations, the rate of decline in FEV_1 is smaller in ex-smokers than in those who continue[7], but the effects of smoking withdrawal have not been fully established in patients who already have obvious emphysema. A study was therefore carried out at King's College Hospital to throw some light on this point[8]. 56 male patients (not of phenotype ZZ or SZ) with radiological evidence[9] of pulmonary emphysema were assesed yearly and the rate of decline in FEV_1 and VC was shown to be significantly less in ex-smokers than in those who continued (Table 3). There was however no difference between the 2 groups in the rate of change in kCO or PaO_2.

Table 3

Annual change in lung function indices
Mean regression coefficients (+ 1 SE) of
variable on time (units per annum)

	Continuing smokers	Ex-smokers	P
FEV_1 (ml/year)	-53 (± 5)	-16 (± 9)	<0.001
VC (ml/year)	-53 (±11)	± 2 (±12)	<0.001
kCO (units/year)	+ 0.007 (±0.007)	± 0.016 (±0.008)	NS
PaO_2 kPa/year	- 0.21 (±0.40)	_ 0.20 (±0.04)	NS

Negative value indicates a decline in the variable, positive value an increase.

Table 4

Annual change in FEV_1 in ml/annum for different
x-ray categories

Mean regression coefficients on time (+ 1 SE)

	Continuing smokers	Ex-smokers
Upper zone	−86 (±10)	−46 (±12)
Lower zone	−38 (±17)	− 3 (±14)
Generalised	−43 (± 8)	− 1 (± 9)
p:UZ vs LZ or G	<0.001	<0.005

RADIOLOGICAL DISTRIBUTION OF EMPHYSEMA

The radiological distribution of emphysema was also found to have a considerable influence on the annual rate of change in lung funtion[10]. The 56 patients were placed in 3 categories; upper zone (UZ), lower zone (LZ) and generalised (G) and the patients in each category were sub-divided into smokers and ex-smokers as before.

With regard to the annual decline in FEV_1 and VC, the following observations were made (Table 4):

1. The smokers declined at a faster rate than the ex-smokers in all 3 radiological categories.

2. Among smokers, the UZ patients deteriorated at a faster rate than the LZ or G patients (who behaved alike).

3. Among ex-smokers in categories LZ and G, FEV_1 and VC underwent no significant change. In contrast, FEV_1 continued to fall in the UZ ex-smokers.

Cigarette smoking is thought to bring about the destruction of the pulmonary elastic tissue either by causing excessive release of proteolytic enzymes or by suppressing the inhibitory effect of AT or by a combination of the two; we might therefore expect the decline in lung function to be halted in ex-smokers as is seen in the LZ and G categories. The continuing decline in FEV_1 in the UZ category even when smoking has ceased, is possibly due to the fact that the mechanical stretching forces are greater in the upper lobes[11], so that stresses like coughing may aggravate the damage already caused by proteolytic action. Thus in smokers both 'mechanical' and 'biochemical' factors combine, whereas in ex-smokers, the biochemical factors are removed, and mechanical factors alone are responsible for the continuing decline in the FEV_1 in UZ cases.

It has already been suggested[12] that the upper and lower zone radiological types correspond respectively to the pathological forms, centrilobular and panlobular emphysema. Our findings confirm that upper zone emphysema behaves in a different way from lower zone or generalised emphysema. Both pathological processes appear to be due to enzymatic destruction of the lung parenchyma; panlobular emphysema could be due to release of elastase from polymorphs, accounting for the predominance of the disease in the lower zones, where the blood flow per unit volume of lung tissue is greater. The centrilobular form, on the other hand, may be caused by enzymes released from alveolar macrophages which accumulate within the respiratory bronchioles. A possible classification system for emphysema appears in Table 5.

UNSOLVED PROBLEMS

Our knowledge of the pathogenisis of emphysema is now extensive, but many questions remain unanswered:

1. In AT deficiency, very large differences in lung function are found, even among non-smokers. What other factors influence the prognosis?

2. Emphysema with normal serum AT. Almost all such patients are smokers, but only a minority of smokers develop emphysema. What is the predisposing factor and what is the explanation for the high male to female ratio?

3. Smoking: how can the toxic agents responsible for emphysema be removed from cigarette smoke?

4. Why do some patients develop centrilobular emphysema and other panlobular?

5. Replacement therapy for AT deficiency: how is this to be achieved?

Table 5 Classification of emphysema

x-ray zone	Serum AT	Pathology	Source of elastase
Upper	Normal	Centrilobular	Macrophage
Lower	a) Deficient b) Normal	Panlobular	Polymorph

REFERENCES

1. C.B. Laurell and S. Eriksson. The electrophoretic alpha$_1$-globulin pattern of serum in alpha$_1$antitrypsin deficiency. Scand. J. Clin. Lab. Invest. 15:132 (1963).
2. S. Eriksson. Studies in alpha$_1$antitrypsin deficiency. Acta. Med. Scand. Suppl. 432 (1965).
3. P. Gross, M.A. Babyak, E. Tolker and M. Kaschak. Enzymatically produced pulmonary emphysema. J. Occup. Med. 6:481 (1964).
4. J. Lieberman, L. Gaidulis, B. Garoutte and C. Mittman. Identification and characteristics of the common alpha$_1$antitrypsin phenotypes. Chest 62 557 (1972).
5. P.J.L. Cook. Genetic aspects of the Pi system. Postgrad. Med. J. 50:362 (1974).
6. D.C.S. Hutchison. British Thoracic Association survey of alpha$_1$antitrypsin deficiency (progress report), Brit. J. Dis. Chest 73:425 (1979).
7. C. Fletcher, R. Peto, C. Tinker and F.E. Speizer. The natural history of chronic bronchitis and emphysema. Oxford University Press, London (1976).
8. J.A. Hughes, D.C.S. Hutchison, D. Bellamy, D.E. Dowd, K.C. Ryan and P. Hugh-Jones. The influence of cigarette smoking and its withdrawal on the annual change of lung function in pulmonary emphysema. Quart. J. Med. In Press (1981).
9. J.W. Laws and B.E. Heard. Emphysema and the chest film: a retrospective radiological and pathological study. Brit. J. Radiol. 35:750 (1962).
10. J.A. Hughes, D.C.S. Hutchison, D. Bellamy, D.E. Dowd, K.C. Ryan and P. Hugh-Jones. Unpublished (1981).
11. J.B. West. Distribution of mechanical stress in the lung, a possible factor in localisation of pulmonary disease. Lancet, 1:33 91971).
12. N.A. Martelli, D.C.S. Hutchison and C.E. Barter. Radiological distribution of pulmonary emphysema. Thorax, 29:81 (1974).

ACKNOWLEDGEMENTS

    Grateful thanks are due to the members of the British Thoracic Association who referred cases to the alpha$_1$antitrypsin deficiency survey: to Mrs Claire Robertson for secretarial assistance: to technical and medical staff in King's College Hospital and elsewhere for their cooperation.

DISCUSSION

LECTURER: Hutchison   CHAIRMAN: Cumming

CUMMING: Thank you Duncan. I think it can be said that you
                trailed your coat adequately in a wide variety of
                areas. I see radiologists and pathologists and
                others putting up their hands. Professor Tricomi
                first.

TRICOMI: I thank the speakers because I think they are
                helping us radiologists to put things straight in
                such a difficult and controversial field as
                emphysema and chronic bronchitis. We radiologists
                have eventually reached an agreement concerning the
                assessment of emphysema, in that we all agree to
                take vascular distribution rather than
                hypertranslucency into account. Emphysema is
                characterised by destruction and therefore
                vascularisation has a tendency to decrease. It
                appears that the decreased vascularity can be more
                limited at the base and greater at the apex, so
                through reading the pulmonary vascularity values we
                can form a closer idea of the physio-pathological
                and anatomical picture of the sort of emphysema we
                are confronted with. The vascular distribution and
                the magnitude of vascularisation enable us to
                distinguish between real emphysema which
                predominantly represents a destructive process and
                the so called surgical emphysema, which is not
                really emphysema. If we surgically remove a lung
                from a patient, the other lung, failing to be
                affected by emphysema, acts as a substitute; now,
                on a radiological basis how can we differentiate
                between these two conditions? Just looking at the
                vascularity state, which is retained in surgical
                emphysema whereas it is reduced in true emphysema.
                Another point I would like to make is: there are
                some types of emphysema mainly associated with
                chronic bronchitis and chronic obstructive
                bronchitis, where there seems to be no destructive
                process from a radiological standpoint, since
                vascularity is retained. These states probably
                represent the early stages of emphysema, in which
                the destructive processes cannot be clinically,

physiopathologically and radiologically perceived, and the airway expansion is still confined to its initial stage where pulmonary vascularity is retained. I am going to end with a question to Dr. Hutchinson concerning the features of heart and aorta in the first type of emphysema you described, the one due to alpha 1 trypsin deficiency. You are probably aware of the work carried out in Sweden by Nordenstrom and colleagues, in which the cardiac-aortic index reversal is attributed to nutritional factors during fetal life. It is customary to say that in the case of the first-type emphysema, the heart shape is small and drop-like, whereas the aorta is comparatively large. Is there any explanation for this phenomenon, frequently observed in this type of emphysema?

HUTCHISON: Thank you for those comments. I am glad to find that you broadly agree with our mode of assessment of emphysema in this condition. You may be aware of a good study by Thurlbeck and Simon, which appeared recently in which they correlated pathological findings with x-rays and found that although you may not see radiologically the earlier form of emphysema, once you see it you are pretty sure that it is there and where it is. To answer your question, I was not aware of this work, the question of the aorta I have not considered at all; the long narrow heart is very characteristic, as you have commented, and I think this simply results from the overexpansion of the lungs which you also see in the very low position of the diagrams.

CUMMING: May I comment about the radiological findings and the work of Thurlbeck and Simon? Firstly, the overall precision of diagnosis in that paper was extraordinarily low, I think about 35%, so in 100 cases with emphysema, proven pathologically, you have a diagnostic technique which will pick up 35 of them. So it is not a very precise technique. I would like to address myself to Professor Tricomi, his discussion hinges very strongly on the word "vascularity". Now vascularity means many things to many people and I rather suspect that to a radiologist it means a different thing than it does to a physiologist. I wonder Professor Tricomi, if you would like to tell us what you mean by vascularity.

TRICOMI: Precisely the lung pattern, consisting of pulmonary
                 arteries and veins. Radiological vascularity
                 corresponds to the anatomical distribution and is
                 completely different from the perfusion of
                 pathologists, as there can obviously be
                 radiologically patent vessels without being
                 perfused. This clearly results from a further step
                 forward; the comparison between x-rays and
                 perfusion scintigraphy on the one hand and
                 ventilation scintigraphy on the other, in that the
                 latter technique enabled us to read chest x-rays
                 with less chance of error and in a functional
                 fashion.

HEATH: Dr. Hutchison, I must say that I have to give a
                 bit more anatomical support to what the Chairman,
                 Professor Cumming, has just said. Any
                 classification such as you gave, which is based on
                 radiology, must be regarded as highly suspect,
                 because I think that, as Professor Cumming said,
                 radiology is only going to pick up quite a small
                 percentage of cases of emphysema and cases in which
                 the degree of emphysema is severe. I did notice
                 that in the nice pictures which you showed, the
                 example of centrilobular emphysema was a very
                 severe confluent form and I am quite certain that
                 this could be picked up radiologically, but I am
                 equally certain that the majority of much more
                 ordinary cases of centrilobular emphysema would be
                 missed radiologically. This bothers me very much,
                 I think that the diagnosis of emphysema is very
                 difficult. I well recall when I worked at the
                 Queen Elizabeth Hospital in Birmingham. When I
                 went there the incidence of centrilobular
                 emphysema, in the hospital, was virtually zero and
                 then we started looking at the lung properly,
                 instead of looking at the routine necropsy.
                 Immediately the incidence of centrilobular
                 emphysema rose. When we lost our reserch interest
                 in that, after about four years, the incidence of
                 centrilobular emphysema in Birmingham suddenly
                 reverted back to a very low level. This is why I
                 think that any radiological classification such as
                 you described is faulty, so that you use, these
                 terms alpha 1 antitrypsin deficiency type emphysema
                 and then you call the others ordinary, but I think
                 what you call the ordinary type of emphysema is a
                 highly selected group consisting of fairly well
                 observable, detectable cases such as could even be

detected at an ordinary necropsy or cases of confluent centrilobular emphysema. What do you think about that criticism?

HUTCHISON: I accept and I have said this already, that radiologically you cannot detect the very early forms of emphysema. In my opening paragraph I said that what I was going to talk about was severe cases of emphysema. The cases you are talking about, who hve minor degrees of centrilobular emphysema at autopsy have never had any respiratory symptoms during life, they may have been cigarette smokers, who didn't have the predisposition to develop emphysema, but it does seem to me that it may be of interest from the pathological point of view, the definition is not of great consequence. What we are dealing with is the very large group of patients who are severely disabled, retiring prematurely, dying a premature death, which we all see in our clinics and it is that group that I am talking about. They must start early, and there is a stage in the evolution of every case of emphysema which is only a biochemical lesion that you can't see by any means at all and then progresses into a pathological lesion which you would recognise and only eventually do you see a radiological lesion. I don't think that this destroys the classification, there must be an evolution of some kind in every disease.

CUMMING: May I make a comment about that? While it is perfectly true that the evolution of the disease must occur in time, the presumption from this is that when the disease is well advanced, then radiological evidence is good. The facts do not support this. Simon and Thurlbeck showed several cases, in just over one thousand cases, I think 20 or 30 have got more than 70% destruction of panacinar emphysema and the three eminent radiologists who read the films said they were normal. I don't want to play down radiology because when emphysema is clearly present you can be certain that it is present, but Donald Heath's point is well taken, it does restrict your study to those cases in which it is radiologically demonstrable and must, by definition, ignore the others. So it is a classification for radiological emphysema but not of all emphysema.

HUTCHISON: But I think that Thurlbeck and Simon's later paper
                demonstrated that once you saw obvious emphysema on
                the x-rays you have a good chance.

CUMMING: Yes, but the point Professor Heath is making is
                that if radiological criteria selects the
                population then your results are on a biassed
                population.

HUTCHISON: No, but he selects the cases he is interested in.

CUMMING: He does indeed.

SOLIMAN: I'd like to ask about the pathological aspects of
                physiological impairment, since emphysema should
                include the function of the small airways. I don't
                think that anyone has covered this point. So I
                have to check function of the small airways. I am
                asking about the pathological reflection of small
                airways, in physiological impairment and I do not
                think anyone would cover the function of the small
                airways from the physiological point of view. I
                would suggest if we had estimated the mid-
                expiratory flow rate or the closing volume, these
                might cover some of the physiological impairment of
                the small airways.

HUTCHISON: Yes, sorry about that. Once emphysema is advanced
                the $FEV_{1.0}$ is quite a reasonable test of the degree
                of disability. It is extremely difficult to carry
                out more detailed tests of small airway functions
                when emphysema is present. The sort of test you
                are thinking of is much more relevant to those
                patients who have relatively normal lung function
                and normal $FEV_{1.0}$ and I don't think they are
                applicable to the patients with very severe degrees
                of emphysema which I am talking about.

DENISON: I am speaking in your defence, Duncan and I want to
                ask Professor Heath a question. Any classification
                must have a name and Duncan's classification was in
                essence setting out two potential mechanisms: one
                air-borne by macrophages, one blood-borne by
                polymorphs as I understood it. He was deliberately
                putting this possible explanation in front of us
                and I think it is accurate to comment on that.
                What do you think about that?

CUMMING: While he is getting the microphone perhaps I could

say that if you make a classification and by some
technique you divide it into two groups and then
you attribute a significance to the two groups and
if your first classification is wrong, then equally
your attribution is wrong.

HEATH: I must say I agree with what the Chairman has just
said. You see, I think what Dr. Hutchinson is
showing seems to me a very highly selected group of
cases, of the disease emphysema, and the morbid
anatomy is not accurately reflected on this system.
I would disagree with a statement that Dr.
Hutchison made in his reply ... Dr. Hutchison says:
well, unless there is a lot of lung destruction the
disease is not going to cause the patient much
harm. I don't think a lot of people in this room
would agree with that, because it is certainly my
experience as a morbid anatomist and it must be
your experience as a physiologist that often the
cases which cause problems are those in which
destruction is relatively slight. Certainly from
the point of view of hypoxemia and the development
of pulmonary hypertension, often the patients who
are most severely disabled are the ones in which
there is relatively little lung destruction and
these would be missed from this classification
altogether. Would you agree with that Duncan.

HUTCHISON: I would agree that they would be missed but I don't
agree with your interpretation of that
classification. The cases you are talking about
are completely different disorders. Many terms
have been applied to this and probably the
commonest is chronic obstructive bronchitis or at
least they might have had that disorder. There are
groups of patients which fit with what you said,
they have a very poor $FEV_{1.0}$, they have disability,
many of them have been smokers, they have little
emphysema on the x-rays, they develop cor pulmonale
and they have pulmonary hypertension. Now that
group is quite different from the ordinary case of
emphysema which we see whether it is AT deficient
or whether it is not. As to what you call that
disease, you can take your choice. They were
classified by the Medical Research Council and were
lumped under the heading of chronic bronchitis, but
chronic bronchitis was also defined by them in
terms of persistent sputum production. A sub-
classification of this, you remember, was chronic

obstructive bronchitis which are the diseases I have just outlined. The trouble is that the moment you get any patient who comes up with any sputum production, they all have chronic bronchitis and therefore they all have chronic obstructive bronchitis. Some people lumped them altogether and called them chronic bronchitis and emphysema as if it were all one word. To come back to the classification if we limit ourselves to cases who have established emphysema, I don't see our classification is effected if we don't include cases who do not have emphysema.

CUMMING: Thank you very much Duncan. I think I would concentrate on this interesting idea that Duncan has put up, that there are two causes for two different types of emphysema, one which is borne by the blood and one which is existing in the airways. I think it's a very interesting idea and should be pursued. My worry is that if you use a technique which is confusing, you may not get a good answer to that classification and I am not attacking the classification by virtue of the idea, but by virtue of the fact that if you use poor definitions you may not get a good answer and you may lose your classification. I think that the argument about radiology is sufficiently dubious to make that not a good first discriminator, I would suggest the use of another discriminator; many are available but to use radiology and to say: I would first divide it into two on this basis may make it worse for you.

FABRI: As regards the last point, I am interested in knowing your opinion about arterial blood gases at rest and on exercise in ZZ phenotype carriers. Moreover I would like to ask you if you have any diagnostic or prognostic experience on haemodynamic investigations on the right heart in this pathology.

HUTCHISON: We have done blood gases over a long period on the bulk of our patients who do not have the deficient phenotypes. In the AT deficient patients this was a multicentere study and I didn't have the opportunity, I felt it was not right to ask for blood gas estimations on other people's patients. The position about blood gases is that in emphysema, whether there is AT deficiency or not,

they are not seriously impaired for the major part
of the course of the disease. They do fall towards
the end of the condition and eventually dioxide
tension in the arterial blood will rise, but only
as a late phenomenon. As regards the second part
of your question, we have done no haemodynamic
investigations though I am aware that there is
quite a bit in the literature.

CUMMING: I will take one last question, from Dr. Rossi.

ROSSI: I think your hypothesis about the difference
                between upper lobe emphysema and lower lobe
                emphysema is fascinating, especially the fact that
                you relate the upper lobe to the macrophage and the
                lower lobe to the neutrophil. My first question
                is: do you have any evidence for that? Secondly,
                do you think the fact that elastase coming from the
                macrophage is not inhibited by alpha-1 antitrypsin
                can play any role in the pathogenesis of upper lobe
                emphysema? Thirdly, we have quite a bit of
                experience of patients with alpha-1 antitrypsin
                deficiency and we perform lavages on them and the
                neutrophils are not in the blood, they are in the
                alveoli. We don't have any data on BAL in
                centrilobular emphysema, but from what I know about
                pathology, they have neutrophils, located not in
                the alveoli but at the junction between the
                terminal bronchioles and the respiratory
                bronchioles.

CUMMING: Would you like to answer those four questions?

HUTCHISON: That's an interesting comment. My point was that
                we have two possible sources for the elastase. On
                the one hand the alveolar macrophages, on the other
                hand the circulating polymorph. We have two types
                of x-ray picture, and two types of pathology and it
                seems that the macrophages are in just the right
                position as they migrate to release the enzymes, as
                I understand they can at that point. They can
                produce a condition which appears to spread from
                the airways into the lobules of the lung. As to
                the question of inhibition by alpha-1 AT, that will
                come up later. I am interested to hear that you
                found a lot of white cells in the bronchial tree,
                people differ considerably in the literature as to
                the number of white cells in the bronchial tree.
                The bulk of the cell population in that area is

macrophages and I have no doubt that neutrophils
can get in when there is frank infection but in
most of our cases with emphysema, they are not in
an infected stage during the majority of their
careers.

CUMMING: I think we have to terminate with this Dr. Rossi.
We have used eight minutes of the next paper, maybe
it arises again when Dr. Baum has presented his
very similar paper.

ANTI-PROTEASES

Harold Baum and Susan Smith

Biochemistry Department
Chelsea College
Manresa Road, London SW3 6LX, UK

INTRODUCTION

Proteases have many important and well authenticated roles to play in mammalian physiological systems. Quite apart from their direct participation in digestion in the G.I. tract, they are involved in processes such as exocrine secretion, hormone release, blood coagulation and host defence as well as controlled turnover of cellular constituents. Clearly the host must be protected against the uncontrolled activity of proteolytic enzymes, otherwise widespread protein degradation would lead to localised necrosis and ultimately to death. Thus, proteases in exocrine secretions are usually in zymogen form and many endogenous proteases such as elastase, collagenase, cathepsin G and neutral proteases are contained by enclosure within intracellular membrane-bound structures (lysosomes), which are particularly numerous in phagocytic cells. However, there are a number of ways in which these potent enzymes are released from within their retaining membrane, the most important ones being damage to the lysosomal membrane itself, cellular degranulation resulting from plasma membrane stimulation, often by immunological stimuli, and lysosomal leakage during phagocytosis (Weissman et al., 1972; Henson, 1971). This latter process is of particular importance in cells involved in protection of the host against microbial attack, namely the polymorphonuclear leukocytes and mononuclear phagocytes. The amounts of protease involved are large. In man, the estimated daily turnover is of the order of 10^{11} PMN leukocytes (Ohlsson, 1979), corresponding to about 1g of proteolytic enzymes.

It is therefore necessary to have a 'second line' of defence against released lysosomal enzymes and any other proteases, for

255

example, those from exocrine secretions which may escape into the circulation, as well as those released by invading microbes. This defence system consists of a number of different protease inhibitors, which are present in blood and other body fluids including bronchial secretions, pancreatic secretions and cervical mucus. There are seven major inhibitors in plasma, but a number of these, antithrombin III, α_2-plasmin inactivator and C_1 inactivator appear to function solely in the coagulation, fibrinolytic and complement pathways respectively and will not be discussed in the present context.

The most important inhibitors in relation to prevention of proteolytic damage to lung and airways are those found in plasma and bronchial secretions. Some data about these antiproteases are presented in Table I. It can be seen that α_1-antitrypsin (α_1AT) is present in the highest molar concentrations in plasma, while the recently characterised antileukoprotease seems to be more important in bronchial secretions. However, as well as the simple concentrations of different inhibitors in various body fluids it is important to note that their specificities vary. There are four major groups of protease; serine, thiol, carboxyl and metal-dependent and not all inhibitors are equally active against all groups or even against all the members of a single group. The specificities of the major protease inhibitors are summarised in Table II and discussed in more detail below. Attention will be focussed on those inhibitors which have attracted the most interest in relation to pulmonary disease, although this is not intended to imply that they are the only inhibitors of importance.

Antileukoprotease

This inhibitor is a polypeptide of approximately 100 residues recently isolated from purulent bronchial secretions by Ohlsson and his co-workers. Very similar compounds have been found in human cervical mucus and seminal plasma (Ohlsson, 1979). These inhibitors have been shown to inactivate trypsin, chymotrypsin and granulocyte, but not pancreatic elastase. With chymotrypsin and elastase, stable 1:1 complexes are formed very rapidly (K_i 9.0 x 10^{-10} M and 2 x 10^{-9} M respectively).

Antileukoprotease has been found to be the major contributor of anti-proteolytic activity in bronchial lavage fluid, the remainder being attributed to α_1AT, α_2-macroglobin (α_2M) and α_1-antichymotrypsin (α_1ACT). Most of the antileukoprotease (ALP) was found to be free, while the remainder appeared to be elastase-bound. The inhibitor has been localised to the tracheal and maxillary sinus mucosa and appears to be locally produced. Although it has not yet been ascribed a definite role, it has been suggested that ALP is important in defence of respiratory mucous membranes against the proteolytic effects of granulocyte enzymes or

TABLE I.   The Major Antiproteases

Antiprotease	Molecular weight	Concentration in plasma[*] g/litre	µM	Concentration in bronchial lavage fluid[+] mg/litre	nM
α_1-antitrypsin (α_1-antiproteinase)	54,000	2.90 (±0.45)	53 (±8)	5.5 (±6.8)	100 (±123)
α_2-macroglobulin	725,000	2.60 (±0.70)	3.6 (±1.0)	9.5 (±10.6)	13 (±14)
α_1-antichymotrypsin	69,000	0.49 (±0.06)	7.1 (±0.9)	2.6 (±2.4)	38 (±34)
antileukoprotease	10,500	?	?	15.0 (±11.5)	1429 (±1095)

[*] After Heimburger (1972) Mean (±SD)

[+] After Ohlsson (1979) Mean (±SD)

TABLE II. Specificities of Major Antiproteases*

Antiprotease	Trypsin	Chymotrypsin	Elastase	Collagenase	Neutral protease	Cathepsin G
α_1-antitrypsin	+	+	++	-	+	(+)
α_1-antichymotrypsin	-	+	-	-	-	+
α_2-macroglobulin	+	+	+	++	++	++
β_1-anticollagenase	-	?	-	+	-	-
antileukoprotease	+	+	++	-	-	+

*After Ohlsson (1979)

bacterial proteases such as those found in purulent sputum.

α_2-macroglobulin

A further inhibitor, which is currently undergoing intensive investigation in our own and other laboratories, is the serum antiprotease α_2M, a large dimeric protein of molecular weight 725,000 daltons. It is not one of the acute-phase reactive proteins, and is found in substantially higher serum concentrations in women than in men. When certain electropheretic techniques are used it displays considerable heterogeneity, in part ascribable to different conformational states.

Unlike the other inhibitors mentioned above, α_2M has a very low specificity and has been shown to inactivate proteases from each of the four major groups (Barrett and Starkey, 1973). The common feature of the enzymes inactivated by α_2M is that they all possess endopeptidase activity; inactive endopeptidase precursors fail to bind to α_2M. Barrett and Starkey (1973) have postulated a steric-hindrance hypothesis whereby the endopeptidase attacks a vulnerable region of the α_2M, analogous to the hinge area of IgG. This alters the conformation of the α_2M, thereby entrapping the protease (see Fig. 1). This would explain a number of other observations made about the antiprotease actions of α_2M. For example the binding of proteases is claimed to be irreversible, so that α_2M can be saturated with one enzyme to prevent the binding of proteases is claimed to be irreversible, so that α_2M can be saturated with one enzyme to prevent the binding of a second. Also proteases in complex with α_2M lose more of their activity against large substrates than small ones. Further evidence that proteases are 'entrapped' rather than bound is provided by the finding that, over a relatively long time period, elastase: α_2M complexes dissociate, as the still active elastase continues to attack the α_2M splitting off a number of inhibitor fragments and eventually releasing itself.

If complex formation is irreversible, unless the inhibitor is destroyed, it should be possible to calculate molar ratios for binding. Numerous groups have done this, most obtaining a 1:1 or 1:2 inhibitor:substrate ratio, but ratios as high as 1:3 have been quoted by some workers. Barrett and Starkey (1973) argue that where a ratio other than 1:1 has been obtained, the calculations have been based on the assumption that the products used are totally pure, whereas this is unlikely to be the case. This is disputed by other workers in the field (Topping and Seilman, 1979).

Since α_2M is extremely large, it cannot pass freely through plasma membranes and therefore may be presumed to have its main site of action in the plasma, except in the case of inflammation when it may escape into the tissues (Barrett and Starkey, 1973).

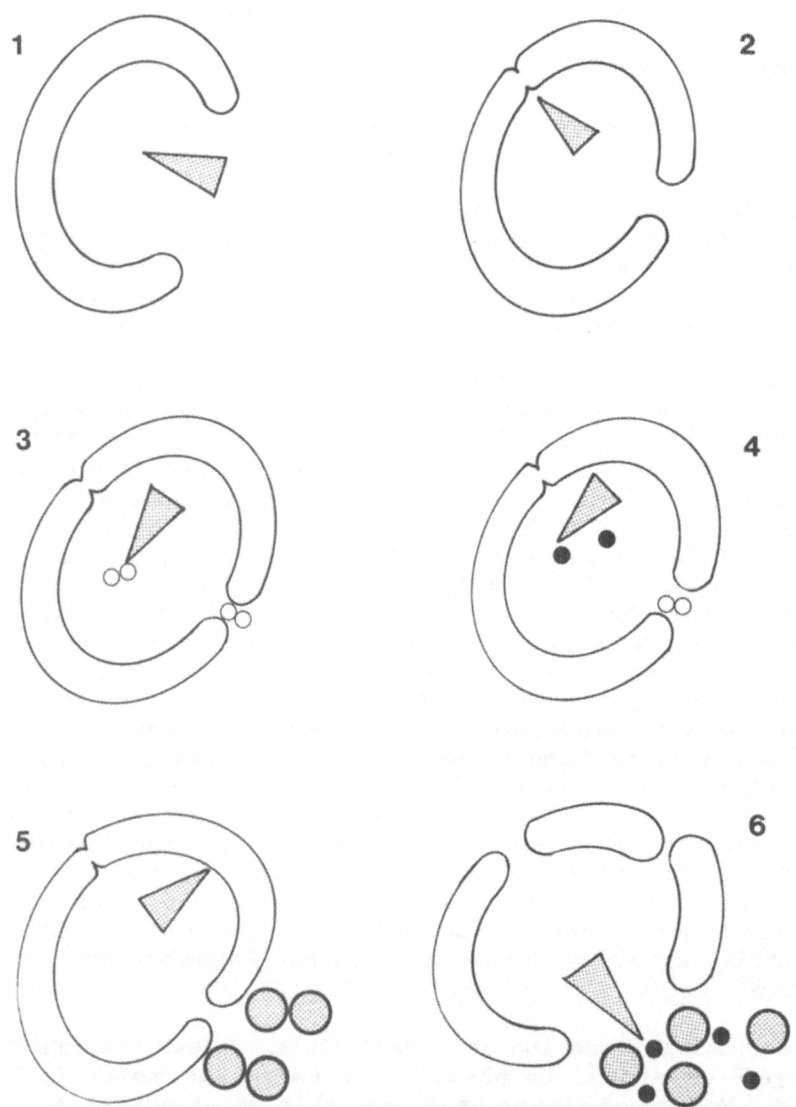

Figure 1. Schematic representation of the mode of action of
          α_2-macroglobulin:

1. Protease (▲) attacks 'hinge' area of α_2M causing
2. a change in α_2M configuration which entraps the protease
3. while still admitting small substrates (∞)
4. which are hydrolysed to products (●) and released.
5. Large substrates (●●) cannot gain access to the protease
6. until it has hydrolysed the α_2M itself.

However, enzyme: α_2M complexes disappear from the circulation very rapidly (Ohlsson and Laurell, 1976) and with no apparent dissociation of the complexes. It is thought that the conformational change induced by the protease binding results in recognition of the complex by the reticulo-endothelial cells of the liver which selectively remove bound inhibitor from the circulation. Also, the enzyme:inhibitor complexes are large enough to provoke phagocytosis. It has been shown that such complexes are rapidly taken up by rabbit alveolar macrophages in vitro and although this situation is unlikely to occur in vivo, complexes could easily be ingested by circulating phagocytic cells. Once within the cell, the active enzyme may be released from the complex (Baumstark, 1970), resulting in intracellular damage and possibly further protease release.

As yet, α_2M has not been clearly implicated (positively or negatively) in the pathogenesis of any disease, although an abnormality of it has been postulated as a contributory factor in cystic fibrosis, and it might also play a role in the development of emphysema, particularly in α_1AT deficient individuals (see below).

α_1-antitrypsin

α_1AT, also known as α-1-proteinase inhibitor, is probably the most extensively investigated of the serum antiproteases. It has been shown to inhibit a number of serine-dependent proteases including elastase, trypsin, chymotrypsin and plasmin. The inhibitor is a glycoprotein with a molecular weight of 54,000, consisting of a single polypeptide chain with several carbohydrate side chains, including 6 or more sialic acid residues. There is convincing evidence that methionine is an essential element of the active site (Johnson and Travis, 1979).

One of the most interesting features of the inhibitor is that it has been shown to exist in a number of genetically determined variant forms, which are known collectively as the 'Pi system'. These variant forms can be separated on the basis of their electrophoretic mobilities, their expression being best explained as under the control of co-dominant alleles (Kueppers and Black, 1974). The frequency of different phenotypes has been investigated and it has been shown that 98% of all neonates in England and Wales carry at least one 'M' gene, while 86% are 'M' homozygotes. There are a number of different subvariants of the M type, but they are probably of little clinical significance. The least common variant is Z, with about 0.03% of the British population being homozygous for this condition, although there is a certain amount of variation in frequency between different ethnic groups (Eriksson, 1965).

Individuals possessing phenotypes other than the normal,

homozygous M condition show a marked reduction in levels of circulating α_1AT (see Table III). Possession of the ZZ phenotype is often accompanied by clinical features affecting the liver and the lung. The condition is often associated with the presence of PAS-positive inclusion bodies in the parenchymal liver cells. The presence of these bodies may be related to an increased incidence of hepatoma or cirrhosis. It is thought that they consist of aberrant α_1AT which is not released from the cells due to a defect in synthesis, perhaps glycosylation (catabolism appears to be normal). The concept of defective glycosylation resulting in reduced secretion from the liver is supported by findings that removal of sialic acid residues from the M protein alter its electrophoretic mobility to that of the Z variant. The overall amino acid and carbohydrate content of the M and Z proteins is similar, with the Z type showing reduced sialic acid levels and the substitution of two glutamic acid residues with one of lysine and one of glutamine. On a molar basis the Z variant has the same inhibitory capacity against proteases as the M form.

A better known and more dramatic clinical feature of α_1AT deficiency than the liver pathology is a predisposition to the development of pulmonary damage, particularly emphysema, a disease characterised by the uncontrolled destruction and disorderly resynthesis of pulmonary elastic tissue (Eriksson, 1965; Laurell and Eriksson, 1963). Protease release from polymorpho-nuclear leukocytes or pulmonary alveolar macrophages has been implicated as a causative agent in emphysema, the condition being mimicked experimentally by intratracheal application of elastase or papain (Snider et al., 1977; Janoff et al., 1977). The high incidence of the disease among antiprotease-deficient individuals is a good indication that disturbance of the protease-antiprotease balance is a significant factor in its development. However, factors other than the simple presence or absence of α_1AT must be involved in the aetiology of the disease since most sufferers are α_1AT normal. Moreover deficient inividuals tend to develop lesions in the lower regions of the lung, but non-deficient sufferers are usually more effected in the upper pulmonary regions (Hutchinson et al., 1972), where the predominant anti-protease might be the antileukoprotease discussed above (Ohlsson, 1979).

Cigarette smoke and the protease:antiprotease balance

Although the hereditary $_1$AT deficiency of the ZZ phenotype is strongly associated with emphysema (Eriksson, 1965) it is not clear whether the heterozygous Z condition is deleterious. In any event, as already stated, the majority of emphysema sufferers are not α_1AT deficient. Most of them, however, are cigarette smokers (Auerbach et al., 1972; Hutchison and Bellamy, unpublished observation). Also, the onset of clinical symptoms is much earlier in ZZ smokers than ZZ non-smokers (Kueppers and Black, 1974). Both

TABLE III.

Concentrations of Circulating α_1-antitrypsin in
Patients of Different Phenotypes

Phenotype	α_1-antitrypsin g/litre Mean ($\pm$SD)	Percentage normal mean concn.
MM	2.86 ($\pm$0.73)	100
MS	2.15 ($\pm$0.47)	75
MZ	1.64 ($\pm$0.44)	57
SS	1.49 ($\pm$0.23)	52
SZ	1.06 ($\pm$0.34)	37
ZZ	0.45 ($\pm$0.08)	16

*After Lieberman et al. (1972)

cigarette smoking and inflammatory disease cause a proliferation of macrophages in the lung, and furthermore, cigarette smoke promotes the release of elastase from these cells (Blue and Janoff, 1978), and also from PMN leukocytes (Hutchison et al., 1980). This enzyme release is probably the result of particle phagocytosis (Weissman et al., 1972).

Under normal circumstances, any proteases lost to the extra-cellular fluid should be inactivated by circulating antiproteases. However, it has been shown that cigarette smoke reduces the elastase inhibitory capacity of serum and bronchial lavage fluid in both man and rat. This inactivation may be due to oxidising agents present in cigarette smoke itself and also to the effects of hydroxyl free radicals which can be released from phagocytosing PMN leukocytes (Carp and Janoff, 1978). Oxidation of the active-site methionine of α_1AT leads to immediate complete loss of protease inhibitory capacity (Johnson and Travis, 1979).

It seems clear that emphysema is the result of a disturbance in the delicate balance between the release of protease (most significantly, elastase), and its inhibition. A bias towards excessive activity can be caused by a hereditary deficiency in inhibitory capacity or by a cigarette smoke induced inhibitory deficiency. It should be borne in mind however, that as well as inhibiting antiproteases, cigarette smoke has been demonstrated to reduce phagocytosis and particle-induced enzyme release and also to reduce the activity of elastase and other lysosomal hydrolases (Ejiofor, Hutchison and Baum, unpublished observation). A further complication is that elastase released from PMN leukocytes in the upper lung may actually be sequestered by alveolar macrophages, and this uptake in turn might be affected by cigarette smoke.

Any multifactorial explanation of this pathological disturbance of the elastase activity:inhibition balance must take into account the fact that many individuals with the MM phenotype smoke without developing emphysema. It would seem that some further hereditary physiological factor may be involved. For example, it has recently been found that in normal non-smoking male subjects the soluble fraction of cigarette smoke inhibited particle-induced lysosomal enzyme release, whereas the cells of emphysematous subjects were apparently not inhibited in this way (Hutchison et al., 1980). These preliminary findings are undergoing further investigation at the present time to determine whether this is a predisposing factor or an adaptive response to smoking. Other predisposing factors could be speculated upon, such as the anti-oxidant capacity of pulmonary secretions; and the significant predisposition of men suggests hormonal involvement.

Interactions of antiproteases

At this point it seems appropriate to note that although we have considered the actions of a number of inhibitors independently in vivo they may well interact with each other, by competing for proteases for which they possess a common specificity. A number of studies of the antiproteolytic capacity of human body fluids have been undertaken to investigate possible interactions, particularly in disease states. For example, it has been observed that when normal human serum is titrated against a protease (pancreatic trypsin), a characteristically shaped inhibition curve is generated (Topping and Seilman, 1979), of a similar shape to Figure 2 below. Topping and his associates have explained this in terms of two binding sites with different affinities for trypsin on α_2M and a third on α_1AT. By resolving the titration curve into four component straight lines, this group can assign α_2M unequivocally into one of seven categories on the basis of the ratio of α and β binding sites, and it is intended to relate this information to disease states.

In contrast, we believe that the titration curve for pancreatic elastase inhibition (Figure 2) by normal serum can be adequately explained on the basis of the enzyme's different affinities for α_1AT and α_2M and different activities of the respective enzyme:inhibitor complexes against the model, low molecular weight substrate employed. Qualitatively, it has not been necessary to postulate more than one α_2M binding site and we are currently investigating these interactions by kinetic analysis and immunoelectrophoretic methods.

A simple qualitative explanation of the curve, consistent with the characteristics of the interaction of the inhibitors with pancreatic elastase as we believe them to be, is as follows: the initial, rapid decline in activity corresponds to enzyme in molar excess over all inhibitors while the minimum activity indicates the point of equivalence between enzyme molecules and enzyme binding sites on all inhibitors present. The subsequent rise in activity in the presence of excess plasma can be explained by preferential, irreversible binding of elastase to α_2M. Since this causes only 60-70% inhibition of the porcine enzyme acting on the low molecular weight substrate employed in these assays, it protects the elastase from total inhibition by α_1AT. This protective effect can be considerably reduced when the serum is treated with methylamine, an inactivator of α_2M (see Figure 2). The titration of pancreatic elastase with fresh serum from an extreme case of α_1AT deficiency (see Figure 3) is consistent with this interpretation both in the presence and absence of methylamine. In such a case α_2M is

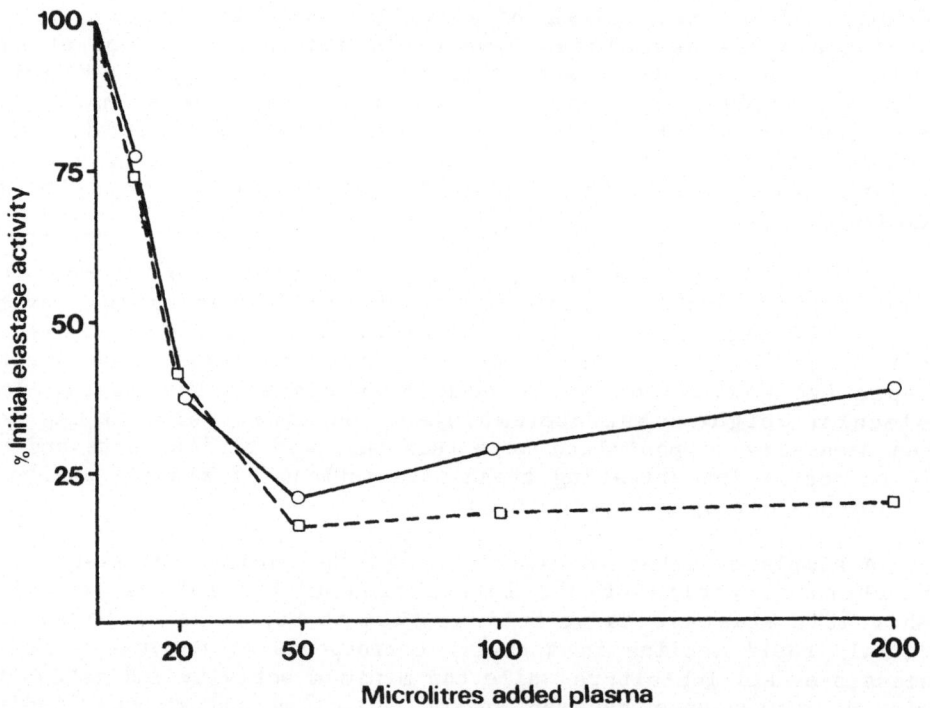

Figure 2. Titration of porcine pancreatic elastase by normal
          human serum before (o—o) and after (□—□) methylamine
          treatment.

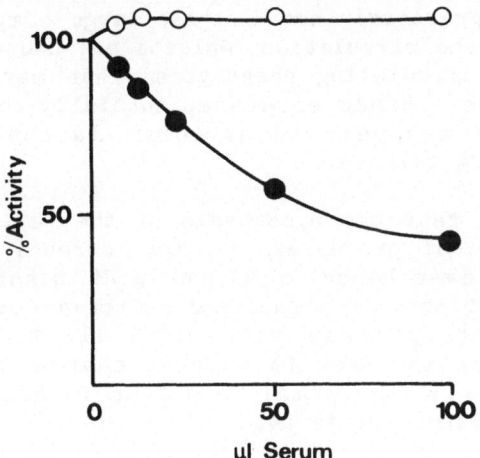

Figure 3. Titration of porcine pancreatic elastase by α₁AT
deficient (ZZ) serum before (●) and after (○)
methylamine treatment.

effectively the only inhibitor present and the removal of this molecule merely reveals the profundity of α_1AT deficiency. However, most ZZ sera do contain significant amounts of α_1AT, so the titration curves are more complicated.

It has been shown (Baum and Sadeghi, 1980) that, unlike the pancreatic enzyme, human leukocytic elastase is preferentially bound to α_1AT. However, this binding is reversible, so any changes in the effective α_1AT: α_2M ratio due to deficiency of one antiprotease or perhaps competition from proteases released by inflammatory processes, may result in significant binding of elastase to α_2M. Such complexes have been shown to be active against substrates of low molecular weight, so they could disturb the equilibrium between elastin breakdown and resynthesis from small precursors, hence causing an emphysema-like condition. Furthermore, as previously noted, these large complexes are rapidly eliminated from the circulation (Ohlsson and Laurell, 1976) and may be scavenged by circulating phagocytes, that may be precursors of lung macrophages. Hence elastase, initially complexed elsewhere with α_2M, may reappear in alveolar macrophages as 'latent elastase' awaiting release.

In order to test the hypothesis of the participation of α_2M in such pathological processes, we are currently surveying ratios of free and protease bound α_1AT and α_2M in normal subjects and emphysematous patients of normal and deficient α_1AT status with a view to correlating these with clinical studies and smoking history. Our initial results suggest that more protease: α_1AT complexes are found in normal sera than in samples from either group of emphysematous patients.

By crossed immunoelectrophoresis we have been able to demonstrate increased inhibitor:elastase complex formation with increasing amount of human leukocyte elastase added to partially purifed α_1AT or α_2M. We are now studying the competitive actions of inhibitors, and our initial results suggest that <u>in vitro</u> at least, α_2M may be able to displace elastase from α_1AT complexes, even when the α_1AT is in molar excess (Longbottom, Baum and Smith, unpublished observation), in confirmation of the conclusion of Baum and Sadeghi, based upon kinetic analysis.

Work is also continuing on the effect of tobacco smoke on phagocytosis, and elastase and inhibitor function. As was stated above, it has been clearly demonstrated that the aqueous phase of cigarette smoke suppresses phagocytosis and associated enzyme release by PMN leukocytes, and also that it inhibits the activities of both human leukocyte elastase and α_1AT. <u>In vitro</u> these inhibitions occur essentially in parallel, so that in normal conditions the net effect on the proteolytic enzyme balance might well be zero. However, recent evidence suggests that when the amount of $\alpha_2 M$ relative to $\alpha_1 AT$ is elevated, $\alpha_2 M$ extends significant protection to the elastase; apparently the complex of the enzyme with $\alpha_2 M$ is less susceptible to attack by aqueous smoke phases, while retaining undiminished esterolytic activity (Ejiofor, Hutchison and Baum, unpublished observation). This vindicates the suggested importance of $\alpha_1 AT$: $\alpha_2 M$ ratios in the aetiology of this disease, and may be of particular significance in $\alpha_1 AT$ deficient patients in whom $\alpha_2 M$ is the major serum antiprotease.

Acknowledgement

We are grateful to the Chest, Heart and Stroke Association for generous support.

References

Auerbach, O., Hammond, E.C., Garfinkel, L., and Benante, C., 1972, Relation of smoking and age to emphysema. Whole-lung section study, N. Engl. J. Med., 286:853.

Barrett, A.J., and Starkey, P.M., 1973, The interaction of α_2-macroglobulin with proteinases, Biochem. J., 133-709.

Baum, H., and Sadeghi, P., 1980, Interactions of elastase with protease inhibitors of human sera. Proc. 13th FEBS Meeting, Jerusalem, Israel, C2-P16.

Baumstark, J.S., 1970, Studies on the elastase-serum protein interaction. II. On the digestion of human α_2-macroglobulin, an elastase inhibitor, by elastase, Biochim. Biophys. Acta, 207:318.

Blue, M-L., and Janoff, A., 1978, Release of elastase from human polymorphonuclear leukocytes by cigarette smoke condensate in vitro, Am. Rev. Respir. Dis., 117-317.

Carp, H., and Janoff, A., 1978, Possible mechanisms of emphysema in smokers: in vitro suppression of serum elastase inhibitory capacity by fresh cigarette smoke and its prevention by anti-oxidants, Am. Rev. Respir. Dis., 118:617.

Eriksson, S., 1965, Studies in α_1-antitrypsin deficiency, Acta Med. Scand. Suppl., 432.

Heimburger, N., 1972, Introductory remark: Proteinase inhibition in human serum: identification, concentration, chemical properties, enzymatic specificity, in: "Pulmonary emphysema and proteolysis", C. Mittman, ed., Academic Press, New York.

Henson, P.M., 1971, Interaction of cells with immune complexes: adherence, release of constituents and tissue injury, J. Exptl. Med., 134: 114s.

Hutchison, D.C.S., Barterm, C.E., Cook, P.J.L., Laws, J.W., Martelli, N.A., and Hugh-Jones, P., 1972, Severe pulmonary emphysema: A comparison of patients with and without $\alpha_1 AT$ deficiency, Quart. J. Med., 41:301.

Hutchison, D.C.S., Desai, R., Bellamy, D., and Baum, H., 1980, The
    induction of lysososmal enzyme release from leucocytes of
    normal and emphysematous subjects and the effects of tobacco
    smoke upon phagocytosis, Clin. Sci. Mol. Med., 58: 403.
Janoff, A., Sloan, B., Weinbaum, G., Damiano, V., Sandhous, R.A.,
    Elias, J., and Kimbel, P., 1977, Experimental emphysema
    induced with purified human neutrophil elastase: Tissue
    localization of the instilled protease, Am. Rev. Respir. Dis.,
    115:461.
Johnson, D., and Travis, J., 1979, The oxidative inactivation of
    human α -1-proteinase inhibitor. Further evidence for
    methionine at the reactive centre, J. Biol. Chem., 254:4022.
Kueppers, F., and Black, L.F., 1974, α_1-antitrypsin and its
    deficiency, Am. Rev. Respir. Dis., 110:176.
Laurell, C-B., and Eriksson, S., 1963, The electrophoretic $_1$-
    globulin pattern of serum in α_1AT deficiency, Scan. J. Clin.
    Lab. Invest., 15:132.
Lieberman, J., Gaidulis, L., Garoutte, B., and Mittman, C., 1972,
    Identification and characteristics of the common $alpha_1$-
    antitrypsin phenotypes, Chest, 62:557.
Ohlsson, K., 1979, The low molecular weight proteinase inhibitor in
    bronchial mucus, INSERM, 84:105.
Ohlsson, K., and Laurell, C-B., 1976, The disappearance of enzyme-
    inhibitor complexes from the circulation of man, Clin. Sci.
    Mol. Med., 51:87.
Snider, G.L., Sherter, C.B., Koo, K.W., Karlinsky, J., Hayes, J.A.
    and Franzblau, 1977, Respiratory mechanics in hamster
    following treatment with endotracheal elastase or collagenase,
    J. Appl. Physiol: Respir. Environ. Exer. Physiol., 42:206.
Topping, R.M., and Seilman, S., 1979, A four-straight-line model
    for the proteinase binding characteristics of human blood
    serum, Biochem. J., 177:493.
Weissmann, G., Zurier, R.B., and Hoffstein, S., 1972, Leukocytic
    proteases and the immunologic release of lysosomal enzymes,
    Am. J. Path., 68:539.

DISCUSSION

LECTURER: Baum   CHAIRMAN: Cumming

CUMMING: Thank you for that fascinating trip through
                biochemical enzymology.

LAURENT: I have two questions. You emphasized that the
                processes that are going on are inevitably a
                balance of synthesis and degradation both of the
                enzymes and the antiprotease. My first question
                is: how are these proteases themselves degraded?
                You mentioned that the macrophage may take up the
                complex of the enzyme and the antiprotease, and
                once it is in the macrophage it may either damage
                it and in turn damage the tissue, or it may be
                released by some other mechanism. The question is:
                how does the macrophage degrade this protease? It
                is a tricky one because they live apparently quite
                happily together in the lysosome. The second
                question relates to the scheme you put up for us,
                it was a scheme within the alveolar space. Here we
                know there is elastin in the interstitium, is it
                necessary to propose membrane damage in the
                epithelium before this sort of process you are
                talking about can occur?

BAUM: I can't give you a firm answer to either question,
                but if we take the first question the alpha 2
                complexes which are very rapidly cleared, may be
                destroyed in one of two ways, either they are
                cleared in a way where they end up inside a
                lysosome, where the lysosome or proteolytic enzyme
                inside alpha 2 would break down the alpha 2 a
                little further so that it falls apart. I know
                nothing of the fate of the alpha 1 proteolytic
                complex, maybe there is somebody else who does.
                The second question I only alluded to when I
                pointed out that one of the factors in the dynamics
                is the question of where the enzyme goes first.
                The only constant is whether the enzyme gets to its
                inhibitor first or somehow gets to the elastin,
                whether it breaks its way out or diffuses out. So
                really the answers to both questions are: I don't
                know.

CORRIN: I wish to refer to Professor Baum's messy eaters,

and of course this is no comment on his behaviour
in the dining-room, but to use his own phraseology
for the release of enzymes from the macrophage, I
think this is a concept that is worth considering a
little further and perhaps ask him whether these
enzymes are released accidentally during
phagocytosis or whether they may be a positive
active secretion of enzymes from the macrophage.
The idea of a secretory activity by mononuclear
phagocytes has been reviewed and there are three
groups of enzymes released by the mononuclear
phagocytes. The first ones are the lysosomal
enzymes, most of which are retained within the cell
for intracellular digestion, a little is released
as you have indicated accidentally. The third
group includes lysosome which is mainly destined
for externalisation from the cell and its release
has nothing to do, and is not modulated by
phagocytosis. I think we must regard it as a true
secretory activity by the phagocytes. Related to
the subject today is the intermediate group which
includes the neutroproteases collagenase and
elastase; most of these are destined for release
by the phagocyte, their release is modulated by
phagocytosis, they are not released during
phagocytosis, they are released after phagocytosis
and I think that these two then should be regarded
as secretory products of the phagocytosis.

BAUM: Yes, I agree entirely; I pointed out at the
beginning there are a number of ways in which
enzymes can be released and you are talking about
the second way I mentioned which was really
triggered degranulation. I was referring to messy
eating, the unintentional release, because the
elastase release closely follows the release of
acid phosphatase and acid glucose aminedase,
enzymes which one normally does not think of as
secretory. In those cases, and this was in
collaboration with Peter Jeffery, we have shown
that we can only qualitatively correlate the amount
of phagocytosis measured directly by electron
microscopy with the amount of enzyme released; one
phenomenon is associated with the other. To that
extent we can't say anything quantitatively, what I
am describing is messy eating associated with
phagocytosis.

LAURENT: I want to make a further comment. There is a man
                called John Gaudy in my department who is
                interested in the control and regulation of alpha 1
                antitrypsin synthesis and in the signal for that
                synthesis to occur. Most of his work has been done
                in rats and mice and in various inflammatory
                conditions, it is quite apparent that the signal
                for synthesis is either distorted or at least the
                effect is different, because in a variety of
                different inflammatory conditions you get release
                by the liver, certainly synthesis by the liver but
                not necessarily release in normal conditions. He
                has also been looking at the turnover of radio-
                labelled alpha 1 antitrypsin in various
                inflammatory conditions, and the level that he
                finds in the circulation does not reflect the route
                of synthesis or the rate at which it is released.
                In this regard when he looks at pneumatoid which
                goes through the lungs and which is even stronger
                than brasiliensis he finds incredibly high turnover
                of alpha 1 antitrypsin but when he looks at the
                serum level it is not changed, and the alpha 1
                antitrypsin is all in the alveolar macrophages.
                That suggests something to me that might support
                some of the things that you have been saying but
                also raises the question as to whether the alpha 1
                antitrypsin is synthesized in sites other than the
                liver, and the alveolar macrophage appears to be
                able to synthesize the alpha 1 antitrypsin as well
                and this raises an additional complication;   a
                local inflammation in the lungs over the above
                already complicated problems that you have alluded
                too.   Do you have any comments on that sort of
                observation?

BAUM: Since it is an acute phase protein, its synthesis
                and release is continually being modulated. Also
                there is the question of weight in addition to the
                suggestion that it may be made by macrophages, I
                have also come across a suggestion that it may be
                made by fibroblasts, which would add yet another
                complication to the system. I agree with you that
                it is a very very complicted system.

KUHN: It seems as though I am going to have nothing more
                to say after coffee break!   That was a superb
                presentation.   Just to make it even more
                complicated, there is now emerging evidence that
                the degradation of elastin is also influenced by

its association with other components of connective tissue, notably the connective tissue glycoproteins. A very elegant study has shown that in order for elastin to be sufficiently degraded by elastase glycoproteins have also been digested. The digestion of glycoproteins can be carried out by other secretory enzymes from the macrophage, notably the so called plasminogen activator secretion so that the control of elastase secretion differs in several respects. There are a whole lot of levels of interaction in this messy system.

HEATH: In emphysema these macrophages are living in a progressively hypoxic environment. I wonder if you would like to make any comments about the role of a low partial pressure of oxygen on the metabolism of macrophages in this complex equilibrium which you have described.

BAUM: That's another very interesting problem that has arisen. We showed some time ago that in the case of neutrophils in a normal subject, phagocytosis and the associated enzyme release is very profoundly inhibited by the vapour-phase constituents of cigarette smoke. In the case of emphysematous subjects in a pilot study there seem to be less susceptibility to such inhibition. Now we didn't know what this was due to, but I thought that it might be a reflection of the fact that these were adapted to a hypoxic environment, a change perhaps of the metabolism to a more glycolytic mode and consequently it would respond differently to some kind of environmental challenge. The macrophage has some very good mitochondria, it could live normally aerobically but as it gets progressively more hypoxic, perhaps it could move to a more glycolytic mode in which case it might very well respond differently to an environmental pollutant challenge. So that's a very useful question and I am glad that you have asked it.

CHAIRMAN: Just before the break for coffee, I'd like to make a small announcement: two years ago the Director of the Centre here, Professor Antonio Zichichi gave a preprandial lecture about the origins of man and ended up by saying "man is thus created". This enthused David Denison with the concept of applying the same idea to the creation of the lung and to do

this he has set up a programme in June of this year
to be held at the Brompton Hospital entitled "First
make a lung" and you will find a sheet with the
preliminary announcement of this course - course is
not the right word really, it is a workshop I
suppose. If you are interested you might like to
attend. And now we have a break for coffee. Come
back at 11.35.

MACROPHAGES AND THE TURNOVER OF CONNECTIVE TISSUE IN THE LUNG

Charles Kuhn

Dept. Medical Biochemistry, University of Manchester (UK)
& Dept. Pathology, Washington University, St. Louis
Mo. (USA)

INTRODUCTION

The function of the alveolar macrophage as an alveolar sentry is well established. As a sentry it has the duty to engage and destroy or capture an intruder such as a bacterium or noxious particle. If the invading force is strong it must call for reinforcements. For this it has a number of secretory products – chemotactic factors, colony stimulating factors, and signals to activate the immune system. The question which this paper will address is whether the macrophage can also serve as a sapper; may it be involved in the restructuring of the lung? Since the pulmonary architecture is determined by its connective tissue, how does the macrophage influence connective tissue?

THE ALVEOLAR MACROPHAGE AND COLLAGEN SYNTHESIS

The idea that macrophages may stimulate collagen synthesis derives from investigations of Heppleston and Styles (1967) into the pathogenesis of silicosis. Since inhaled fibrogenic silica is taken up first by macrophages, which then undergo necrosis releasing their silica for another round of phagocytosis, these investigators tested the ability of extracts of macrophages treated with silica to stimulate collagen synthesis. They reported that lysed rat macrophages which had been treated with silica in vitro, contained a soluble material which stimulated the conversion of radiolabelled proline to hydroxyproline by chick fibroblasts several fold. Other investigators have performed similar experiments; most have obtained some stimulation of collagen synthesis but some have found suppression (Harrington et al, 1973). In addition, there has been no uniformity of results with controls, i.e. the products of lysed normal macrophages: Heppleston and Styles observed minimal stimulation of collagen synthesis, Harrington et al. significant depression, and Aalto et al. (1976) slight depression of collagen synthesis. Hence, there is ambiguity about the proper baseline. Aalto et al. considered their results to show that silica-treated macrophages did stimulate fibrosis, since they found extracts of silica-treated macrophages to be stimulatory compared to extracts of normal macrophages, although the "stimulated" collagen synthesis was only comparable to that by untreated fibroblasts.

Even in instances where authors felt their results supported the Heppleston and Styles system, they have differed in detail. For example, Heppleston and Styles found that killing macrophages by freezing and thawing prevented the production of the collagen-stimulating factor, which seemingly required viable macrophages, while Aalto et al., produced activity by exposing isolated subcellular organelles to silica. The diversity of results probably is caused in part by the use of complicated and impure systems and in part by variations in experimental design from laboratory to laboratoy. Studies have varied in many details – species, type of macrophage (alveolar vs peritoneal, resident vs elicited), means of exposure to silica (in vivo vs in vitro) type of silica (quartz vs tridymite and cristobolite) collagen synthesizing system (fibroblast, sponge granuloma, cell-free translation). Fibroblast targets have varied as to species, age, maturity, and phase of growth. Recently Aalto et al. have reported an extensive series of investigations which illustrate some of these problems. In their system, extracts of silica-treated resident peritoneal cells were stimulatory compared to untreated macrophages, whereas extracts of thioglycollate-elicited macrophages were maximally stimulatory without silica. Fibroblasts from granulomas in rats responded to the extracts, while chick tendon fibroblasts did not, (Aalto et al., 1976, 1979). In general the effects on collagen synthesis were small and conclusions could vary depending on the baseline selected. Recently, these investigators have turned their attention to purifying the active principle in medium from silica-treated macrophages. Stimulatory activity was associated with a non dialyzeable protein, M_r 14,300 (Aalto and Kulonen, 1979). With the availability of purified systems, perhaps order can be brought to this confused field.

There is also evidence that macrophages may be able to regulate fibroblast growth and migration. Leibovich and Ross (1975) observed that suppression of the macrophage influx into healing wounds also prevented fibroblastic proliferation. Investigating this further, they found that fibroblasts proliferate in the presence of serum prepared from whole blood. When they used serum prepared from plasma depleted of platelets, fibroblasts failed to proliferate although the medium was nutritionally adequate. Medium from cultures of macrophages in serum from platelet-poor plasma could replace the missing platelet-derived factor. This fibroblast mitogen may be the result of macrophage action on a plasma protein rather than a direct secretory product of macrophages (Leibovich, 1978).

The fibroblast mitogen appears to be clearly different from putative collagen-stimulating factor. It requires intact macrophages, is critically dependent on the presence of platelet poor plasma serum and does not require perturbation of the macrophages.

MACROPHAGES AND COLLAGEN DEGRADATION

Macrophages have been implicated in rapid collagen breakdown in other organs, of which bone resorption and the involution of the post partem uterus are striking examples. The idea that macrophages are involved in the breakdown of collagen in lung disease is supported by considerable indirect evidence.

The first step in collagen breakdown is thought to be the cleavage by the enzyme, collagenase of the triple helical collagen monomer at a single bond 3/4 of the way from the amino-terminal end of each alpha chain (Harris and Krane, 1974; Weiss, 1976). The products of this cleavage are unstable and undergo denaturation at body temperature. Once denatured, they are susceptible to digestion by non-specific proteases, with which macrophages are well-endowed. Macrophages cultured in serum-free medium secrete an enzyme with the typical properties of mammalian collagenases, a metalloproteinase capable of cleaving collagens of the major types (types I, II and III). Part of the secreted enzyme is not immediately active but can be activated by neutral proteases, including some also produced by macrophages (Horowitz et al., 1976). Some of the factors which regulate the secretion of this activity have been investigated. In at least two species, mouse and guinea pig, peritoneal macrophages do not produce collagenase unless stimulated. (Wahl et al., 1974a; Werb and Gordon, 1975a). Endotoxin, lymphokines, prostaglandin E and the ingestion of particles all provoke collagenase secretion. Rabbit alveolar macrophages secrete easily measureable amounts of collagenase and can be further stimulated by particles or M. Butyricum (Horowitz and Crystal, 1976). Those from BCG-treated animals secrete still higher levels. Although it is probable that alveolar macrophages are indeed more active than peritoneal macrophages in respect to collagenase-secretion, the direct comparison in the same species has not been reported.

It is reasonable to assume that macrophage collagenase contributes to the turnover of collagen in lung disease. Collagenase activity appears in lung lavage fluid from patients

with idiopathic pulmonary fibrosis (Gadek et al., 1979). Although
the source of this activity is not known in the human disease,
collagenase also appears in the lavage fluid from rats with
experimental pulmonary fibrosis induced by bleomycin. Collagenase
secretion by macrophages washed from the lungs of these animals
rises at the appropriate time to be contributing to this activity
(Marom et al., 1980).

ALVEOLAR MACROPHAGES AND THE DEGRADATION OF ELASTIN

The degradation of elastic fibres is believed to be critical
to the pathogenesis of emphysema (Kuhn and Senior, 1978). The
association of macrophages with the early lesions of emphysema and
with the characteristic lesions of cigarette smoking has focused
attention on their possible function in elastin turnover.
Macrophages produce a variety of products which may influence
elastin turnover in complex ways.

Macrophages from many species contain esterases which
hydrolyze synthetic substrates often used to measure elastases
(Levine et al., 1976). In general, it has not been possible to
detect apppreciable activity degrading insoluble elastin in
macrophage extracts. When cultured in serum-free medium, however,
animal macrophges do secrete an enzyme which degrades elasin (Werb
and Gordon, 1975; White et al., 1977). This enzyme has recently
been purified from cultures of mouse macrophages (White et al.,
1980; Banda and Werb, 1981). It is a metalloproteinase with a
molecular weight estimated by polyacrylamide gel electrophoresis
and gel filtration to be in the range 21 to 28 kilodaltons. It is
resistant to inhibitors of serine proteases including serum alpha-
1-protease inhibitor (α-1-PI). Like most so-called elastases, it
is a non-specific protease active against other proteins as well as
elastin. Macrophage elastase preferentially hydrolyzes peptide
bonds on the amino side of leucyl residues and hence does not
hydrolyze synthetic substrates such as succinyl trialanyl-p-
nitroanilide (SLAPN) used to measure pancreatic and granulocytic
elastases (Kettner et al., 1981).

There is some information about the regulation of elastase-
secretion. As was the case with collagenase, resident peritoneal
cells secrete negligible amounts of elastase, but can be stimulated
to secrete by phagocytosis of particles. Thioglycollate-elicited
macrophages are more active secretors of elastase and respond to
stimulation by phagocytosis by increased output. Resident alveolar
macrophages secrete elastase as actively as elicited peritoneal
macrophages which have been given a phagocytic stimulus. In
contrast to peritoneal macrophages, phagocytosis by alveolar
macrophages is not associated with elastase secretion above the
already high baseline (White et al., 1977; White and Kuhn, 1980).
Although the explanation for this refractoriness is unclear, it is

not because alveolar macrophages are maximally stimulated: exposure in vitro to vinblastine, colchicine, or cytochalasin B, increases elastase secretion (White et al., 1981c). The mechanism for these pharmacologic effects is unclear.

Cyclic nucleotides, sympathetic and parasympathetic agonists have small effects if any on elastase secretion by mouse macrophages. Corticosteroids are strongly inhibitory. In vivo, tobacco smoke increases elastase secretion by alveolar macrophages, but in vitro cigarette smoke is ineffective suggesting that the effect in vivo may be indirect (White et al., 1979).

While the secretion of an elastase by some animal macrophges is firmly established, the data for human alveolar macrophages are conflicting. DeCremoux et al., (1978) detected an enzyme with properties similar to those of mouse macrophage elastase, while others have detected only low levels of activity in human macrophage cultures and such activity as they did detect had properties of a serine protease and was able to hydrolyze SLAPN (Rodriquez et al., 1977; Green et al., 1979; Hinman et al., 1980). After a careful reappraisal of the available data, Hinman et al., (1980) concluded that the low levels of activity found in human macrophages were unlikely to be of pathologic significance and could be explained as granulocyte-derived (vide infra).

Elastin degradation will only occur in vivo when the amount of free enzyme exceeds the capacity of the available inhibitors. Macrophages can indirectly influence the amount of free elastase in tissue both by the production inhibitors and by their ability to selectively bind and internalize certain glycoproteins including granulocyte elastase.

The two major elastase inhibitors of plasma ae alpha-1-protease inhibitor and alpha-2-macroglobulin. Alpha-1-PI is readily measureable in lung lymph and bronchoalveolar lavage fluid, and because of its small size, probably comes mainly from plasma. Alpha-2-macroglobulin is present in lung lymph in nearly half its concentration in serum (Ganrot et al., 1970). Because of its large size, it is not able to cross the alveolar epithelium to the alveolar space, and is found only inconstantly in alveolar lavage fluid. Alveolar macrophages have been shown to synthesize both alpha-1-P.I and alpha-2-macroglobulin in culture (White et al., 1980; 1981a and b). How quantitatively important the synthesis of these inhibitos is in lung compared to the passive diffusion of plasma inhibitors is uncertain. It is notable that Banda and Werb (1981) obtained an 8-fold increase in total elastase activity during its purification from culture fluid, suggesting that in serum-free culture fluid much elastase is in inhibited form.

While macrophages can apparently produce inhibitors, they can also inactivate inhibitors under certain circumstances. Stimulated mononuclear phagocytes reduce oxygen to several products including superoxide anion, hydrogen peroxide, and hydroxyl radicals. The elastase inhibitory capacity of alpha-1-P.I. can be reversibly destroyed by oxidation of a methionine residue at the active inhibitory site (Johnson and Travis, 1979). The oxygen reduction products released by stimulated monocytes and alveolar macrophages inactivate the elastase inhibitory capacity of alpha-1-P.I. and this effect can be reversed by reducing agents (Carp and Janoff 1980).

In the context of emphysema, it is notable that when stimulated, alveolar macrophages from smokers produce more superoxide anion than those from non-smokers.

Another way in which macrophages can indirectly influence the degradation of elastin is by the binding of extracellular leukocyte elastase and enzyme-inhibitor complexes. Macrophages selectively recognize and bind extracellular granulocyte elastase and rapidly incorporate it into their lysosomal granules (Campbell et al., 1979). Other proteases including human pancreatic elastase are taken up much more slowly. Granulocyte elastase inhibited with phenyl methyl sulfonyl fluoride is bound similarly to the active enzyme but enzyme inhibited with serum or alpha-1-P.I. is taken up only slowly. The receptor which recognizes elastase also binds another protein from the granulocyte lysosome, lactoferrin.

Complexes between alpha-2-macroglobulin and proteases retain someenzymatic activity, usually for low molecular weight substrates. In the case of complexes between alpha-2-M and granulocyte elastase, there is also activity toward tropoelastin (Galdston et al., 1979). Complexes between alpha-2-M and proteases have only a short half-life in vivo and in vitro are rapidly bound and internalized by macrophages and the protease degraded (Kaplan and Nielsen, 1979 a and b). The receptor involved is clearly different from that which recognizes free granulocyte elastase, since it is trypsin sensitive and requires divalent cautions, while the elastase receptor is insensitive to trypsin and does not require cations.

THE DEGRADATION OF COMPLEX SUBSTRATES

In vivo, connective tissue proteins occur as part of a complex matrix, and the interaction of the various matrix components may alter the susceptibility of individual components to degradation. Werb and her colleagues (Jones and Werb, 1980; Werb et al., 1980 a, b) have investigated this in an elegant in vitro system. Smooth muscle cells or other mesenchymal cells were cultured and allowed to form a labelled extracellular matrix. The muscle cells and soluble matrix components were removed by washing with NH_4OH and macrophages plated on the insoluble matrix and the released products monitored. These studies showed that the association with glycoproteins protects the elastin from degradation by macrophage elastase. In the presence of plasminogen, the secretion of plasminogen activator by macrophages leads to the generation of plasmin, degradation of the glycoproteins, and enhanced elastolysis. Morphologic studies (Werb et al., 1980b) showed that the catabolism of matrix fibres was initiated extracellularly and was accelerated in the immediate vicinity of the macrophages. The final breakdown to amino acids was intralysosomal.

MACROPHAGES AND FIBRONECTIN

Fibronectins are a family of closely related glycoproteins which are found both in soluble form in plasma and in insoluble form on the surface of a variety of cells, in fibres of the extracellular matrix of some cultured cells including lung. Plasma fibronectin is formed of two subunits, M_r 220,000 joined by a single disulfide bond near their carboxyl ends. Insoluble fibronectin is of similar structure and cross-reacts immunologically with the plasma form. Some authors have found a small difference in molecular weight between the two forms. Both forms of fibronectin bind to collagen, fibrin, heparin and fibroblasts and both can be substrates for plasma transglutaminase (factor XIII). Fibronectins have been implicated in a variety of functions including cell-cell interaction, malignant transformation, the attachment and spreading of cells on extracellular matrices, and the recognition and uptake of certain particles (Mosher, 1980; Mosesson, 1980). Cultured blood monocytes synthesize fibronectin (Alitalo et al., 1980), and recently cultured human alveolar macrophages have also been found to synthesize fibronectin. Metabolic labelling studies of cultured macrophages show incorporation of radiolabelled precursors into a molecule of molecular weight 440,000 unreduced, 220,000 under reducing conditions, which binds to denatured collagen and reacts with antibodies to plasma fibronectin. Although found associated with cells 15 minutes after pulse labelling, the molecule is detectable in the culture medium by 30 minutes and by eight hours 80% of the label is extracellular. Immunoperoxidase staining shows the presence of fibronectin only with the Golgi apparatus and endoplasmic reticulum (Fig. 1). The plasma membrane is devoid of fibronectin, except rarely at points of intercellular contact (Villiger et al., 1981). Thus, in contrast to fibronectin synthesized by fibroblasts, that synthesized by macrophages is soluble, and not associated with the membrane or extracellular matrix.

Fibronectin synthesis may be particulary important for the function of alveolar macrophages. In other sites, macrophages probably have access to plasma fibronectin, but because of its

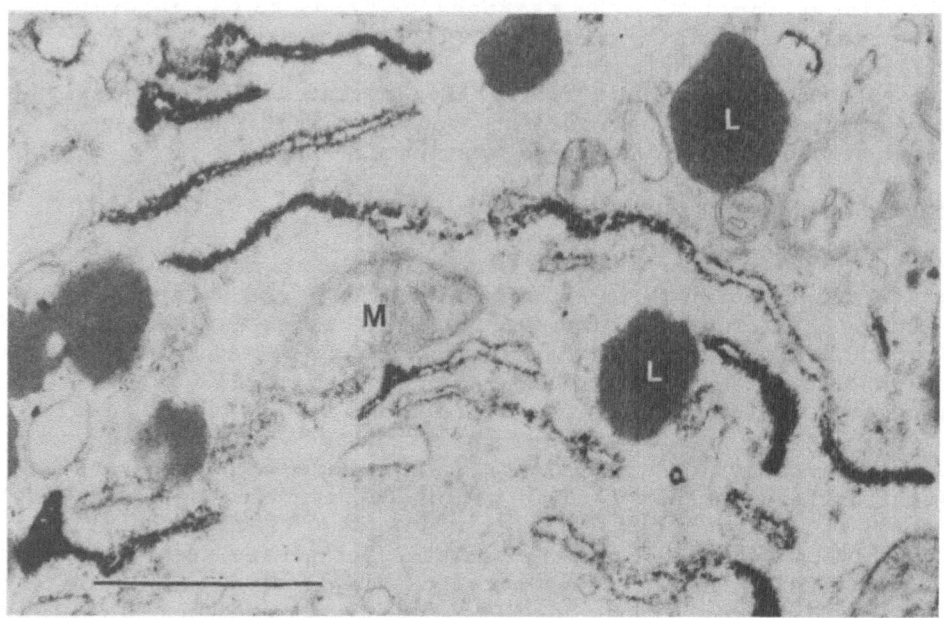

Figure 1:   Human alveolar macrophage fixed with periodate-lysine-
paraformaldehyde, treated with 0.1% (per cent) triton X-100 to
increase its permeability and stained with F(ab) fragments of
antifibronectin immunoglobulin conjugated to horseradish
peroxidase.   The black reaction product is localized to the
endoplasmic reticulum.
M = mitochondrion;   L= lysosome;   Bar = 1 micron.

large size, it is improbable that plasma fibronectin has access to
the alveolar spaces.   Therefore macrophage-derived fibronectin in
alveoli may be needed for the opsonization of bacteria (Kuusela
1978).   In addition, although the degradation of extracellular
matrix substrates is started in the extracellular space, it is
completed in lysosomes (Werb et al., 1980b).   Fibronectin promotes
the uptake of particles coated with denatured collagen, (Doran,
1980) (Fig. 2) and is likely to serve to promote the binding and
uptake of denatured collagen or the denatured fragment produced by
collagenase.   Thus, macrophage-derived fibronectin may also
contribute to the remodelling of lung tissue.

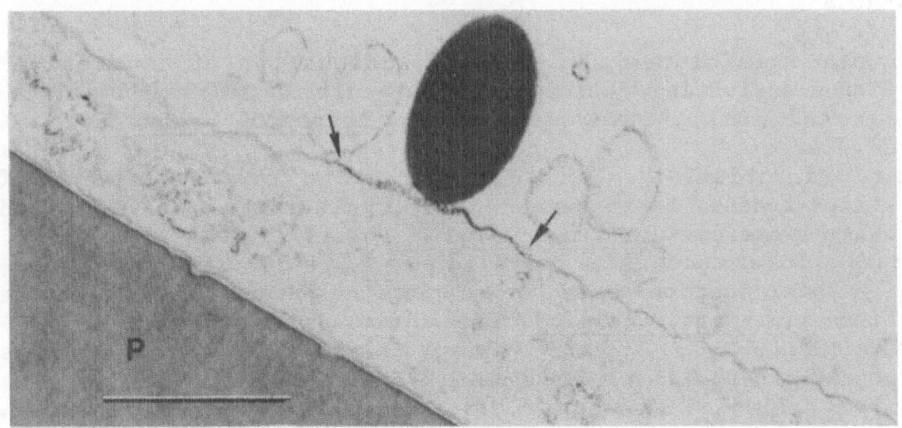

Figure 2:   Binding of gelatin-coated latex particle to alveolar macrophage.   Fibronectin (between arrows) is localized by the peroxidase-labelled F(ab) fragment of antifibronectin immunoglobulin to the plasma membrane of the macrophage just beneath the particle.   The remainder of the plasma membrane lacks fibronectin.   No fibronectin is stained between the macrophage and the plastic culture dish (P) to which is is adherent.   Bar = 1 micron.

CONCLUSIONS

1. Despite much investigation, the importance of the macrophage as a regulator of connective tissue synthesis remains uncertain.

2. The macrophage has the enzymatic machinery to degrade the major connective tissue components, collagen, elastin, and glycoproteins.

3. The macrophage has a capacity for controlling connective tissue-degrading proteases through soluble inhibitors and membrane receptors.

4. The interaction of enzyme secretion and the defensive measures will be a useful field for future investigation.

5. Macrophages can synthesize and utilized fibronectin.   In lung, fibronectin from macrophages may function as an intraalveolar opsonin.

Acknowledgement: The author is indebted to Dr. Richard White for making available several of his manuscripts prior to publication.

REFERENCES

Aalto, M., and Kulonen, E., 1979, Fractionation of connective-tissue-activating factors from the culture medium of silica-treated macrophages, Acta Path. Microbiol. Scand. Sect. C. 87:271.

Aalto, M., Potila, M., and Kulonen, E., 1976, The effect of silica-treated macrophages on the synthesis of collagen and other proteins in vitro, Exp. Cell Res. 97:193.

Aalto, M., Turakainen, H., and Kulonen, E., 1979, Effect of SiO_2-liberated macrophage factor on protein synthesis in connective tissue in vitro, Scand. J. Lab. Invest. 39:205.

Alitalo, K., Hovi, T., and Vaheri, A., 1980, Fibronectin is produced by human macrophages, J. Exp. Med. 151:602.

Banda, M.J., and Werb, Z., 1981, Mouse macrophage elastase. Purification and characterization as a metalloproteinase, Biochem. J. 193:589

Campbell, E.J., White, R.R., Senior, R.M., Rodriquez, R.R., and Kuhn, C., 1979, Receptor-mediated binding and internalization of leukocyte elastase by alveolr macrophages in vitro, J. Clin. Invest. 64:824.

Carp, H., and Janoff, A., 1980, Potential mediator of inflammation Phagocyte-derived oxidants suppress the elastase-inhibitory capacity of $alpha_1$ proteinase inhibitor in vitro, J. Clin. Invest. 66:987.

DeCremoux, H., Hornbeck, W., Jaurand, M.C., Bignon, J., and Robert, L., 1978, Partial characterization of an elastase-like enzyme secreted by human and monkey alveolar macrophages, J. Pathol. 125:171.

Doran, J.E., 1980, Cold insoluble globulin-enhanced phagocytosis of gelatinized targets by macrophage monolayers, J.R.E.S., 27:471.S

Gadek, J.E., Kelman, J.A., Fells, G.A., Weinberger, S.E., Horwitz, A.L., Reynolds, H.Y., Fulmer, J.A. and Crystal, R. G., 1979, Collagenase in the lower respiratory tract of patients with ifiopathic pulmonary fibrosis, New Engl. J. Med. 301:737.

Galdston, M., Levytska, V., Liener, I.E., and Twumasi, D.Y., 1979, Degradation of tropoelastin and elastin substrates by human neutrophil elastase free and bound to alpha-2-macroglobulin in serum of the M and Z (Pi) phenotypes for alpha-1-antitrypsin, Am. Rev. Respir. Dis. 119:435.

Ganrot, P.O., Laurell, C.B., and Ohlsson, K., 1970, Concentration of trypsin inhibitors of different molecular size and of albumin and haptoglobin in blood and lymph of various organs in the dog, Acta Physion. Scand. 79:280.

Green, M.R., Lin, J.s., Berman, L.B., Osman, M.M., Cerreta, J.M., Mandly, I., and Turino, G.M., 1979, Elastolytic activity of alveolar macrophages in normal dogs and human subjects, J. Lab. Clin. Med. 94:549.

Harrington, J.S., Ritchie, M., King, P.C., and Miller, K., 1973, The in vitro effects of silica-treated hamster macrophages on collagen production by hamster fibroblasts, J. Pathol., 109:21.

Harris, E.D., and Krane, S.M., 1974, Collagenase, New Engl. J. Med. 291:557, 605, 652.

Heppleston, A.G., and Styles, J.A., 1967, Activity of a macrophage factor in cllagen formation by silica, Nature 214:521.

Hinman, L.M., Stevens, C.A., Matthay R.A., and Gee, J.B.L., 1980, Elastase and lysozyme activities in human alveolar macrophages:effects of cigarette smoking, Am. Rev. Respir. Dis. 21:263.

Horwitz, A.L., and Crystal, R.G., 1976, Collagenase from rabbit pulmonary alveolar macrophages, Biochem. Biophys. Res. Commun. 69:296.

Horwitz, A.L., Kelman, J.A., and Crystal, R.G., 1976, Activation of alveolar macrophage collagenase by a neutral protease secreted by the same cell, Nature 264:772.

Johnson, D., and Travis, J., 1979, The oxidative inactivation of human alpha-1-proteinase ihibitor: further evidence for methionine at the reactive centre, J. Biol. Chem. 254:4022.

Jones, P.A., and Werv, Z., 1980, Degradation of connective tissue matrices by macrophges. II. Influence of matrix composition on proteolysis of glycoproteins elasin and collagen by macrophages in culture, J. Exp. Med. 152:1527.

Kaplan, J., and Nielsen, M.L., 1979, Analysis of macrophage surface receptors. I. Binding of -macroglobulin-protease complexes to rabbir alveolar macrophages, J. Biol. Chem. 254:7323.

Kaplan, J., and Nielsen, M.L., 1979, Analysis of macrophage surface receptors. II. Internalization of -macroglobulin-trypsin complexes by rabbit alveolar macrophages, J. Biol. Chem. 254:7329.

Kettner, C., Shaw, E., White, R., and Janoff, A., 1981, The specificity of macrophage elastase on the insulin B-chain, Biochem. J. (in press).

Kuhn, C., and Senior, R.M., 1978, The role of elastases in the development of emphysema, Lung 155:185.

Kuusela, P., 1978, Fibronectin binds to staphylococcus aureus, Nature 276:718.

Levine, E.H., Senior, R.M., and Butler, J.V., 1976, The elastase activity of alveolar macrophages: measurements using synthetic substrates and elastin, Am. Rev. Respir. Dis. 113:25.

Leibovich, S.J., 1978, Production of macrophage-dependant fibroblast stimulating activity (M-FSA) by murine macrophages. Effect on BALBC 3T3 fibroblasts, Exp. Cell Res. 113:47.

Leibovich, S.J., and Ross, R., 1975, The role of the macrophage in wound repair, Am. J. Pathol. 78:141.

Marom, Z., Weinber, K.S., and Fanburg, B.L., 1980, Effect of bleomycin on collagenolytic activity of the rat alveolar macrophage, Am. Rev. Respir. Dis. 121:859.

Mosesson, M.W., and Amrani, D.L., 1980, The structure and bilogical activities of plasma fibronectin, Blood 56:145.

Mosher, D.F., 1980, Fibronectin, Prog. Hemostasis Thromb. 5:111.

Rodriguez, R.J., White, R.R., Senior, R.M., and Levine, E.A., 1977, Elastase release from human alveolar macrophages: a comparison between smokers and non-smokers, Science 198:313.

Villiger, B., Kelley, D.G., Engleman, W., Kuhn, C., and McDonald, J.A., 1981, Human alveolar macrophage fibronectin: synthesis, secretion and ultrastructural localization in vitro. (Submitted for publication).

Wahl, L., Wahl, S.M., Mergenhagen, S.E., and Martin, G.E., 1974, Collagenase production by endotoxin activated macrophages, Proc. Nat. Acad. Sci, USA 71:3598.

Wahl, L.M., Wahl, S.M., Mergenhagen, S.E. and Martin, G.R., 1975, Collagenase production by lymphokine activated macrophages, Science 187:261.

Weiss, J.B., 1976, Enzymic degradation of collagen, Int. Rev. Connect. Tissue Res. 7:102.

Werb, Z., Banda, M.J., and Jones, P.A., 1980a, Degradation of connective tissue matrices by macrophages. 1. Proteolysis of elastin, glycoproteins and collagen by proteinases isolated from macrophages, J. Exp. Med. 152:1340.

Werb, Z., Bainton D.F., and Jones, P.A., 1980b, Degradation of connective tissue matrices by macrophages III. Morphological and biochemical studies on extracellular, pericellular and intracellular events in matrix proteolysis by macrophages in culture, J. Exp. Med. 152:1537.

Werb, Z., and Gordon S., 1975a, Secretion of a specific collagenase by stimulated macrophages, J. Exp. Med. 142:346.

Werb, Z., and Gordon, S., 1975b, Elastase secretion by stimulated macrophages. Characterization and regulation, J. Exp. Med. 142:361.

White, R., Habicht, G.S., Godfrey, P. Janoff, A., Barton, E., and Fox, C., 1981a, Secretion of elastase and alpha-2-macroglobulin by cultured peritoneal exudative macrophages: studies on their interction, J. Lab. Clin. Med. (in press).

White, R., Janoff, A., and Godfrey, H.P., 1980a, Secretion of alpha-2-macroglobulin by human alveolar macrophages, Lung, 158:9.

White, R., and Kuhn, C., 1980b Effects of phagocytosis of mineral dusts on elastase secretion by alveolar and peritoneal exudative macrophages, Arch. Environ. Health 35:106.

White, R., Lee, D., Habicht, G.S., and Janoff, A., 1981b, Secretion of alpha-1-proteinase inhibitor by cultured rat alveolar macrophages, Am. Rev. Respir. Dis. (in press).

White, R., Leon, I. and Kuhn, C., 1981c, Effect of colchicine, vinblastine, D_2O and cytochalasin B on elastase secretion, protein synthesis and fine structure of mouse alveolar macrophages, J. Reticuloendoth. Soc. (in press).

White, R., Lin, H.S., and Kuhn, C., 1977, Elastase secretion by peritoneal exudative and alveolar macrophages, J. Exp. Med., 146:802.

White, R.R., Norby, D., Janoff, A., and Dearing, R., 1980c, Partial purification and characterization of mouse peritoneal exudative macrophage elastase, Biochim. Biophys. Acta, 612:233.

White, R., White, J., and Janoff, A., 1979, Effects of cigarette smoke on elastase secretion by murine macrophages, J. Lab. Clin. Med. 94:489.

DISCUSSION

LECTURER: Kuhn   CHAIRMAN: Cumming

CUMMING: Thank you for a fascinating glimpse into the
                future. I find that, as your Chairman, I have
                created somewhat of a difficulty, because we have
                already consumed 18 minutes of your lunch hour and
                yet you may wish to continue the discussion of this
                interesting topic. So I would use the technique
                used by the ancient Roman emperors at the
                gladiatorial contests: thumbs up please all those
                who would like to go to lunch; thumbs down all
                those who would like a discussion period. The
                discussion period has it. I will take questions
                from the floor.

BAUM: A brief comment and a short question. The brief
                comment is that to the extent that there may have
                been any discrepancies between what he said and
                what I said - he is right. The brief question is:
                do you know whether the mouse macrophage elastase
                is inhibited by the antileukoprotease?

KUHN: The answer, no, I don't know.

HUTCHINSON: Returning to the question of emphysema in animals,
                if you don't mind. I believe that it's the case
                collagenase doesn't play much part in this but
                elastase is the main determinant of emphysema,
                certainly in the experimental situation, Snider and
                colleagues and I am sure yourselve have shown this.
                Would you like to comment on that point.

KUHN: The evidence really is not as overpowering as one
                might suppose because it is true that you cannot
                produce experimental emphysema using bacterial
                collagenases. The problem is that bacterial
                collagenases attack the basement membrane and the
                animal soon dies of haemorrhage before one degrades
                very much of the straight banded collagen fibres.
                To my knowledge nobody has ever injected a
                mammalian collagenase which would really be
                interesting because it would not attack the
                basement membrane and therefore you could really
                see what happens if you selectively destroy banded
                collagen fibres in vivo. But there is no

commercially available mammalian collagenases and it is not abundant in any form. Certainly all the experimental evidence we have points to elastin but collagen has not been eliminated rigorously.

KAY: Could you tell us whether any congenital or acquired disorders of fibronectin have been described and if so, what the clinical features are.

KUHN: The only thing that I am aware of is that there has been a case report of a family in which the fibronectin concentrations in plasma were normal but in which there was a defect in platelet function; (platelet granules contain fibronectin). Incidentally, it is really very remarkable how diverse are the function attributed to fibronectin. These patients had a poor-clot formation but the addition of exogenous plasma fibronectin corrected the defect. Presumably it was an undefined molecular abnormality but not a deficiency.

SORS: Do you have information about the activity of non-steroidal anti-inflammatory drugs upon protease secretion by alveolar macrophages?

KUHN: No, I don't think I do, I am trying to remember. There have been studies done some years ago but they were done using a synthetic esterase substrate and I don't think we know that they were measuring the true elestase activity.

LAURENT: I want to ask a question, but first I should make a statement to clear up the business of the stimulation of fibroblast collagens in the models. You high-lighted the confusion in the literature on that. Carrington was at Brompton two months ago and was discussing his reasearch which was the main area of conflict. He has now repeated that work using silica particles of different types and sizes and by choosing the right type and size, he now repeats the Heppleston results. So I don't think the confusion is as great now.

KUHN: Thank you very much. That makes it unanimous that everybody who has tried it got some stimulation although it certainly differs a great deal in its properties and magnitude.

CUMMING: There appears to be a falling off in the enthusiasm
             for questions.   So I will make one brief
             announcement. Dr. Tricomi from Rome is giving a
             seminar this evening in Trapani and he is giving it
             at 7.30, so to permit those of the members of this
             Course who would wish to attend his symposium, I
             would like to finish a little earlier and may we
             therefore begin this afternoon's session not at
             3.30 as was advertised yesterday, but at 3.00
             o'clock. Particularly for Madame Lafuma, since she
             is the first speaker - is that possible. Thank you
             very much.

ELASTIN AND STRUCTURAL GLYCOPROTEINS IN NORMAL AND PATHOLOGICAL

LUNG: BIOSYNTHESIS AND DEGRADATION

Chantal Lafuma and Ladislas Robert

Lab. Biochimie du Tissu Conjonctif (GR CNRS N° 40), Fac.
Medecine, Univ. Paris-Val de Marne, 8 rue du General
Sarrail, 94010 CRETEIL CEDEX (France)

INTRODUCTION

Lung parenchyma is composed of cells and intercellular matrix.
Intercellular matrix plays a crucial role in the maintenance of
lung geometry and in the facilitation of its main physiological
role which is the optimisation of the conditions for the rapid
exchange of gases between the inhaled air and the circulating
blood. This is obtained by the maximisation of the surface of
contact between air and capillary walls. Therefore the qualitative
and quantitative aspects of lung intercellular matrix play an
important role in the maintenance of physiological function.

As with other intercellular matrices (connective tissues),
lung intercellular matix is also composed of macromolecules
belonging to what we have called the four families of intercellular
matrix macromolecules (for a review, see Hance and Crystal, 1976;
Horwitz et al., 1976; Robert, 1980). These are the collagens,
elastin, proteoglycans and structural glycoproteins. In this
report, we shall only present the structure and function of elastin
and structural glycoproteins and touch only shortly on
proteoglycans.

As a result of the studies of several laboratories over the
last decade, the biosynthesis of such a precise microarchitectural
arrangement as the lung parenchyma can be envisaged as the result
of the coordinated biosynthesis of macromolecules belonging to the
above four classes.

The rate of synthesis of the pulmonary intercellular matrix
molecules is different for each and changes with the age of the

293

individual. We do not know yet what determines these relative
rates of synthesis, but many factors play a role in the regulation
of elastin and glycoproteins biosynthesis. Among the factors which
are of paramount importance in determining these relative rates of
biosynthesis of macromolecules, are certainly the number of cells
which participate in any given biosynthetic process and the amount
of total biosynthetic activity devoted by every cell to the
synthesis and excretion of every macromolecule. On the other hand
there are the degradative mechanisms which may become of importance
in such pathological conditions as emphysema. These may be
insufficient in some other pathological conditions to remove the
excess of connective tissue macromolecules synthesised (Bienkowski
et al., 1978) and this seems to be the case in pulmonary fibrosis.

Moreover, these relative rates of synthesis appeared to be
regulated by a genetically determined program in such a fashion
that their relative intensity changes with age which results in an
age dependent modification of the structure and microarchitecture
of pulmonary tissue.

STRUCTURE AND FUNCTION OF ELASTIN

Elastic fibers have been shown to be synthesised in mammalian
tissue such as aorta, tendon, ligamentum nuchae, on a
microfibrillar network. It is highly probable that this synthetic
process is the same in the lung. The microfibrillar network
appears to be composed of structural glycoproteins. Figure 1 shows
schematically this process whereby a mesenchymal cell (which can be
a smooth muscle cell or a fibroblast) will synthesise the
components of these microfibrils as well as the soluble precursor
of elastic fibre which was recently shown to be tropoelastin
(Sandberg, 1980). Tropoelastin is a soluble, relatively globular
protein of a molecular weight of about 74 000, rich in lysine,
about 48 out of 1000 total residues. It is probable the through
this positive charge conferred upon it by the high lysine content
it may interact electrostatically with the negatively charged
acidic glycoprotein microfibrils (Robert, 1980b).

Early observations with the electron microscope demonstrated
that the elastic fibers were heterogeneous, both amorphous and
fibrillar components being visualized (Ross, 1973). The
description of elementary elastic units was proposed by Daria Haust
(1965) who realised that the amorphous structures which interact
with the microfibrillar aggregates are the elementary morphological
units of elastic tissues. This contention was confirmend by
several investigators (B. Robert et al., 1970; Ross and Bornstein,
1969).

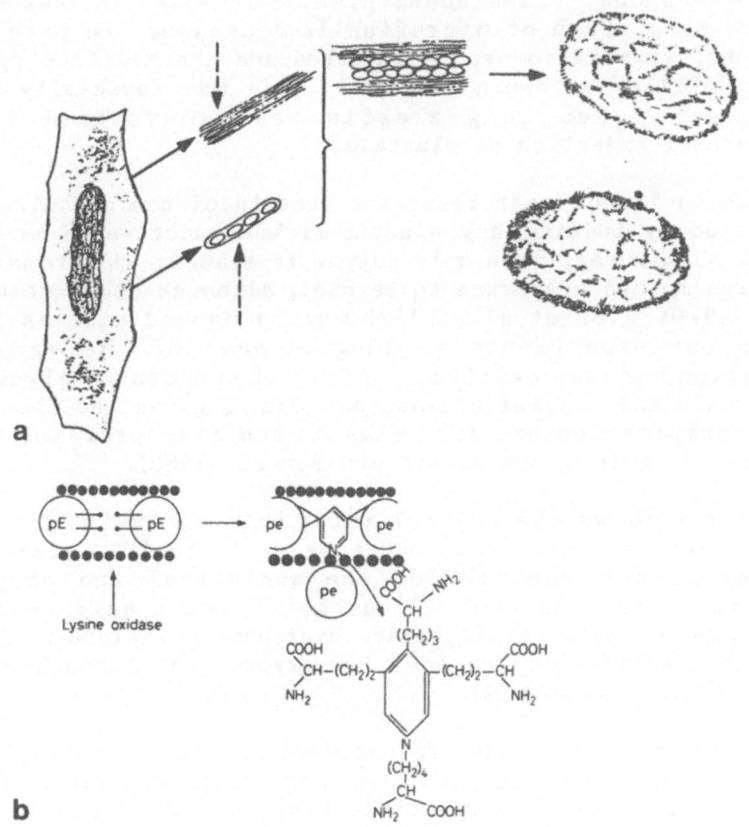

Figure 1 – Schematic representation of elastogenesis. (a) The smooth muscle cell synthesizes structural glycoprotein-microfibrils and tropoelastin or proelastin. These two of molecules interact to form first the young elastic fiber and then the mature elastic fiber. The former is rich in microfibrils, the latter is poor in microfibrils. (b) Details are shown on this procedure: between two rows of microfibrils, two molecules of proelastin face each other with two and two lysine residues opposing each other. After the action of lysyl-oxidase, a desmosine link is formed between the two pairs of lysine residues, giving rise to a covalent desmosine cross-link. This would be the mechanism of dimer and oligomer formation between vicinal tropoelastin subunits. The microfibrillar structures could correspond, according to Cotta-Pereira (1979) to the oxytalan fibers, young elastic fibers rich in microfibrils, to the elaunine fibers, and the mature elastic fibers would acquire orceinophilia and the other histochemical characteristics of mature elastic fibers.

Later, during maturation of the elastic tissue, the proportion
of the amorphous, translucent elastic lamellae increases and the
relative proportion of microfibrils decreses. In pathological
conditions, this ratio can be inversed and the relative proportion
of microfibrils can again increase. This was especially shown to
be the case when lung elastin was neosynthesized after
intratracheal injection of elastase.

Elastin is the most resistant protein of the organism and can
be purified by boiling any elastin-rich connective tissue in 0.1N
NaOH for 45 minutes, when only polymeric elastin will remain. This
very harsh method has often to be used, although other methods (Paz
et al., 1976; John et al., 1972) were proposed such as those we
used in our experiments on lung because of the very strong
association between calcium, lipids and structural glycoprotein
microfibrils and polymeric elastin.   Fig. 2 gives the flow diagram
for the preparation of lung elastin and glycoproteins (Lafuma,
1980) (for a review, see Robert and Robert, 1980).

Table 1 shows the typical amino acid analysis of pulmonary
elastin.   Bovine ligamentum nuchae being 70 to 80% elastin it is
often used as a control for the analytical and preparative
procedure.   More than 80% of the total amino acid residues of
elastin are composed of aliphatic, hydrophobic residues.   It can be
seen that elastin contains some hydroxyproline although much less
than collagen (about 1.5%).

The polar amino acids are composed of some dicarboxylic amino
acids and some basic ones and the most characteristic amino acids
are the cross-linked amino acids isolated and characterised by
Partridge (1964) and by Franzblau (1970), desmosines and
lysinonorleucine.   These amino acids are synthesised from two four
lysine residues respectively as shown on the Fig. 2.   The formation
of cross-links renders elastin insoluble and assures its
elasticity.   The oxidation is catalyzed by lysyl-oxidase.   The work
of Brody et al. (1979) on lung lysyl-oxidase activity in rabbit
lungs, demonstrated that this activity remained high during the
first weeks after birth and then decreased by about 50% to a stable
level thereafter.

Elastin has interesting physicochemical properties in that it
it is hydrophobic and can interact with lipids (Fig. 3).

When elastin in vitro is swollen by immersion in
dimethylformamide and mixed with lipids, these will strongly
interact with elastin and even after washing, a part of these added
lipids adhere to the elastic fibers.   When elastin is then
hydrolysed by elastase, a turbid solution is obtained as contrasted
to the control elastin not mixed with lipids, which yields a
perfectly clear solution (Chaudière et al., 1980).   The speed of

Entire lung freed from circulating blood

Bronchoalveolar lavage
_____

Lung dissected free of trachea   Bronchial mucins, surfactant
major bronchia and bronchioles   alveolar proteins...

Incubation of lung slices with
^{14}C-glucosamine (5 hrs)

1M NaCl + protease inhibitors
          2x
_____

1M NaCl extract   Residue
                               2M MgCl$_2$
          _____

          2M MgCl$_2$ extract   Residue

                               Delipidation
                               (CHCl$_3$ MeOH 2/1)
                    _____

                    Residue   Lipids

               5M guanidinium‐
               chloride buffer
               + DTT
          _____

          Guanidine   Residue
          Extract I
     _____

     STRUCTURAL GLYCOPROTEINS   + collagenase

                    Residue

               5M guanidinium‐   Collagenase
               chloride buffer   extract
               + DTT
          _____

          Guanidine   Residue elastin
          Extract II
     _____

     MICROFIBRILLAR COMPONENT

Figure 2 - Flow sheet for the preparation of lung elastin and
structural glycoproteins.

Table 1.   Amino acids composition of parenchymal elastins and
           structural glycoproteins.   Results expressed as
           residues/1000 amino acids residues.
           * from Paz et al. (1976)   ** from Francis et al. (1974)

	Elastin (*)	Glycoproteins (**)
Hyo	15.0	0
Asp	10.0	103.5
Thr	7.0	50.5
Ser	8.0	65.0
Glu	20.9	159.0
Pro	110	47.2
Gly	313	69.2
Ala	220	80.5
Cys	0	12.8
Val	135	52.5
Met	0	16.8
Ile	25.9	42.0
Leu	63.3	100.8
Tyr	8.5	28.5
Phe	37.1	37.2
Iso-Des	1.4	0
Des	3.9	0
Lys	8.8	14.3
His	3.5	50.0
Arg	7.8	69.4

elastolysis was enhanced in vitro by many of the lipid ligands
ligands introduced in to elastin. This simple model experiment
shows that relatively important amounts of lipids can accumulated
in elastic fibers.   Dr. Szemenyi (1980) showed that the auto-
fluorescence of pulmonary elastin increases with age because of the
interaction of lipids, auto-oxidatain of the lipids and the
formation of fluorescent pigments on elastin. This progressive
interaction with lipids and calcium salts is one of the important
factors determining the ageing of elastin and the decrease of
elasticity, and also explains the increased elastolysis noticed
with age in elastic tissue such as aorta, skin and the lung.

This is considered as part of a more or less normal ageing
process, and may be the reason which determines the marked decrease
with age of the maximal breathing capacity of the lung: most
physiological functions decrease with age but one of the most
marked age-dependent decrease was observed in the maximal breathing
capacity (Shock, 1959).

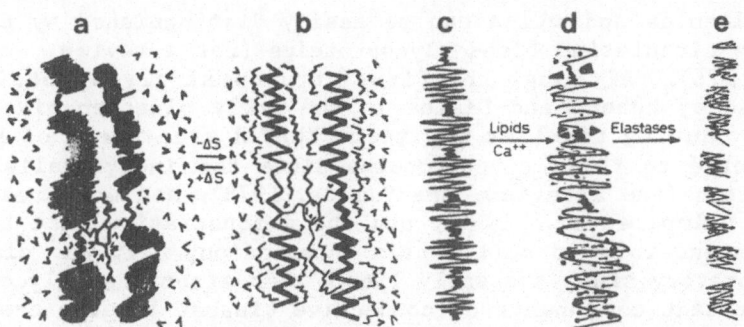

Figure 3 - Schematic representation of the entropic nature of elasticity of elastin (a, b) and the effect of the saturation by lipids of the elastin peptide backbone (c, d). The elastin peptide backbone is stabilized by hydrophobic interactions with the exclusion of water molecules (v) from the vicinity of these hydrophobic regions. The hydrophilic regions containing the desmosine cross-links are represented as branched cyclic structures. By stretching the fiber, the hydrophobically stabilized regions are distorted and water has to take a more ordered conformation around the exposed hydrophobic residues, this process being accompanied by a decrease of entropy. This is the basis of the entropic nature of the elasticity of elastin (Robert and Robert, 1980). The return to the normal length after the release of stretching forces would be due to the disorganization of the water solvent around the reformed hydrophobic regions. This entropy-driven structural change would be abolished if lipids and calicium salts were inserted into the peptide backbone. This same saturation by lipids accelerates the action of elastolytic enzymes which could more easily attack the peptide bonds and could readily be absorbed on the polar, negatively charged side chains introduced by fatty acids in the lipid-saturated elastin.

There are specific stains which can be used to stain elastin selectively, they are resorcinol iron basic (Weigert) or orcein iron hematoxylin (Verhoeff). The so-called oxytalan and elaunine (ref. Fig. 1) fibers are not stained by orcein with oxidation whereas the fully fledged major elastic fibers are orcein positive without oxidation.

STRUCTURAL CLYCOPROTEINS

Generalities

It was shown years ago that connective tissues contain besides collagen, elastin and proteoglycans, glycoprotein components which are synthesised by the mesenchymal cells at the time as the other

macromolecules and which can be easily distinguished by this fact
from the circulating blood glycoproteins (for a review, see Robert
et a., 1976). Although the first structural glycoproteins were
described by Robert and Dische in the early nineteen sixties, it
was only during the last few years that a great deal of interest
was devoted to these glycoproteins since the individualisation of
fibronectin (fro a review, see Yamads, 1978), and most recently of
laminin (Timpl et al., 1980), and the demonstration that they play
an important role in cell matrix interactions. It was claimed by
our laboratory since the early 1960's that structural glycoproteins
are important components of connective tissues because they ensure
the specificity of cell matrix interaction and play an important
role in the oriented biosynthesis of connective tissues.

Lung structural glycoproteins

Over the past few years we have studied the structural
glycoproteins of lung parenchyma. For this purpose organ cultures
of the lung of several species such as rat, hamster, and baboon
were incubated with ^{14}C-glucosamine in sterile organ culture
conditions followed by the sequential extraction of the
macromolecular components (as shown on the flow sheet (Fig. 2))
followed by the characterisation of these glycoproteins. Fig. 4
shows that several distinct glycoproteins could be identified in
the lung parenchyma which actively incorporate ^{14}C-glycosamine and
show up as periodic acid Schiff positive bands on polyacrylamide
gel electrophoresis (Lafuma et al., 1978, 1981). Their relative
proportions and their relative labelling intensities are also
characteristic of the lung parenchyma as compared to other
connective tissues such as aorta, or cornea which were studied in
great detail in our laboratory (Moczar, 1978). The amino acid
composition of these glycoproteins (Table 1) is very similar to the
general pattern which was described for structural glycoproteins
that is a relatively high dicarboxylic amino acid content, a high
aliphatic amino acid (alanine, valine, leucine content and a higher
aromatic amino acid content than in the other matrix components
(collagen, elastin or proteoglycans). They are often rich in
disulphide linkages which certainly contribute to their low
solubility. Contrasting with their low solubility, they have
usually a very active metabolism being rapidly synthesised in most
connective tissues which were studied (Robert et al., 1976).

Fibronectin was detected immunologically in connective tissue
fractions from adult human lung by Anderson Bray (1978).
Fibronectin antigen could be solubilized from the parenchyma and
from the alveolar basement membrane by collagenase digestion
indicating that fibronectin occurs in lung connective tissue in
association with collagen. Lung fibronectin plays probably a
similar role as in other tissues, in assuring a specific contact
between some of the pulmonary cells and the intercellular matrix,

and may probably influence the function of alveoli. Nevertheless, fibronectin in our previously described experiments turned out to be a minor component and that several other glycoproteins are certainly quantitatively more important. We know from recent studies on cartilage and on cornea that other glycoproteins similar to fibronectin but quite distinct immunochemically, may be of importance in assuring the cell matrix interaction in tissues. It is not impossible that besides fibronectin some of the glycoproteins we studied are of a major importance in cell matrix interaction and in morphogenesis in the lung.

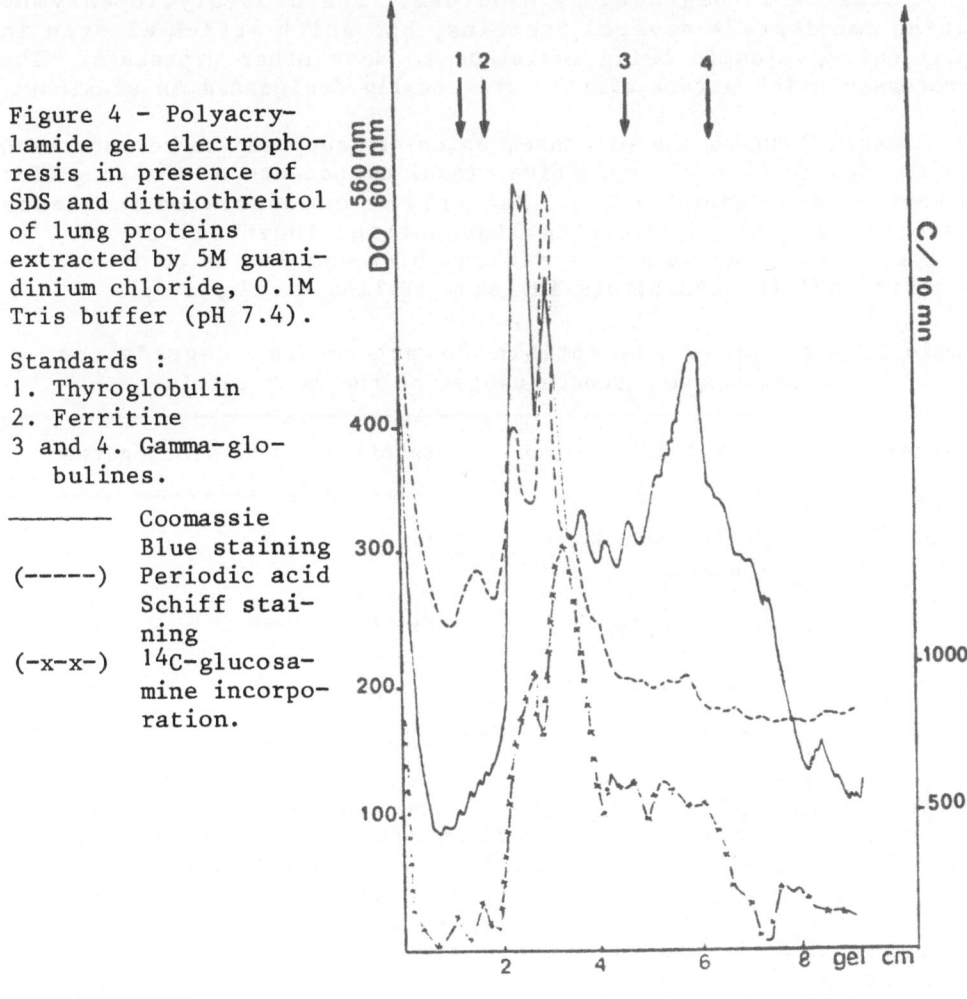

Figure 4 - Polyacrylamide gel electrophoresis in presence of SDS and dithiothreitol of lung proteins extracted by 5M guanidinium chloride, 0.1M Tris buffer (pH 7.4).

Standards :
1. Thyroglobulin
2. Ferritine
3 and 4. Gamma-globulines.

———— Coomassie Blue staining
(-----) Periodic acid Schiff staining
(-x-x-) ^{14}C-glucosamine incorporation.

PROTEOGLYCANS

These consist of a central protein core to which long polysaccharide chains, the glycosaminoglycans, are attached such as chondroitin-sulphate (4 and 6-sulphate), dermatan-sulphate, heparan-sulphate. The protein portion is then linked to other similar proteoglycans through hyaluronic acid and link glycoproteins. More details about their structure and their composition and ultrastructural localisation in the lunk can be found in the following references: Horwitz et al. (1976), Sampson et al. (1979) and Vaccaro et al. (1979).

DEGRADATION OF ELASTIN AND GLYCOPROTEINS IN NORMAL AND PATHOLOGICAL LUNGS

Elastin is degraded by elastases, the proteolytic enzymes which can degrade several proteins, but which attack elastin in particular, elastin being resistant to most other proteases. The proteases which attack elastin are usually designated as elastases.

Table 2 shows the elastases which are supposed to be important in the destruction of connective tissue components of the lung (for a review, see Gadek, 1980a). Recently we demonstrated elastases associated with circulating lipoproteins (Hornbeck et al., in prepartation) and we have a method which enables us to determine elastase and its inhibitors in serum (Bellon et al., 1978).

Table 2. Proteases susceptible to act on the degradation of elastin and glycoconjugates of the lung matrix.

Enzyme	Cellular source	Elastin	Glycoconjugates
Elastase	Alveolar macro-phage	++	0
	Leucocyte	++++++	++
Cathepsin G	Leucocyte	+	++

Elastases are important enzymes because they were shown to be of importance (Kuhn et al., 1980) in the experimental production of emphysema in hamsters and they probably play a role in human emphysema also (Gądek et al., 1980a). It is now generally accepted that the integrity of human alveolar structures is partly dependent on maintenance of a balance between inflammatory cell derived proteases and the anti-proteases which have access to the lower respiratory tree. The only enzyme which correlated with emphysema and was demonstrated in bronchoalveolar lavage fluids is leucocyte elastase. This enzyme is inhibited by alpha$_1$-antitrypsin, alpha$_2$-macroglobulin and bronchial mucus inhibitor (so-called antileukoprotease) (for a review, see Hochtrasser, 1980; Ohlsson,

1980 and Franken, 1980). It was shown by Janoff (1980a) and Travis (1980) that these inhibitors are oxidised by tobacco smoke components and functionally depressed. In our laboratory, Hornebeck et al. demonstrated that there was a strong decrease with age of elastase inhibitors in the serum of smokers (Robert et al., 1980). No such decrease could be demonstrated in the serum of non smokers (Hornebeck et al., in preparation). This fact may be of importance in explaining increasing elastolysis with age in the pathogenesis of emphysema.

Carp and Janoff (1980) showed that the oxidative inactivation of alpha$_1$-proteinase inhibitor by the reactive oxygen species released in the microenvironment of lung inflammatory cells at sites of acute or chronic inflammation may allow proteases from these cells to damage adjacent connective tissue components more readily.

When lung elastin is attacked by elastolytic enzymes, elastin peptides are released (Kucich, 1980), and are susceptible to act on monocytes, neutrophils or macrophages (Senior, 1980), which in their turn can secrete elastolytic enzymes and may influence the progress of emphysema.

Although elastin is resynthesised in the animals which received intracheal elastolytic enzymes, the structure of this elastin is not identical to the one found in normal lung (for reviews, see Kuhn, 1980). This elastin seems much richer in microfibrillar material than normal elastin (Szemenyei, 1980). Elastolytic proteases from leukocytes and also from other sources can attack not only elastin but also other proteins such as proteoglycans or structural glycoproteins (Table 2). We showed that the highly purified elastolytic protease obtained from skin fibroblast cultures, when injected in the dermis of rabbits degrades elastic fibers and also the microfibrils (Godeau et al., 1981). Our recent experiments (Lafuma, 1980; Moczar, 1980) on elastase induced emphysema in hamsters demonstrated the alterations of metabolism of glycosamino-glycans and microfibrillar components. It appears therefore that the proteolytic attack on glycoproteins which ensure the contact between cells and intercellular matrix is probably of important in the disorganisation of the specific microarchitecture of lung parenchyma. It was shown in the laboratory of Prof. Crystal (McDonald, 1979, 1980) that leucocyte elastase attacks fibronectin much faster than other matrix components. The release of biologically active fibronectin fragments by leucocyte elastase suggests a new potential mechanism for alteration of cell connective tissue interactions at sites of inflammation in vivo, and this reaction is probably of importance in the production of the emphysematous lesion.

Therefore, several laboratories actively study replacement

therapy in emphysema providing artificially inhibitors of elastases
in order to diminish the degradation of intercellular matrix macromolecules (for a review, see Gadek et al., 1980b; Janoff, 1980b).

CONCLUSIONS

We conclude that lung intercellular matrix is composed of
macromolecules belonging to four classes of matrix macromolecules.
Besides collagen, elastin and structural glycoproteins are major
components of the lung parenchyma. Proteoglycans are also present,
but in a much lower amount. The normal function of the lung
depends not only on the relative proportions, of these
macromolecules, but also on their specific arrangement. It appears
that this specific arrangement is deranged or modified in
pathological conditions such as emphysema. Several important
aspects of the biosynthesis and degradation of these macromolecules
have been elucidated over the last few years but many are not yet
understood. For instance, we do not know yet which cells are
responsible at every period of development and aging for the
biosynthesis and the removal of elastic fibrils and if they have a
turnover at all in lung parenchyma.

Finally, although we demonstrated the presence of several
structural glycoproteins, we still have to assign them a specific
function and location in normal and pathological lung tissues. We
can expect however, that thanks to the development of modern
methodology, rapid progress will be accomplished in the
understanding of these problems and also in the molecular
pharmacology of lung disease.

ACKNOWLEDGEMENTS

The original part of this lecture was supported by grants from
the C.N.R.S. (GR N⁰ 40), I.N.S.E.R.M., Fondation pour la Recherche
Médicale Francaise and Commission de Recherche of the Université
Paris-Val de Marne.

Helpful discussions with Mrs Madeleine MOCZAR, Chargee de
Recherche au CNRS, member of this laboratory and with Professor
BIGNON (Départment de Physiopathologie, Hôpital Intercommunal,
Université Paris-Val de Marne) are thankfully acknowledged.

REFERENCES

Bellon, G., Hornebeck, W., Derouette, J.C., and Robert, L., 1978,
        Simplified methods for the quantification of elastase and its
        inhibitor in physiologic fluids, Path. Biol., 26:515.
Bienkowski, R., Baum, B., and Crystal, R.G., 1978, Fibroblasts
        degrade newly synthesised collagen within the cell before
        secretion, Nature, 276:413.

Bieth, J., Natural and synthetic inhibitors of pancreatic and leucocyte elastases, in: "Frontiers of Matrix Biology" L. Robert, ef., S. Karger, Basel, Vol. 8, pp. 216-224 (1980).

Bray, B.A., 1979, Cold insoluble globulin (fibronectin) in connective tissues of adult human lung and in trophoblast basement memnbrane, J. Clin. Invest., 62:745.

Brody, J.S., Kagan, H., and Manalo, a., 1979, Lung lysyl oxidase activity: relation to lung growth, Ann. Rev. Resp. Dis., 120:1289.

Cantor, J.O., Keller, S., Parshley, M.S., Darnule, T.Y., Darnule, A.T., Ceretta, J.M., Turino, G.M., and Mandl, I., 1980, Synthesis of cross linked elastin by an endothelial cell culture, Biochem. Biophys. Res, Com., 95:1381.

Carp, H., and Janoff, A., 1980, Potential mediator of inflammation, J. Clin. Invest., 66:987.

Chaudiere, J., Derouette, J.C., Mendy, F., Jacotot, B., and Robert, L., 1980, In vitro preparation of elastin-triglyceride complexes. Fatty acid uptake and modification of the susceptibility to elastase action, Atherosclerosis, 36:183.

Cotta-Pereira, G., Kattenbach, W.M., and Guerra-Rodrigo, F., Elastic related fibers in basement membranes, in: "Frontiers of Matrix Biology" L. Robert, ed., S. Karger, Basel, Vol. 7, pp. 90-100 (1979).

Francis, G., and Thomas, J., 1975, Isolation and characterisation of acific structural glycoproteins in pulmonary tissues, Biochem. J., 145:299.

Franken, C., Kramps, J.A., Meyer, C.J.L.M., and Dijkman, J.H., 1980, Localisation of a low molecular weight protease inhibitor in the respiratory tract, Clin. Resp. Physio., 16:231.

Franzblau, C., Faris, B., Lent, R.W., Salcedo, L.L., Smith, B., Jaffe, R., and Crombie, G., Chemistry and biosynthesis of cross-links in elastin, in: "Chemistry and Molecular Biology of the Intercellular Matrix" E.A. Balazs, ed., Academic Press, New York, Vol. 1, pp. 617-639 (1970).

Gadek, J.E., Hunninghake, Fells, G.a., Zimmerman, R.L., Keogh, B.A., and Crystal, R.G., 1980a, Evaluation of the protease anti-protease theory of human destructive lung disease, Clin. Resp. Physiol., 16:27.

Gadek, J.E., Keoga, B.A., and Crystal, R.G., 1980b, Opportunities for the specific therapy of destructive lung disease, Clin. Resp. Physiol., 16:389.

Godeau, G., Frances, C., Hornebeck, W., Rouges, O., Brechemier, B., and Robert, L., 1981, submitted to Invest. Dermatol.

Hance, A.J., and Crystal, R.g., 1976, "Collagen in lung biology in health and disease" R.G. Crystal, ed., Vol. 2:214.

Haust, M.D., 1965, Fine fibrils of extracellular space (microfibrils). Their structure and role in connective tissue organization, Am. J. Path., 47:1113.

Hochstrasser, K., 1980, The acid-stable proteinase inhibitors of the respiratory tract. Chemistry and function. <u>Clin. Resp. Physiol.</u>, 16:223.

Horwitz, A.L., Elseon, N.A., and Crysta, R.G., Proteoglycans and elastic fibers, <u>in</u>: "Lung biology in health and disease" R.G. Crystal, ed., Vol. 2, pp. 273-311 (1976).

Janoff, A., Carp, H., and Lee, D.K., 1980a, Inactivation of alpha$_1$-proteinase-inhibitor and bronchial mucous proteinase inhibitor by cigarette smoke in vitro and in vivo, <u>Clin. Resp. Physiol.</u>, 16:321.

Janoff, A., and Dearing, R., 1980b, Prevention of elastase induced experimental emphysema by a synthetic elastase inhibitor administered orally, <u>Clin. Resp. Physiol.</u>, 16:399.

Kucich, U., Christner, P., Weinbaum, G., and Rosenbloom, J., 1980, Immunological identification of elastin derived peptides in the serums of dogs with experimental emphysema, <u>Am. Rev. Resp. Dis.</u>, 122:461.

Lafuma, C., Moczar, M., and Robert, L., 1979, Non collagen proteins of pulmonary interstitial matrix in rats. <u>Clin. Resp. Physiol.</u>, 15:38.

Lafuma, C., Lange, F., Moczar, M., Bignon, J., and Robert, L., 1980, Modifications of ^{14}C-glucosamine incorporation into hamster lung glycoconjugates in elastase induced emphysema, <u>Clin. Resp. Physiol.</u>, 16:91.

Lafuma, C., Moczar, M., and Robert, L., 1981, Isolation and characterization of lung connective tissue glycoproteins, submitted.

McDonald, J.A., Baum, B.J., Rosenberg, D.M., Kelman, J.A., Brin, S.C. and Crystal, R.G. 1979. Destruction of a major extracellular adhesive glycoprotein (fibronectin) of human fibroblasts by neutral proteases from polymorphonuclear leukocyte granules, <u>Lab. Invest.</u>, 40:350.

McDonald, J.A., and Kelley, D., 1980, Degradation of fibronectin by human leucocyte elastase, <u>J. Biol. Chem.</u>, 255:8848.

Moczar, M., and Moczar, E., 1978, Glycoproteins from the aorta, <u>Path. Biol.</u>, 26:63.

Moczar, M., Lafuma, C., Lange, F., Bignon, J., and Robert, L., 1980, Glycosaminoglycans in elastase induced emphysema, <u>Bull. Europ. Physiopath. Resp.</u>, 16:99.

Partridge, S.M., Eldsen, D.F., Thomas, J., Dorfman, A., Telser, A. and Ho, P.L., 1964, Biosynthesis of the desmosine and isodesmosine cross bridges in elastin. <u>Biochem. J.</u>, 93:30.

Paz, M.a., Keith, D., Travers, O.H., and Gallop, P.M., 1976, Isolation, purification and cross linking profiles of elastin from lung and aorta, <u>Biochem.</u>, 15:4912.

Reilly, C.F., and Travis, J., 1980, The degradation of human lung elastin by neutrophil proteinases, <u>Biochim. Biophys, Acta.</u>, 621:147.

Robert, A.M., Robert, B., and Robert, L., Chemical and physical properties of structural glycoproteins, <u>in</u>: "Chemistry and Molecular Biology of the Intercellular Mtrix", E.A. Balazs, ed., Academic Press, New York, Vol. 1, pp. 237-242 (1970).

Robert, L., Junqua, S., and Moczar, M., Structural glycoproteins of the intercellular matrix, in: :Fronteirs of Matrix Biology", L. Robert, ed., S. Karger, Basel, Vol. 3, pp. 113-142 (1976).

Robert, L., 1980, Nouveaux concepts de la pathologie du conjonctif pulmonaire, Bull. Europ. Physiopath. Resp., 16:421.

Robert, L., and Robert, A.M., Elastin, elastase and arteriosclerosis, in: "Frontiers of Matrix Biology", L. Robert, ed., S. Karger, Basel, Vol. 8, pp. 130-173 (1980).

Robert, L., Bellon, G., and Hornebeck, W., 1980, Characterization of different elastases. Their possible role in the genesis of emphysema, Bull. Europ. Physiopath. Resp., 16:199.

Ross, R., and Bornstein, P., 1969, Elastic fiber, J. Cell Biol., 40:366.

Ross, R., 1973, The elastic fiber. A review, J. Histochem. Cytochem., 21:199.

Sandberg, L.b., Leach, C.t., Leslie, J.g., Torres, R.A., and Alvarex, V.L., Structural studies of porcine aortic tropoelastin, in: "Fronteirs of Matrix Biology" L. Robert, ed., S. Karger, Basel, Vol. 8, pp. 69-78 (1980).

Sampson, P.M., Jimenez, S.A., and Bashey, R., 1979, Isolation and partial characterization of proteoglycans from sheep lung parenchyme, Biochim. Biophys. Acta., 558:129.

Shock, N.W., "Biological Aspects of Aging" 2nd ed., Academic Press, New York and London (1962).

Senior, R.M., Griffin, G.L., and Mecham, R.P., 1980, Chemotactic activity of elastin derived peptides. J. Clin. Invest., 66:859.

Szemenyei, K., 1980, Papain induziertes Emphysem als Versuchsmodell des panobularen Emphysems, Ergebn. exp. Med., 35:553.

Timpl, R., and Rhode, 1979, Characterization of laminin, a major glycoprotein of basement membrane. J. Biol. Chem., 254:9933.

Travis, J., Beatty, K., Wong, P.S., and Matheson N.R., 1980, Oxydation of alpha$_1$-proteinase inhibitor as a major contributing factor in the development of pulmonary emphysema. Clin. Resp. Phys., 16:341.

Vaccaro, C.A., Brody, J.S., 1979, Ultrasturctural localization and characterization of proteoglycans in the pulmonary alveoles Am. Rev. Resp. Dis., 129:901.

Yamada, K.M., and Olden, K., 1978, Fibronectins - adhesive glycoproteins of cell surface and blood, Naure, 275:179.

DISCUSSION

LECTURER: Lafuma   CHAIRMAN: Cumming

BAUM: I was extremely interested in your comment about
                the interaction between elastin and lipid which
                rendered it much more susceptible to degradation by
                elastase.   I wonder whether there was any
                specificity in the kind of lipid that binds the two
                elastins, and in particular whether steroids may
                interact because this may relate to a possibility
                that the interaction between female sex hormones
                and elastin may somehow protect it against
                elastolasis.

LAFUMA: I think that lipids that are mixed with elastin are
                lipids of a different group. I don't know if there
                is any correlation between the nature of lipids and
                their reaction with elastin.

DENISON: Madame Lafuma, I will ask my question in English
                first and then in French. In your first or second
                slide you showed that a central step in the
                formation of structural proteins was taken by the
                enzyme nysin oxidase. This means that the enzyme
                has a need for oxygen. Have you any idea what the
                Michaelis constant for oxygen is?   In the
                discussion of oxidases and oxygenases one
                frequently sees quoted the rate constant for the
                proteins they are manipulating, but you very rarely
                see mentioned their need for oxygen and how
                sensitive they are. Quite often their Michaelis
                constants for oxygen are very high, of the order of
                5 to 250 mm of mercury which means that many of
                these proteases are really quite sensitive to local
                oxygen pressure.

CUMMING: Perhaps I should explain to the audience who may
                not be familiar with Michaelis constants that this
                refers to the pressure of oxygen at which the
                enzymic activity is appropriate.

WILLIAMS: I will make my comment and question in three
                languages, first in Welsh, then in French, then in
                English in case you do not understand either. Do
                you recognise collastin? You tell us that elastin
                comes from fibroblasts and from small muscle,
                collagen comes from fibroblasts, is there a

substance that is both collagen and elastin?

COHEN: May I ask you or Dr. Kuhn whether the fibronectin
            you are talking about is the same molecule as the
            fibronectin that Dr. Kuhn was talking about and if
            so why should its subcellular localisation be very
            different.

LAFUMA: If the fibronectin is bound to the same membrane of
            the fibroblast it is the same, the fibronectin of
            the macrophage is the same.

COHEN: Is it the same molecule?

LAFUMA: I think so because I don't think the distribution
            is different.

CUMMING: Does that answer your question?

KUHN: It's an excellent question to which the answer is
            not entirely known. There are described some
            differences between soluble fibronectin and the
            insoluble cell surface associated fibronectin.
            These differences are very small differences in
            molecular weight and differences in the state of
            aggregation which are not understood, but cell
            surface fibronectin is probably a polymer of much
            larger than two million. The biochemical basis for
            these differences is not clear, there are many many
            similarities including immunologic reactivity.
            Many biological properties such as binding
            collagen, heparin and fibrin aminoacid composition
            are virtually identical. So we are not entirely
            sure what is the basis for these small differences.

CUMMING: We should now say to Madame Lafuma that she has
            made a very valiant attempt to stand in for Dr.
            Robert, made an excellent job and we should give
            her a round of applause. Now I am going to ask
            Geoff Laurent to present his paper.

COLLAGEN IN NORMAL LUNG AND DURING PULMONARY FIBROSIS

Geoffrey J. Laurent

Biochemistry Laboratory, Department of Thoracic Medicine
Cardiothoracic Institute, Brompton Hospital
Fulham Road, London   SW3 6HP

Collagen is the most abundant protein of lung connective tissue and as such will influence normal lung function. Thus alteration in its structure, metabolism or distribution may impair lung function – in particular that related to gas exchange. In crypto-genic fibrosing alveolitis (also termed idiopathic pulmonary fibrosis) there is morphological evidence for excessive and disorded collagen deposition. For this reason, evidence has been sought to explain the disease on the basis of alterations in collagen biochemistry. In this paper, some aspects of collagen biochemistry as it relates to the lung will be discussed and the concept of whether or not pulmonary fibrosis is a disease of altered collagen metabolism examined.

Collagen structure and its synthetic and degradative pathways are now known in some detail and these aspects have been the subject of several excellent reviews (more recently, Prockop et al, 1979 a, b; Bornstein and Sage, 1980). The collagen molecule is currently thought to exist in at least 5 different forms (called types I, II, III, IV, and V). Each type consists of three separate polypeptide chains (known as α-chains) which are intertwined in a right-handed helix. Type I collagen contains two identical α-chains in association with another different α-chain so that the formula is usually written as $[\alpha 1(I)]_2 \alpha 2$. The two α-chains of type I collagen have different amino acid sequences and are therefore coded by distinct genes. The chain compositions of types II and III collagens are shown in Table 1. Type II collagen, which is found only in cartilage tissue, is a trimer of only one type of α-chain. This is also the case for Type III collagen, which is found in most tissues usually associated with type I collagen. Types IV and V collagens, which are usually associated with

311

basement membranes, have uncertain chain compositions and it is
possible that these chains represent more than two collagen types.

There is now evidence that all of the collagen types are
present in lung (see Table 1). This comes from both chemical
extraction methods (Bradley et al, 1974; Madri and Furthmayr, 1979;
Laurent et al, 1981) as well as immunochemical techniques using
collagen type specific antibodies (Madri and Furthmayr, 1979, 1980;
Bateman et al, 1981).

Table 1   Collagen Types and their Location in Lung Tissue

Type	Chains	Molecular Composition	Location in Lung
I	$\alpha 1, \alpha 2$	$[\alpha 1(I)]_2 \alpha 2(I)$	Bronchi, blood vessels, pleura interstitium
II	$\alpha 1$	$[\alpha 1(II)]_3$	Trachea, bronchi
III	$\alpha 1$	$[\alpha 1(III)]_3$	Interstitium blood vessels
IV	$\alpha 1, \alpha 2,$	Uncertain	Alveolar and capillary basement membranes
V	$\alpha 1, \alpha 2, \alpha 3$	Uncertain	Alveolar and capillary basement membranes, interstitium

The chemical extraction of collagen has been made difficult by
the extreme insolubility of the protein. One approach to this
problem has been to subject tissue collagens to long digestions in
the presence of pepsin. This enzyme, by cleaving the short non-
helical extensions (telopeptides), solubilised a fraction of the
total collagen pool (usually less than one-third in the case of
lung). Once this has been accomplished, the different collagen
types can be separated based on their solubilities in NaCl
solutions of varying concentrations and pH. Figure 1 shows the
electro-phoretic patterns for types I to V collagen separated from
rabbit lung by such methods. Collagen α-chains have been separated
from each other as well as the dimers (β-chains), trimers (δ-
chains) and very high molecular weight polymeric collagen by
polyacrylamide gel electrophoresis. The patterns are similar to
those obtained previously using similar extraction methods on other
tissues. Further evidence for the identity and purity of these
collagens can be obtained from amino acid analyses as well as the
electrophoretic patterns obtained following CNBr treatment (Laurent
et al, 1981). Fig. 1 indicates the presence of 9 distinct α-chains

which indicates that there are at least this number of genes being actively expressed in the various cells of the lung.

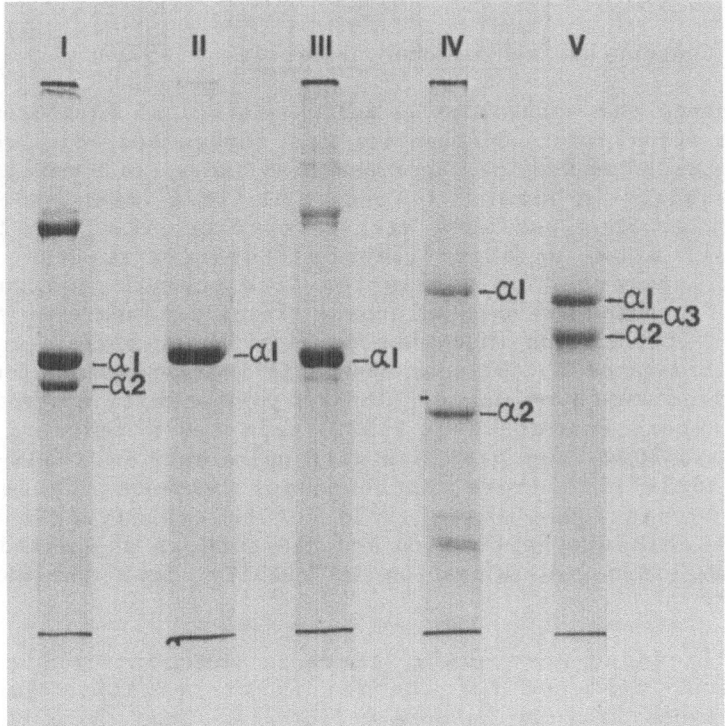

Fig. 1.   Polyacrylamide gel electrophoresis of lung collagens. All
          samples have been reduced with mercaptoethanol prior to
          electrophoresis in the presence of sodium dodecyl
          sulphate. The collagen type is designated at the top of
          each track and the α-chain, identified according to
          Bornstein and Sage, 1980, is designated for each collagen
          type.

     The concept of polymorphism of lung collagens has important implications in the consideration of lung function. If each collagen has its own distinctive physico-chemical properties, shifts in relative amounts of the different collagen types might be expected to alter lung function. Evidence exists that such a shift occurs for types I and III collagen in human pulmonary fibrosis. These collagens, which together represent the bulk of total lung collagen, are normally present in a ratio of 1.5–2.5 : 1 (Type I :

Type III) but analysis of post-mortem lung samples from human patients with pulmonary fibrosis suggests this ratio increases to about 4 : 1 (Seyer et al, 1976; Madri and Furthmayr, 1980) and it was suggested that such a change may be responsible for the alteration of lung compliance in this disease.

Collagen Content During Pulmonary Fibrosis

Although the morphological and physiological manifestations of this form of pulmonary disease are well documented, only relatively recently has a biochemical approach been made. In an initial study of this subject in humans, Fulmer et al (1976) examined biopsied lung samples from patients with idiopathic pulmonary fibrosis. They could show no significant differences in the collagen contents, expressed as a fraction of either dry weight or DNA content, for patients with pulmonary fibrosis compared with normal controls. Furthermore there was no significant difference between the in vitro rates of collagen synthesis in explants of lung tissue from patients with pulmonary fibrosis compared to controls. In a further report Fulmer et al (1980) reported a rearrangement of collagen in lung from patients with pulmonary fibrosis with the interstitial fibres more thickened and twisted. These results suggested that the disease could not be explained in terms of excessive collagen deposition and the authors suggested that it more likely involved alteration in "quality, form and location of collagen".

As discussed previously, there is immunohistochemical and biochemical evidence for changes in the relative amounts of collagen types in human pulmonary fibrosis. Thus there is evidence for increases in the amounts of Type I and V compared to Type III collagens (Seyer et al, 1976; Madri and Furthmayr, 1980). Furthermore, Bateman et al, 1981, using immunofluorescent techniques, observed that in areas described as "active" in fibrosis, such as sarcoid nodules or areas characterized by large numbers of inflammatory cells and fibroblasts, there was an increased amount of fluorescence associated with type III collagen. These results suggest that there may be a temporal change in the deposition of different collagen types with excessive type III collagen production in the early stages with a switch to type I collagen later. Certainly there is evidence for such a shift in skin wound formation where only type III collagen is produced in the first 48 hours after which there is a rapid increase in type I collagen (Gay et al, 1978). Clearly these reports suggest that changes in production of different collagen types is of importance in the pathogenesis of pulmonary fibrosis. However, further measurements are still required preferably involving chemical quantitation of lung cyanogen bromide extracts (Laurent et al,

1981) rather than the more subjective estimates necessitated by immuno-fluorescent methods or estimates on pepsin solubilized collagen which may represent less than 10% of the total lung collagen (Madri and Furthmayr, 1980).

The observations of Fulmer et al (1976) that neither collagen content nor synthesis rate alter in pulmonary disease is important since these results suggest that pulmonary fibrosis is not a disease of excessive collagen deposition. However, this conclusion is equivocal and there are alternative explanations. Lung collagen content is commonly expressed in two ways: either as concentration - with respect to another lung component or components (i.e. protein, DNA or dry weight); or alternatively as total collagen content per lung or set of lungs. Because collagen concentration is determined by two parameters, changes in either of these will determine the value obtained. Thus, for example, if one expresses collagen concentration in terms of DNA content and its content changes concomitantly with that of collagen (i.e. due to infiltration of inflammatory cells) the concentration will not change. The possibility that such a phenomenon may operate has received support from several studies of more acute forms of pulmonary disease in both humans and experimental animals.

Zapol et al (1979) have examined both total lung collagen content and concentration, with respect to dry weight, in post-mortem lungs from 12 patients who died within 25 days following acute respiratory failure. Although there was an apparent decrease in the collagen concentration the total lung collagen content increased. Similar observations have been made in studies of bleomycin-induced pulmonary fibrosis in hamsters (Goldstein et al, 1979) and baboons (McCullogh et al, 1978). This phenomenon is also demonstrated in Figure 2 which shows the time-course of changes in total lung collagen compared to collagen concentration following an intratracheal instillation of bleomycin into rabbits. At no time was there a significant increase in collagen concentration, when injected rabbits were compared to untreated controls, but the collagen content was significantly increased 2 days after instillation and was more than doubled after 14 days. In this model for pulmonary fibrosis morphological changes were also investigated. At all times inflammatory cells were prominent but there was no histochemical evidence of increased collagen deposition until 8 weeks (Laurent et al, 1981). These data demonstrated firstly, that compared to biochemistry, histological methods are relatively insensitive in detecting the early increases in collagen content; and secondly, that measurements of collagen concentration are inappropriate for detection of changes in collagen content. Measurement of total lung collagen content clearly demonstrates increased collagen deposition whereas changes in collagen concentration do not.

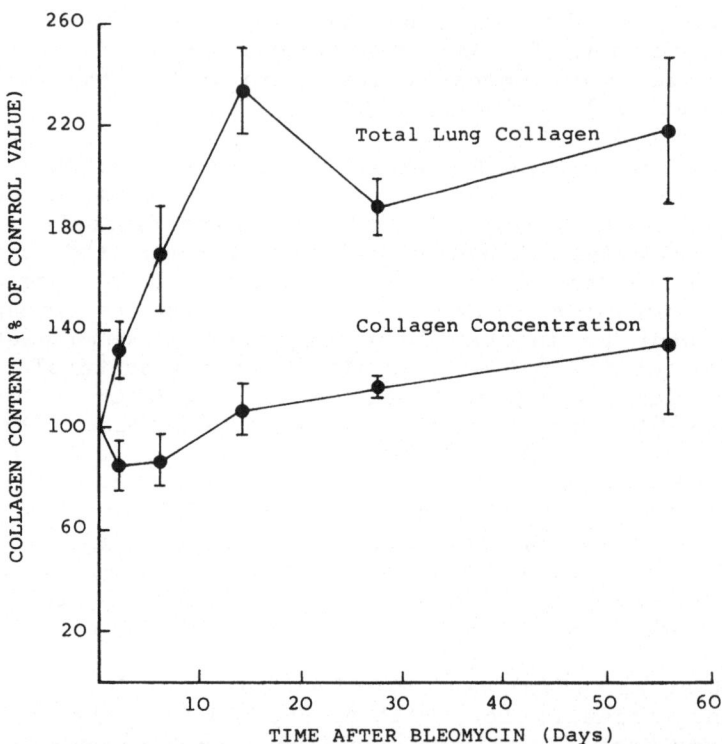

Fig. 2.   Changes in collagen content in bleomycin induced pulmonary
          fibrosis.   Adult rabbits received an intra-tracheal
          instillation of 10 mg/kg body weight of bleomycin sulphate
          and were sacrificed in groups of at least 4 animals at
          various times between 2 and 56 days following.   Lung
          collagen was then determined from hydroxyproline contents
          (Laurent et al, 1981) and is expressed as both a
          concentration (with respect to wet weight) and as total
          content per set of lungs.   The values shown are changes
          relative to age matched controls (which equal 100%) $\pm$
          S.E.M.

Collagen Metabolism in Pulmonary Fibrosis and the Use of "Anti-Collagen" Drugs

The importance of changes in collagen synthesis during
pulmonary fibrosis is, like changes in content, still equivocal.
As mentioned previously, results of Fulmer et al (1976) suggest

that there was no change in collagen synthesis rate for lung biopsy explants assessed by in vitro uptake of ^{14}C-proline into collagen hydroxyproline. However, in laboratory animals using a similar technique large increases in collagen synthesis have been reported (Phan et al, 1980; Clarke et al, 1980). As with collagen content, the method of expressing results may be important. Clarke et al, 1980 expressing synthesis rates as hydroxyproline incorporation per lung unit time, observed a 10-fold increase in synthesis rate. Phan et al (1980) expressed uptake with respect to DNA (as did Fulmer et al, 1976) protein and dry weight and showed in all cases about a doubling in the synthesis rate within 4 days of the intratrachal injection. Preliminary studies from the author's laboratory have also suggested an increase in synthesis rate, measured by in vivo methods, for rabbits with bleomycin-induced pulmonary fibrosis (Table 2). Six days after instillation of the drug, there was almost a doubling in the rate of collagen synthesis.

Table 2   In vivo Lung Collagen Synthesis for Rabbits with
            Bleomycin-Induced Pulmonary Fibrosis[a]

Collagen Synthesis Rate (%/day)[b]

CONTROL	10.1 ± 0.3
BLEOMYCIN	17.4 ± 2.0

[a]Unpublished data from Laurent and McAnulty

[b]Synthesis rates were calculated from uptake of ^{3}H-proline which was administered with a high dose of unlabelled proline. Animals were sacrificed after 180 minutes and synthesis rates calculated from the relative specific radioactivities of ^{3}H-proline in the tissue-free and protein pools according to previously described methods (McNurlan et al, 1979). A synthesis rate of 12.6%/day means that an amount of collagen equivalent to 12.6% of the total collagen pool is being synthesized each day.

Clearly there are differences in the reported changes in collagen synthesis in pulmonary fibrosis. The human studies have not shown increased synthesis rates whereas the animal studies of more acute forms of pulmonary disease have. It is uncertain at present whether these discrepant results occur because of methodological differences or more fundamental differences in the two forms of pulmonary disease.

The question of whether or not there is increased collagen deposition in human pulmonary fibrosis is of vital interest for without evidence for such an increase, it would be difficult to justify the use of drugs which inhibit its deposition. Current drug therapy in the treatment of fibrotic lung disorders is directed against the inflammatory and immune processes. Most success has been attained with corticosteroids which have been shown to be most effective in patients with inflammatory cell infiltrates rather than an advanced fibrosis as assessed by histological criteria (Carrington et al, 1978). Some patients, however, do not respond to steroid therapy and for these patients, the prognosis is poor, often with a rapid deterioration of lung function. For such patients there may be a role for drugs which interfere with collagen metabolism. Theoretically a wide range of drugs, interfering with the metabolism at various steps in collagen synthesis and processing could be used (see Chvapel, 1975 for review). In general, however, these types of drugs have not been employed in the treatment of human pulmonary fibrosis. An exception is penicillamine – an inhibitor of cross-linking and thus polymerization of collagen molecules. This drug has been employed in both humans (Goodman et al, 1981) and laboratory animals with experimentally induced fibrosis (Fedullo et al, 1978) but has been unsuccessful in improving lung mechanics. Recently, however, Kelly et al, 1980 have used proline analogues in the treatment of rats with pulmonary fibrosis induced by oxygen toxicity and bleomycin respectively. These agents, by replacing proline and hydroxyproline in the newly synthesized collagen polypeptide, inhibit post-translational events required for collagen cross-linking and fibril formation. In both these studies the drugs inhibited the increased collagen levels associated with the fibrosis and also prevented the lung mechanical changes associated with the disease. Thus there is interest in the possible use of such drugs in human patients.

In conclusion, the diversity of collagen types in lung is now well established and there is evidence to suggest that changes in the relative amounts of these different types plays a role in the pathogenesis of pulmonary fibrosis in man. Whether or not there is excessive collagen deposition in human fibrotic disorders is at present uncertain mainly because of the difficulties associated with the method of expressing collagen content. The use of collagen concentration, with respect to other lung constituents, which has by necessity been used for lung biopsy sample, is clearly inappropriate. This issue should be resolved because if excessive collagen deposition does occur in human pulmonary fibrosis a role may be envisaged for drugs which inhibit collagen deposition – some of which have been used with success in experimental models for pulmonary fibrosis in experimental animals.

Acknowledgements

The author is grateful to Mr. R. McAnulty and Mr. P. Cockerill for their contributions to the work reported in this paper; also to Professor M. Turner-Warwick for her encouragement and support. Finally, I wish to thank Miss H. Rolls for the preparation and typing of the manuscript.

References

Bateman, E.D., Turner-Warwick, M. and Adelmann-Grill, B.C., 1981. Immunohistochemical study of collagen types in human foetal lung and fibrotic lung disease. Thorax (in press).

Bornstein, P. and Sage, H., 1980. Structurally distinct collagen types. Ann. Rev. Biochem. 49: 957.

Bradley, K., McConnell-Breul, S. and Crystal, R.G., 1974. Lung collagen heterogenity. Proc. Natl. Acad. Sci. (USA), 71: 2828.

Carrington, C.B., Gaensler, E.A., Coutin, R.E., Fitzgerald, M.X., and Gupta, R.G., 1978. Natural history and treated course of usual and desquamative interstitial pneumonia. New Engl. J. Med. 298: 801.

Chvapil, M. 1975. Pharmacology of fibrosis: definitions, limits and perspectives. Life Sciences, 16: 1345.

Clarke, J.G., Overton, J.E., Marino, B.A., Vitto, J. and Starcher, B.C. 1980. Collagen biosynthesis in bleomycin-induced pulmonary fibrosis in hamsters. J. Lab. Clin. Med. 96: 943.

Fedullo, A.J., Lucey, E.C., Karlinsky, J.B., Goldstein, R.H., and Snider, G.L., 1978. Effect of penicillamine on bleomycin pulmonary fibrosis in hamsters. Clin. Res. 24: 384A.

Fulmer, J.D., Bienkowski, R.S., Cowan, M.J., Bradley, K.H., Roberts, N.C. and Crystal, R.G. 1976. Comparison of collagen concentration, distribution and synthesis in fibrotic and normal lungs. Clin. Res. 24: 384A.

Fulmer, J.D., Bienkowski, R.S., Cowan, M.J., Breul, S.D., Bradley, K.M., Ferrans, V.J., Roberts, W.C. and Crystal, R.G., 1980. Collagen concentration and rates of synthesis in idiopathic pulmonary fibrosis. Am. Rev. Respir. Dis. 122: 289.

Gay, S., Viljanto, J., Raekallio, J. and Penttinen, R. 1978. Collagen types in early phases of wound healing in children. Acta. Chir. Scand., 144: 205.

Goldstein, R.H., Lucey, C.F., Franzbalu, C. and Snider, G.L., 1979. Failure of mechanical properties to parallel changes in lung connective tissue composition in bleomycin-induced pulmonary fibrosis in hamsters. Am. Rev. Respir. Dis. 120: 67.

Goodman, M., Knight, R.K. and Turner-Warwick, M., 1981. Pilot study of penicillamine therapy in steroid failure patients with interstitial lung disease, in "Modulation of Autoimmunity and Disease: The Penicillamine experience", R.N. Maini and H. Berry, eds. Holt-Saunders Ltd. Eastbourne. (in press).

Kelley, J., Newman, R.A. and Evans, J.N., 1980. Bleomycin-induced pulmonary fibrosis in the rat. Prevention with an inhibitor of collagen synthesis. J. Lab. clin. Med. 96: 954.

Laurent, G.J., Cockerill, D., McAnulty, R.J. and Hastings, J.R.B., 1981. A simplified method for quantitation of the relative amounts of type I and type III collagen in small tissue samples. Analyt. Biochem. (in press).

Laurent, G.J., McAnulty, R.J., Corrin, B., and Cockerill, D. Biochemical and histological changes in pulmonary fibrosis induced in adult rabbits with intratracheal bleomycin. Eur. J. Clin. Invest. (submitted for publication).

Madri, J.A. and Furthymayr, H. 1980. Collagen polymorphism in the lung. Hum. Pathology. 11: 353.

McCullough, B., Collins, J.F., Johanson, N.G. and Grover, F.L. 1978. Bleomycin-induced diffuse interstitial pulmonary fibrosis in baboons. J. Clin. Invest. 61: 79.

McNurlan, M.A., Tomkins, A.M., and Garlick, P.J. 1979. The effect of starvation on the rate of protein synthesis in rat liver and small intestine. Biochem. J. 178: 373.

Phan, S.H., Thrall, R.S. and Ward, P., 1980. Bleomycin-induced pulmonary fibrosis in rats: biochemical demonstration of increased rate of collagen synthesis. Am. Rev. Respir. Dis., 121: 501.

Prockop, D.J., Kivirikko, K.I., Tuderman, L., and Guzman, N.A. 1979a. The biosynthesis of collagen and its disorders (I). New Engl. J. Med. 301: 13.

Prockop, D.J., Kivirikko, K.I., Tuderman, L., and Guzman, N.A. 1979b. The biosynthesis of collagen and its disorders (II). New Engl. J. Med. 301: 77.

Riley, D.J., Berg, R.A., Edelman, N.H., and Prockop, D.J., 1980. Prevention of collagen deposition following pulmonary oxygen toxicity in the rat by cis-4-hydroxy-L-proline. J. Clin. Ivest. 65: 643.

Seyer, J.M., Hutcheson, E.T. and Kang, A.H. Collagen polymorphism in idiopathic pulmonary fibrosis. J.Clin. Invest. 57: 1498.

Zapol, W.M., Trelstad, R.L., Coffey, J.W., Tsai, I and Salvador, R.A. 1979. Pulmonary fibrosis in severe acute respiratory failure. Am. Rev. Respir. Dis. 119: 547.

DISCUSSION

LECTURER: Laurent   CHAIRMAN: Cumming

CUMMING: Thank you Geoff, your paper is now open for
                  discussion. May I begin by asking one question
                  myself?   Since the total collagen content is
                  increased but the concentration is not, what is the
                  diluent?

LAURENT: Can I answer by drawing a diagram? We have made
                  measurements not only of collagen but several other
                  constituents in the model for fibrosis and I can
                  express changes above controls.   100% increase
                  would denote that changes are doubling in the
                  amount of the constituents and this is going on for
                  three weeks.   I showed the changes in collagen
                  content which look something like that and also
                  made measurements of changes in several other
                  constituents. To answer your question, Gordon, the
                  changes occur for almost everything we measure and
                  you can see the changes in total collagen and total
                  protein and this would represent about 90% of the
                  change.   The total protein in this case represents
                  everything except the collagen and with that sort
                  of effect very early, one can imagine why you have
                  a dilution effect on the collagen.

LAFUMA :   I was a little surprised by the turnover of the
                   collagen that you measured.   To study the rate of
                  synthesis of DNA is a very difficult problem and
                  when one works on the whole lung it maybe really
                  difficult. We compared the turnover of elastin
                  which is very low and the turnover of the collagen.
                  The result is very striking, it means that collagen
                  changes completely every ten days. Did you take
                  into account the labelled hydroxylproline in the
                  soluble fraction because this labelled
                  hydroxylproline may represent not the newly
                  synthetized but degraded protein.

LAURENT: In measurements of synthesis rate for any protein
                  it is important to extract the total protein pool.
                  So we tritiated and acid precipitated all proteins
                  and measured the hydroxylproline specific
                  radioactivity. So we are measuring the specific
                  radioactivity of total collagen including the newly

synthetised pool as well as the collagen that has been present for a long time. This is a total synthesis rate and we can't differentiate between different collagen pools.

LAFUMA: It may appear that when you look at the turnover of collagen you take into account some collagen in which the synthesis begins but stops somewhere or is very rapidly degraded.

LAURENT: All one can quote is the average mean rate. Almost certainly the newly synthetised collagen is turning over at a very rapid rate, perhaps 300 or 400% a day, whereas the mature collagen fibre may have a rate of less than 1% a day. There is evidence that about 30% of the collagen is degraded before it even leaves the cell. I'd like to think of this as a wastage pathway but I think there may be very good reasons for that process.

ROSSI: You mentioned patients in acute respiratory failure. Were those patients on respirators?

LAURENT: I read the method section in that paper quite carefully but I don't remember that statement. I couldn't be certain but it's probable.

ROSSI: I am a little shocked by the results, it is possible that by looking only at collagen concentration we are losing the total lung collagen, but the other possibility you mentioned is that we are looking at two different models of disease. One possible explanation for this is that if these patients were on a respirator then you can assume that most of the lung damage was induced by oxygen and that's the same mechanism which is supposed to be at the basis of induced lung fibrosis. So another possibility is: since there is maybe some evidence that lung fibroblasts are not represented by a single population, but different populations, we can act in a different way with a different synthesis rate, with a different secretion of collagen of one type compared to another. This is one possible explanation for the difference in our results and may be based on all those question marks.

LAURENT: Yes, I don't think there is much I can add to your comment. It seems to me that the obvious experiment, and I am absolutely amazed it has bever been done in the past, is to have measured total collagen content of the lungs from the CFA patients. NIH is at the moment trying to collect some lungs, and it is an absolutely vital question whether there is a change in total collagen content.   I think without evidence for such a change it's quite difficult to justify anticollagen drugs.

ROSSI: Do you have any explanation for the fact that the total content of protein in the lung in experiments is going up and then down. In the last diagram you pointed out that the first curve was the total protein was varying up and down.   What's the explanation for this curve?

LAURENT: I can't give any explanation for that, I am just giving you my thoughts about it. It represents two classes of protein; first it was an inflammatory cell complement and we know that there are lots of inflammatory cells entering the lungs; I showed you the morphology and these may represent a great part of that increase in protein content. There is also hyperplasia of existing structures and increased size of cells. These are resident cells of the lung and so they will contribute as well. I didn't show you changes in the collagen synthesis rates but they also go up so that I think that we have the explanation, at least partially.

SOLIMAN: First of all I would like to thank you for this marvellous lecture but I would like to tell you my own observation. It was a part of my thesis for MD about ten years ago and I obtained an experimental model of lung fibrosis in guinea-pigs by giving high doses of oestrogenic hormones. The observations were that after two weeks of oestrogenic hormones there was a very severe type II cell proliferation and a stage of transient pulmonary oedema. After one month there was collagen deposition; after two months this collagen increased; after three months there was massive collagenosis associated with endothelial changes and endothelial degeneration with marked pinocytosis of the endothelium, rupture of vesicles and loss of the integrity of the endothelium. Would you like to comment on these observations.

LAURENT: Thank you very much.   It's nice to hear about
              models for pulmonary fibrosis. I presume you are
              making these assessments on morphological grounds?

SOLIMAN: Part of this was done by light microscopy and part
              by electron microscopy. it was therefore based on
              a morphological description.

LAURENT: The sort of changes you have talked about seem to
              be fairly common for quite a wide variety of
              injuries induced to the lung with oxygen toxicity
              and a variety of other drugs such as paraquat.

SOLIMAN: You mentioned that there is no change in collagen
              content and no change in collagen synthesis. We
              observed on a morphological basis a progressive
              increase in collagen formation in the guinea-pigs
              we studied.

LAURENT: Are you saying that you have evidence in that model
              for no change in collagen?

SOLIMAN: <u>Increased</u> densities of collagen.

LAURENT: Well, these changes are consistent with the changes
              that we are seeing.

SOLIMAN: I think you mentioned that there is no change in
              either collagen content or collagen synthesis.
              Does this apply to experiments or not?

LAURENT: I hope the point I was making was whether one sees
              changes in those parameters depends on the method
              you use. I think it is important to use good
              techniques, preferably in vivo techniques, for
              assessing synthesis. For a collagen content one
              must measure total lung collagen, if you don't I am
              am careful about both these things, I think it is
              quite possible not to observe changes.

WILLIAMS: Just a brief point, in that part of my questions
              have already been answered and that is: I am
              worried about the old pathologists of long ago
              turning over in their graves at the moment outside
              in the mist, in that you said that the maximum
              amount of collagenosis from the whole
              interpretation and measurement was there, at the
              stage of histology where you were showing
              presumably proliferative round cells in the broad

sense. Could you tell us whether at that stage you
were seeing fibroblasts in the exudate because it
would be quite early. I was worried about the fact
that all the proteins come at 8 weeks, because we
still see that stuff there and it stays there on ur
definition. So I think perhaps your measurements
are different from what I mean by collagenosis
coming from fibrosis.

LAURENT: In answer to the first question I could see
evidence for increased fibroblasts but there are
difficulties in distinguishing between the
monocytes and the fibroblasts at the light
microscopy level. So, although it was tempting to
say that such an increase occurred, one cannot be
certain. As to the second part of the question, do
you want me to hand that over to Dr. Rossi?

SOME INTERSTITIAL CELL DISEASES OF THE LUNG

Bryan Corrin

Department of Lung Pathology
Cardiothoracic Institute, Brompton Hospital
London   SW3 6HP

This paper will describe three unusual diseases of the lung, each of which is characterised by an abnormal proliferation of one particular cell type. They represent three different forms of interstitial lung disease.

EOSINOPHILIC GRANULOMA

Eosinophilic granuloma was first described in bone (Lichtenstein and Jaffe, 1940; Otani and Ehrlich, 1940) and almost immediately linked with the previously described Hand-Schuller-Christian and Letterer-Siwe diseases (Farber, 1941) the term Histiocytosis X(HX) being introduced by Lichtenstein in 1953 to embrace all three. Letterer-Siwe and Hand-Schuller-Christian diseases represent the acute and chronic disseminated forms respectively, and eosinophilic granuloma the localised form of HX. Although these three diseases behave quite differently and carry different prognoses, their essential unity has subsequently been supported by the electron microscopic identification of a distinctive cytoplasmic marker organelle in all forms of HX (Basset and Nezelof, 1969). All three diseases are now thought to represent abnormal proliferations of the Langerhans' cell (Corrin and Basset, 1979). Letterer-Siwe disease behaves as a high grade malignant neoplasia of Langerhans' cells but the other two forms of HX are often self-limiting. However they may induce sufficient damage to involved organs to be life-threatening.

Langerhans' Cells

Langerhans' Cells are constantly found in the epidermis and, contrary to earlier views, are unrelated to melanocytes or Merkel

327

cells (Birbeck et al, 1961). They are believed to be of mesodermal origin and related to the mononuclear macrophage system. They have been described in the dermis, lymph nodes, thymus and tonsils but not, so far, in normal lung. They have however been observed in idiopathic interstitial fibrosis of the lung (Basset et al., 1976[a]) and within a bronchiolo-alveolar tumour (Basset et al., 1974).

Langerhans' cells possess surface receptors for Fc and C3 (Nezeloff et al., 1977; Stingly et al., 1977) and Ia surface antigens (Rowden et al., 1977; Klareskog et al., 1977). An increase in dermal Langerhans' cells is found in cell mediated hypersensitivity reactions, when close contact between Langerhans' cells and lymphocytes has been observed. Ferritin labelling suggests that Langerhans' cells are involved in antigen transport to lymph nodes (Silberberg et al., 1975, 1976; Shelley and Huhlin, 1976; Silberberg-Sinakin et al., 1977; Symposium, 1978). In this respect Langerhans' cells resemble macrophages but they differ in their low phagocytic potential, scanty lysosomes and paucity of acid phosphatase and non-specific esterase (Elleder et al., 1977; Elema and Poppema, 1978).

Langerhans' cells have an indented nucleus with a fine chromatin pattern, and a moderate amount of lightly eosinophilic cytoplasm. Ultrastructurally they contain small elongated bodies which are scattered throughout the cytoplasm but are most numerous near the cell membrane. These have a pentalaminar structure and measure 40-45 nm is width with a constant longitudinal periodicity (10nm) to their central lamina (Birbeck et al., 1961; Wolff, 1967). They often terminate in a small dilatation. Sometimes the pentalaminar bodies fuse with the cell membrane. Desmosomal connections are rare and lysosomal dense bodies are scanty.

Eosinophilic granuloma of the lung

Pulmonary involvement in the disseminated forms of HX was noted by Weinstein and colleagues and by Ackerman in 1947 but it was not until 1951 that Farinacci and co-workers first described a form of the disease limited to the lungs. This is comparable to the solitary osseous lesions and is termed either eosinophilic granuloma of the lung or primary pulmonary HX. Although it is not common, several large series have now been described (Auld, 1957; Anderson and Foraker, 1959; Basset et al., 1978).

Eosinophilic granuloma of the lung affects the young and elderly alike with a mean age of about 30 years. Men are more commonly affected than women, in a proportion of about four to one. Presenting features include cough, shortness of breath, pneumothorax, weight loss and fever. Chest radiographs typically show bilateral reticulonodular opacities. The opacities are widespread but most marked in the mid-zones, with initial sparing

of the cost-phrenic angles. As the disease advances the opacities
increase in size and number, cystic changes appear and all parts of
the lungs become involved. Functional tests show a restrictive
pattern of disease. Blood eosinophilia is not a feature of the
disease.

Ultrastructural recognition of Langerhans'-type cells in
lavage fluid supports the diagnosis but biopsy is often required.
To study the important topographical distribution of the lesions
within the lung acinus, open lung biopsy is indicated.
Microscopically, eosinophilic granuloma of the lung is
characterised by a focal interstitial infiltration centred upon
small blood vessels and airways. Active lesions consist of nodular
collections of histiocytes and lymphocytes are ordinary macrophages
and as the lesions age these acquire Perls' and periodic acid-
Schiff positive brown pigment. Other mononuclear cells lack pigment
and have a particularly fine nuclear chromatin pattern and an
indented nuclear outline: these represent the Langerhans'-type
cells which are thought to be fundamental to the disease process.
Varying numbers of lymphocytes, plasma cells and eosinophils are
also found. The last may suggest the diagnosis, and their
interstitial localisation will help to distinguish the condition
from eosinophilic pneumonia. The macrophages and Langerhans'-type
cells enter the air spaces and large numbers may suggest a
diagnosis of desquamative interstitial pneumonia, but this error
will be avoided if attention is paid to the characteristic
interstitial foci. As the lesions age, fibrosis develops and
stellate scars appear. These cause irregular scar emphysema and
bronchiolectasis. The end stage is a non-specific honeycomb lung.

The natural history of the disease is very variable. Some
patients pursue a rapidly downhill course and die early. Others
improve spontaneously and become symptom-free with total
radiologial remission. Others experience spontaneous remissions
and relapses. The disease may arrest at any stage. Unfavourable
prognostic features include multiple pneumothoraces, widespread
radiological involvement of the lungs, a low diffusing capacity and
prolonged constitutional disturbances such as fever and weight
loss. There is unfortunately no effective remedy. Steroids have
often been employed and occasionally cytotoxic drugs, but without
obvious success. The effectiveness of these drugs is difficult to
assess with such a variable natural history but cytotoxic drugs
have been shown to be beneficial in the rapidly progressive
disseminated form of HX, Letterer-Siwe disease (Starling et al.,
1972; Lahey, 1975).

LYMPHANGIOLEIOMYOMATOSIS

This rare disease may, like HX, be confined to the lungs,
combine pulmonary and extrapulmonary lesions or affect only

extrapulmonary sites, notably lymph nodes and lymphatics. The pulmonary lesions of lymphangioleiomyomatosis are identical to those of tuberous sclerosis (epiloia, Bournevilles disease) but there are intriguing differences in the sex distribution and family history of these two conditions. A large series of pulmonary lymphangioleiomyamatosis (PLAM) was described by Corrin and colleagues (1975).

PLAM is confined to women in the reproductive years, most of whom die of the disease before reaching the menopause. In a few cases the disease appears to have arrested at the menopause, and ovariectomy has recently been claimed to be successful in this respect (Kitzsteiner and Mallen, 1980). Such a seemingly drastic measure is fully justified in the face of a uniformly fatal disease for which there is no other effective therapy. Androgens were unhelpful in the one case reported by Bush and associates (1969) but they were only employed late in the course of the disease.

The essential feature of PLAM is a proliferation of short spindle shaped interstitial cels. The nature of these is often not immediately obvious to the histologist but electron microscopy shows that they have all the features of smooth muscl cells (Basset et al., 1976b). Their derivation from pulmonary lymphatics is largely presumed from the association of similar lesions along extrapulmonary lymphatics and within lymph nodes.

Clinically the patients develop a progressive shortness of breath, and three notable complications: haemoptyses, chylous effusions and repeated pneumothoraces, all of which can be explained by the pathological distribution of the disease. Obstruction of small veins explains the haemoptyses and the haemosiderosis evident microscopically. Obstruction of lymphatics explains the chylothorax and chylous ascites which are often evident, whilst valvular obstruction of small airways leads to air cysts, rupture of which explains the frequent pneumothoraces. The lungs are finally reduced to a mass of emphysematous bullae and PLAM is yet another case of honeycombing.

Whereas PLAM is confined to adult women and shows no familial pattern, tuberous sclerosis affects both sexes, is evident from birth and has a strong familial pattern of inheritance. This seemingly clear distinction becomes somewhat blurred, however, when it is realised that patients with a family history of tuberous sclerosis who show pulmonary involvement are almost exclusively female and that clinical status is dominated by their lung disease; the neurological stigmata of tuberous sclerosis are seldom prominent insuch patients (Dwyer et al., 1971; Valensi, 1973). The pathology of the lung is identical in these two conditions (Valensi, 1973; Corrin et al., 1975) and features such as renal angiomyolipomas and extrapulmonary lymphangiomas may be found in either. It has therefore been suggested that PLAM represents a forme fruste of tuberous sclerosis (Valensi, 1973). The exclusively female sex distribution of PLAM suggest however that hormonal factors may be uppermost in importance. In this connection it is interesting to note that fibromyomas have been induced in guinea pigs with oestradiol (Lipshutz and Vargas, 1939).

HYALINISING INTRAVASCULAR "BRONCHIOLOALVEOLAR" TUMOUR

This rare condition is to be regarded as a multifocal neoplasm of the lungs but it presents clinically as a restrictive form of lung defect. It has a wide age range and affects both sexes, but most patients are young adult women.

It was first recognised by Liebow. So far there are few reports in the literature. A single case was reported by Farinacci and colleagues (1973) as representing deciduosis of the lung, although other cases have involved men. Liebow's 20 cases have so far only been reported in abstract from (Dail and Liebow, 1975). Corrin and associates (1979) described three cases with emphasis on fine structural identification of the cells involved. A further case has been reported by Ferrer-Roca (1980).

The pulmonary tumours are multiple and quite characteristic histologically. They have a poorly cellular hyaline core and a more cellular periphery. They grow into adjacent alveoli in a micropolypoid manner and centrally obliterate blood vessels. Cells clothing the peripheral intra-alveolar extensions have the appearances of type II pneumocytes at the periphery from tumour cells proper. The latter lack lamellar inclusions but possess abundant cytoplasmic filaments. The appearances are those of mesechymal rather than epithelial cells. Some of the tumour cells contain abundant Weibel-Palade bodies indicating endothelial differentiation and it is suggested that the tumours are derived from vasoformative reserve cells (Corrin et al., 1979)

The course of the disease is variable but generally the tumours increase in number and size only slowly. Typically the disease is progressive and the patients die of restrictive lung disease after an illnes of 2 to 12 years. Occasionally however the course of the disease is so slow as to suggest that it has arrested spontaneously. At death the tumours are usually confined to the lungs but occasional metastases are found in distant organs.

REFERENCES

Ackerman, A. J., 1947, Eosinophilic granuloma of bones associated
     with involvement of the lungs and diaphragm.
     Am. J. Roentgen., 58, 733.
Anderson, A. E., and Foraker, A. G., 1959, Eosinophilic granuloma

of lung.   Clinical features and connective tissue patterns.
Arch. Intern. Med., 103, 966.

Auld, D., 1957, Pathology of eosinophilic granuloma of the lung.
Arch. Pathol., 63, 113.

Basset, F., Corrin, B., Spencer, H., Lacronique, J., Roth, C.,
Soler, P., Battesti, J. P., Georges, R., and Chretien, J.,
1978, Pulmonary Histiocytosis X.
Am. Rev. Resp. Dis., 118, 811.

Basset, F., and Nezelof, C., 1969, L'Histiocytose X: Microscopie
electronique, culture in vitro et histoenzymologie.
Discussion a propos de 21 cas.
Rev. Fr. d'Et Clin. Biol., 14, 31.

Basset, F., Soler, P., Wyllie, L., Abelauet, R., Le Charpentier,
M., Kreis, B., and Breathnach, A. S., 1974, Langerhans' cells
in a bronchiolalveolar tumour of lung,
Virchow's Arch. A, 362, 325.

Basset, F., Soler, P., Wyllie, L., Mazin, F., and Turiaf, J., 1976,
Langerhans' cells and lung interstitium.
Ann. N.Y. Acad. Sci., 278, 599.

Birbeck, M. S., Breathnach, A. S., and Everall, J. D., 1961, An
electron microscopic study of basal melanocytes and high level
clear cells (Langerhans' cell) in vitiligo.
J. Invest. Derm., 37, 51.

Bush, J. K., McLean, R. L., and Sieker, H. O., 1969, Diffuse lung
disease due to lymphangiomyoma.
Am. J. Med., 46, 645.

Corrin, B., Liebow, A. A., and Friedman, P. J., 1975, Pulmonary
lymphangiomyomatosis.
Am. J. Path., 79, 348.

Corrin, B., Manners, B., Millard, M., and Weaver, L., 1979,
Histogenesis of the so-called "Intravascular
bronchioloalveolar tumour".
J. Path., 128, 163.

Corrin, B., and Basset, F., 1979, A review of histiocytosis X with
particular reference to eosinophilic granuloma of the lung.
Invest. Cell Pathol., 2, 137.

Dail, D., and Liebow, A., 1975, Intravascular bronchiol-alveolar
tumour.
Amer. J. Path., 78, 6a.

Dwyer, J. M., Hickie, J. B., and Garvan, J., 1971, Pulmonary
tuberous sclerosis. Report of three patients and a review of
the literature.
Quart. J. Med., 40, 115.

Elema, J. D., and Poppema, S., 1978, Infantile histiocytosis X
(Letterer-Siwe Disease).   Investigations with
enzymehistochemical and sheep erythrocyte rosetting
techniques.
Cancer, 42, 555.

Elleder, M., Povysil, C., Rozkovocova, J., Cihula, J., 1977,   -D-
annosidase activity in histocytosis X.

Virchow's Arch. B, 26, 139.

Farber, S, 1941, The nature of 'solitary or eosinophilic granuloma' of bone.
Amer. J. Path., 17, 625.

Farinacci, C. J., Jeffery, H. C., and Lackey, R. W., 1951, Eosinophilic granuloma of the lung. Report of 2 cases.
US Armed Forces Med. J., 2, 1085.

Ferrer-Roca, O., 1980, Intravascular and sclerosing bronchio-alveolar tumour.
Amer. J. Surg. Path., 4, 375.

Kitzsteiner, K. A., and Malen, R. G., 1980, Pulmonary lymphangiomyomatosis: treatment with castration.
Cancer, 6, 24.

Klareskog, L., Tjernlund, U. M., Forsum, U., and Peterson, P. A., 1977, Epidermal Langerhans' cells express Ia antigens.
Nature, 268, 248.

Lahey, M. E., 1975, Histiocytosis X – comparison of three treatment regimes.
Paeiat., 87, 179.

Lichtenstein, L., 1953, Histiocytosis X. Integration of eosinophilic granuloma of bone, Letterer-Siwe disease and Hand-Schuller-Christian disease as related manifestations of a single nosological entity.
Arch. Path., 56, 84.

Lichtenstein, L., and Jaffe, H. L., 1940, Eosinophilic granuloma of bone.
Am. J. Path., 16, 595.

Lipschutz, A., and Vargas, L., 1939, Experimental tumorigenesis with subcutaneous tablets of oestradiol.
Lance, 236, 1313.

Nezelof, C., Diebold, N., Rousseau-Merck, M. F., 1977, Ig surface receptors and erythrophagocytic activity of Histiocytosis X cells in vitro.
J. Path., 122, 105.

Otani, S., and Ehrlich, J. C., 1940, Slitary granuloma of bone simulating primary neoplasm.
Am. J. Path., 16, 479.

Rowden, G., Lewis, M. G., and Sullivan, A. K., 1977, Ia antigen expression of human epidermal Langerhans' cells.
Nature, 268, 247.

Shelley, W. B., and Huhlin, L., 1976, Langerhans' cells form a reticulo-epithelial trap for external contact antigens.
Nature, 261, 46.

Silberberg, L., Baer, R. L., Rosenthal, S. A., Thorbecke, G. J., and Berezowsky, V., 1975, Dermal and intravascular Langerhans' cells at sites of passively induced contact sensitivity.
Cell. Immunol., 18, 435.

Silberberg-Sinalin, L., Fedorko, M. E., baer, R. L., Rosenthal, S. A., Berezowsky, V., and Thorbecke, G. J., 1977, Langerhans' cells: target cells in immune complex reactions.

Cell. Immunol., 32, 400.

Starling, K. A., Donaldson, M. H., Haggard, M. E., Vietti, T. J., and Sutow, W. W., 1972, Therapy of Histiocytosis X with vincristine, vinblastine and cyclophosphamide.
Am. J. Dis. Child., 123, 105

Stingl, G., Wolff-Schreiner, E. C., Pichler, W. J., Gschnbait, F., and Knapp, W., 1977, Epidermal Langerhans' cells bear Fc and C3 receptors.
Nature, 268, 245.

Symposium, 1978, The Langerhans' cell and contact dermatitis.
Acta Dermato Venereologica, 58, Suppl. 79.

Valensi, Q. J., 1973, Pulmonary lymphangiomyoma, a probable forme fruste of tuberous sclerosis - A case report and survey of the literature.
Am. Rev. Resp. Dis., 108, 1411.

Weinstein, A., Francis, H. C., and Sprofkin, B. F., 1947, Eosiniphilic granuloma of bone; report of a case with multiple lesions of bone and pulmonary infiltration.
Arch. Int. Med., 79, 176.

Wolff, K., 1972, The Langerhans' cell.
Current Problems in Derm., 4, 79.

DISCUSSION

LECTURER: Corrin   CHAIRMAN: Cumming

CUMMING: This paper is now open for discussion.

TRICOMI: I'd like to draw your attention to the histocytosis
                pictures which you showed. Each of them is a
                completely non-specific picture protraying the
                natural history of the disease at that time. This
                implies that the radiologist looking at
                interstitial fibrosis can at most recognise the
                stage the disease has reached without being able to
                diagnose what it is. His diagnostic suspicions are
                strengthened should histocytosis be accompanied by
                bone changes in the frontal bone. In all the
                different cases x-rays reflected the clinical and
                anatomic stages the disease had passed through. As
                in all fibrosis there is a component related to the
                interstitium and a component related to cyst
                formation, which radiologically account for honey-
                comb lung or even a parafibrotic emphysema. As for
                the extremely rare, chronic diseases, I'd like to
                remind you of the existence of a cystic
                lymphangioma, the diagnosis of which involves a
                relapsing chylothorax. The diagnosis of this
                disease which is likely to be similar to the one
                you described is arrived at by means of
                lymphography such as we are performing in Rome.

CORRIN: Thank you very much. Certainly if histocytosis is
                generalised and there are lytic lesions in bone,
                the radiologists can often suggest the diagnosis.
                It seems in those cases where the disease is
                confined to the lungs that the radiological
                diagnosis will prove more difficult. If the
                lesions are at the costal-phrenic angles, as it
                often is in the early stages of eosinophilic
                granuloma, this may be helpful in distinguishing
                the disease from interstitial fibrosis which
                typically starts at the bases and even at an early
                stage involves the costal-phrenic angles. Your
                observations about the relationship of
                lymphangioleiomyomatosis to another disease – I am
                afraid I cannot comment on this, because I can't
                really make out whether that is a separate disease

or if it is yet another example of the same disease. What would be the nature of those effusions? I think if it is perhaps a chylothorax and if the patient is a woman in the reproductive years, a biopsy may well show lymphangioleiomyomatosis.

CANNIN: I would like to learn from Professor Corrin about the prevalence of clinical features like crepitations, clubbing, cyanosis or at least what are the main differentiating features from cryptogenic fibrosing alveolitis.

CORRIN: I am glad you have asked me about the clinical and radiological features of this disease because it is an opportunity to pay a tribute to the French Commission who made the analysis of the clinical and radiological features. Professor Chretian headed this group at Laennec and also at Bichat Hospitals. I really am a histopathologist and not capable of comments on the value of crepitations, but the clinical features are summarised on the diagram with the ones which they felt were most helpful and particular combinations of them most suggestive of the diagnosis clinically. As a pathologist I'd recommend getting some tissue and establishing a tissue diagnosis.

CANNIN: You did not mention crepitations or clubbing in these cases. Is it because you did not concentrate on these signs.

CORRIN: These were not present in the cases I autopsied personally.

SOLIMAN: Thank you very much Professor Corrin for this useful edition of the three diseases. Regarding the second disease we had a case in Alexandria. She was a female of 23 years with chylous effusion and we exhausted all the investigations to find out a cause. We aspirated the fluid, we added some steroids but two months later the patient died without any explanation of the cause of death and it is unlucky that we didn't have any post-mortem because she died at home. Could this be explained on the basis of the second disease?

CORRIN: Certainly it could be explained on the basis that a very rare disease could occur in the absence of any other pathological evidence by metastatic tumours in the halo lymph node or a lymphoma destroying the halo lymph nodes or any other cause of lymphatic obstruction. There are many causes of chylous effusions. She does seem to be of the right sex and age of lymphangioleiomyomatosis.

SOLIMAN: All these causes have been excluded, all the manifest causes that could lead to an obstruction have been excluded. I would like to think of the second disease.

CORRIN: Well, I thought I showed fairly well a tumour of the thoracic duct in one particular case, and similar lymphangiomatous tumours develop in the halo lymph nodes and many of the lymph nodes in the body. This obstructs the lymphatic pathway and subsequent rupture of lymphatic vessels leading to chylous pleural effusion.

SOLIMAN: That's an acquired real tumour of large size.

CORRIN: That is part of the disease complex. The disease may be confined to the lung or there may be the extrapulmonary manifestations with lymphatics, lymph nodes and lymphangiomyomas.

BIENENSTOCK: Recently Langerhans' cells in the skin have acquired considerable interest from the immunological point of view, since they appear to be capable of processing and presenting antigens in immuno-responses. I was interested to see that they were found in sites other than the skin. Could you tell us what the characteristics are? Is it simply those ultra-structural tennis rackets? Do you know anything more about how would they be recognised in tonsils for example?

CORRIN: The ultra-structural features are the presence of the characteristic tennis racket granules, positive dense bodies in cells which one might otherwise assume from their structure to be macrophages; if one could study the surface characteristics and perhaps recognise the FC and S3 sites and the IA antigens and if one could test their function one would observe that they are poorly phagocytic.

BIENENSTOCK: Have you seen them in the circulation? Do you know if they are ever found in the thoracic duct?

CORRIN: I don't really know the answer to that.

KUHN: There has been one case, a case that was otherwise morphologically monocytic leukemia. I want to ask you a question about the last disease, the HIVBAT. If that's a basal formative cell why is the predominant site of growth infra-alveolar?

CORRIN: It is a poorly malignant, but nevertheless malignant tumour which is infiltrating and protruding into neighbouring alveolar lumen as any tumour might do by process of expansion and infiltration. I'd like to picture them beginning in the interstitium.

KAY: Would you agree that it is curious in a way because eosinophils in the skin are remarkably uncommon, in fact I can't think of any other condition where eosinophils are a marked feature in the skin.

CORRIN: It is not constant and there are cases where there is a very marked proliferation of the mononuclear cells one would take to be Langerhans' cells by the light microscope with no eosinophils to be seen. This is one of the most difficult cases to diagnose histologically and I think that eosinophils arrive later with lymphocytes, with the pigment-loaded macrophages as a late reaction in the scarring stage.

KAY   Can I follow this with a quick semantic point. You referred to the eosinophils as a reactive cell and I know that this term is used a lot by pathologists. I wonder if you could tell us actually what you mean by that and whether you feel it is a useful term.

CORRIN: By reaction I mean a response to a stimulus which in the case of bacterial infection we recognise, and in case of many diseases we do not. In the case of this disease there is very little information of what triggers this pathological proliferation of Langerhans' cells, but there are a few reports of immunological disturbances in histocytosis, particularly pulmonary histocytosis. The related paper published by King and colleagues in 1979 has reported on the presence of circulating

immune complexes in pulmonary histocytosis and the
detection of immunoglobulin of various classes and
complement by immunofluorescence microscopy in the
lung tissue.   Perhaps the stimulus is an
immunological one.

WILLIAMS: To go back to the Langerhans' cells there is a very
nice article and I regret I can't remember the name
of the author in a recent book on mononuclear
phagocytes;   he does say, as far as I can
recollect, these cells are also non-specific
esterase positive. Did you find this because if so
thats additional evidence that we are thinking of
monocyte macrophage series. The second point I'd
like to ask, please, do you think therefore from
your discussion that these are the essential tumour
cells?   They don't look malignant in my own
experience on them, but I would be glad to have
your view.   If I may ask a third question about
HIVBAT presentation, do you think those cells look
at all like the cells some others have seen in
cardiac myxoma?

CORRIN: The third question related to a non-specific
esterase I can't answer from my own experience.
The second question is whether I regard the
Langerhans' cells as the primary cells of this
tumour.   I didn't say I regarded this as a
neoplastic disease, it is self-limiting in many
cases.   I do regard this as the essential cell of
the disease. The disease represents a pathological
proliferation of Langerhans' cells, other cells are
regarded, if Barry Kay permits me to use the term,
as reactive.   But many of the cases are self-
limiting, in fact the disease often gets better, so
I hesitate to regard it as a malignant condition.
The third question related to the possibility, (I
borrowed the term vaso motive) of vaso formative
reserve cell from Stain's studies of cardiac
myxoma, he seems to describe just such a cell in
cardiac myxoma and that's the name he gives to it.
Professor Heath will you be referring to the little
pimples on the heart valve?   Yes, the same cells
are found there and again I think they regard
these as some form of primative vaso formative
cells.

CUMMING: Thank you very much Brian.   I'll draw the
discussion to a close at that point and make one

brief announcement. Tomorrow, you'll remember, we
are all going on our outing to Mazio and to have
dinner in the vineyard in Marsala. We wish to
leave on the coach at 1.15 p.m. It will be, I
think, impossible for those who wish to go on the
coach to go to a restaurant and have a normal
lunch. However, I have arranged that in the dining
room of San Rocco there will be available
sandwiches, fruit and drinks from 12.00 o'clock
onwards, so you may take a light lunch and join the
bus at 1.15. Thank you very much.

PROSTAGLANDINS AND THE LUNG

H. Sors and P. Even

Hôpital Laennec
42, rue de Sèvres
75007 Paris

GENERAL PROPERTIES AND METABOLISM OF PROSTAGLANDINS

Prostaglandins (PGs) are a group of lipidic compounds derived from C-20 polyunsaturated fatty acids, mainly arachidonic acid (AA). Biosynthesis of these compounds is an ubiquitous process, occuring in the cell membrane under various stimuli; synthesis is intermittant, short lasting and self limited and is followed by an immediate intra and/or extra-cellular release; PGs exert local action on cells by interaction with specific receptors and are then subjected to a spontaneous or enzymatic intra-cellular catabolism. As a consequence, these substances are local mediators rather than classical hormones or transmitters and, with the possible exception of prostacylcin (PGI2), there is no circulating PGs.

The first step of PG biosynthesis is the hydrolysis of membrane bound phospholipids by phospholipase(s) (mainly phospholipase A2) which gives rise to free AA; this is a rate-limiting step regulated by various mechanical and hormonal stimuli and inhibited by corticosteroid drugs. Free AA can be converted either by the PG endoperoxide synthetase or by lipoxygenase(s). The PG endoperoxide synthetase displays two distinct activities i.e. a cyclodioxygenase activity, converting AA to the endoperoxide PGG2 and a peroxidase activity converting PGG2 to PGH2. The enzyme, which have three binding sites for AA, molecular O2 and the endoperoxide, has very particular autoregulating properties with a positive feedback by the product and a self-catalysed irreversible destruction by free radicals leading to a time limited and explosive synthesis of endoperoxides; the cyclodioxygenase is inhibited by non-steroidal antiinflammatory drugs (NSAI) such as aspirin and indomethacin. PG endoperoxides are unstable compounds

341

(t 1/2 = 5 mn) which share important biological activites upon platelet aggregation, contraction of smooth muscle and inflammatory and anaphylactic processes.

Depending on the enzymatic equipment of each cell type and probably other unknown factors, PG endoperoxides can be converted to primary PGs (E2, F2α and D2), thromboxane A2 (TX A2) and prostacyclin; TXA2 is a short-lived compound (t 1/2 = 30 sec) with very powerful activities, inducing platelet aggregation and contraction of smooth muscle; it is subjected to a rapid spontaneous catabolism to the stable thromboxane B2 (TXB2). Prostacyclin is also an unstable compound (t 1/2 = 3 mn) whose biological activities are opposite to those of TXA2; PG12 synthetised by vascular endothelial cells has a major role in inhibiting platelet aggregation and relaxating vascular smooth muscle; PG12 is rapidly catabolized to the stable 6-keto-PGF1α (6 K-PGF1α). As PG12 is continuously sunthestised and released by the lung and is not removed by the pulmonary circulation, it could act as a circulating PG. The conversion of AA into hydroxylated fatty acids (HETE's) represents another pathway not directly related to PG synthesis but which now appears of central importance; 12-HETE was the first monohydroxy acid isolated in platelets, synthestised by a lipoxygenase, via a 12-hydroperoxy precursor (12-HPETE). More recently other monohydroxylated derivatives of AA (i.e. 5, 8, 9 and 11 HETE) have been isolated. 5-HETE was recently reported to be a major product of AA conversion in polymorphonuclear of several species including man. For the first time in 1979 Samuelsson et al. described a new biosynthetic pathway leading from AA (by the action of a 5-lipoxygenase) to an unstable epoxide intermediate named leukotriene A4 (LTA4) and then by the introduction of gluthatione to LTC 4 which can be further transformed by a glutamyl-transpeptidase to LTD 4; LTC 4 and LTD 4 are now known to be the main componants of the slow-reacting substance of anaphylaxis (SRS-A). In several species including man hydrolysis of LTA 4 could also lead to 5, 12 di-HETE (LTB 4) which was previously isolated from polymorphonuclear leukocytes.

Since lipoxygenases are not inhibited by NSAI, the lipoxygenase pathway of AA metabolism is rather enhanced when PG cyclooxygenase is inhibited by these drugs of such a manner that the immunological release of SRS-A is potentiated by indomethacin but reversed by inhibitors of both lipoxygenase and cyclooxygenase such as nordihydroguaiaretic acid, eicosatetraynoic acid and 1-phenyl-3-pyrazolidone (phenidone).

Enzymatic catabolism of PGs is a cytoplasmic process involving successive enzymatic reaction i.e. 15-hydroxyl-dehydrogenation and 13, 14-reduction leading to the biologically inactive 13, 14-dihydro-15-keto metabolites subsequently catabolized via β and ω oxidation processes. Interconversion between E and Fα PGs has been reported in vitro via the action of 9-keto-reductase and 9-hydroxy-dehydrogenase but the significance of these processes in vivo is still controversial.

BIOSYNTHESIS AND CATABOLISM OF PROSTAGLANDINS IN THE LUNG

Like seminal vesicles, kidney and platelets, mammalian lung is able to convert very actively AA into PG derivatives. However, at present, a comprehensive review of PG biosynthesis by the lung is difficult for the following reasons: i) the extreme heterogeneity of the pulmonary cell populations, each of them having probably different sensitivities and biosynthetic profiles to specific stimuli, the overall organ synthesis; being necessary and averaged and imprecise reflect of biosynthesis for technical reasons only few investigation of PG biosynthesis by isolated lung cells in culture have been performed (e.g. type II alveolar epithelial cells and lung fibroblasts) and further investigations are required in this field; ii) before 1975-1978 and still now, most studies were dealing with primary PGs and did not include PG metabolites, TXs, PG12 and HETE's measurements; since lung synthesized preferentially these last compounds, these studies could hardly open to any interpretation. Furthermore, some methodological difficulties (unreliability of some bioassay and. radioimmunoassay measurements) have contributed to some discrepancies; iii) finally, lung PG biosynthesis is very dependent qualitatively and quantitatively on a number of conditions such as specie, age, experimental conditions (i.e. type of material), physiological situation of the lung at the time of the experiment (PO2, pH, perfusion flow ventilation) and nature of the stimulus.

As in all other cells, the lung seems to release PGs upon a large number of stimulating procedures all of them having in common an alteration of the cell membrane. Thus PG synthesis has been detected after: i) mechanical distorsions of the cells elicited by various physical conditions but also bronchial constriction, pulmonary vasculary pressure increase, and hyperventilation; ii) phagocytosis of particles; iii) interactions with specific membrane receptors induced by mediators such as histamine, C5a, phospholipase A2 and its activators (bradykinin, angiotensin II, RACS-RF); iv) immunological challenge of the lung; v) incubation with or perfusion of PG precursors.

By contrast to its biosynthetic activity, it is well established that the lung is very efficient in removing circulating PG's. In most species, including man, 70 to 95% of intravenously injected PGs of the E and Fα types are removed from the circulation during a single passage through the pulmonary circulation. This disappearence is not due to any storage but reflects an intracellular catabolism of PG's by some undetermined lung cells. Current evidence suggests that a saturable carrier-mediated specific transport process ensures the uptake of PGs from the circulation into the lung and is mainly responsible for the high removal of PGs.   Structural requirements of PG molecules essential for uptake have recently been reported as well as several inhibitors (bromocresol green, diphloretin phosphate).   In contrast to PGs of the E and Fα types, other PGs (i.e. PGA, PG12 and probably PGD) are not affected by their passage through the pulmonary circulation, either because a lack of affinity for the transport system or bec{use they are not substrates of 15-PGDH; it was recently suggested that the PG12 generated by blood vessels of the lung and probably in other organs could therefore remain in the circulation and act as a ciruclating hormone.

PHARMACOLOGICAL ACTIONS OF PROSTAGLANDINS ON THE LUNG

Pharmacological actions on the airways

Evaluation of the pharmacological actions of PGs on the lung and the airways has led to conflicting result mainly explained by the designs of experiments and variability of the action of PG's depending on a number of factors.

Important differences have been observed according to the specie and the natural basal tone of the airway; as an example, PGF2 contracts bronchial smooth muscle in vitro in guinea-pig and man but has no effect on cat and rat airways; PGE relax isolated bronchial muscle of species with a high natural basal tone such as guinea pig, sheep and pig but have no effect in cat, monkey and rabbit, except if constriction is induced by histamine or acetylcholine.

The route of administration is an essential determinant of the pharmacological properties of PGs.   By comparison with I.V. infusion, the respiratory effects of aerosolised PGs are delayed but more potent and long-lasting.   These differences have been attributed to the rapid and important (90%) inactivation of circulating PG's while PG's perfused via the airways are less extensively catabolized (10-40%) and/or an easier access of aerosolised PG's to the receptor sites.

As a further point, recent studies have emphatized that it was of importance to delineate the PGs effects on central and peripheral airways.   Central airways (trachea and large bronchi) are easily accessible to aerosolised PG's but less by I.V. route and can be explored by the measurement of FEV1, airway conductance, while peripheral airways (respiratory bronchioles, alveolar ducts and alveoli) are more easily accessible to PG's delivered by I.V. route and can be explored by the static pressure volume curve and the density dependence of the MEFV curve.

One of the most striking points is that, over a wide range of doses, all PG's exhibit to varying degrees contractant, relaxant and irritant activities; in addition, one given PG, at the same concentration, may induce either contraction or relaxation; on human isolated bronchial muscle in vitro PGF2α , PGE2, PGI2 and PGE1 contract 95, 75, 22 and 2% of the preparations respectively. Furthermore irritative properties are independant of constrictor or dilator effects. It has therefore been hypothesized that these effects could be mediated through separate receptors subserving contraction, relaxation and irritation.

PGF2α is a potent constrictor of human bronchial muscle in vitro. Inhalation of 10-1000 of PGF2 in normal subjects elicits a log-dose related decrease in SGaw (with important inter-individual variations), an increase of respiratory frequency and tidal volume and may induce cough, wheezing, tracheal and pharyngeal soreness. In pregnant women I.V. or intra-amniotic administration of PGF2α elicit some decrease of FEV1, FVC, MMEF, alterations of MEFV curve and a slight hypoxaemia. Although 15-methyl-PGF2α given intramuscularly in human was shown to have a predominant effect on the peripheral airways other experiments have concluded that the site of the constriction induced by PGF2α varied among individuals.

PGE, usually considered to be dilators, relax the human isolated bronchial muscle, PGE1 being more active that PGE2 which acts as a mild bronchoconstrictor on most preparations; at high concentrations a paradoxical constrictor effect of PGE1 with crossed tachyphylaxis with PGF2α has been reported (suggesting a crossed stimulation of PGE1 and PGF2α, receptors). By aerosol route, PGE is a weak bronchodilator in human (PGE1/isoprenaline = 1), eliciting a small increase in SGAW without effect on FEV1 probably because the respiratory smooth muscle is almost fully relaxed in healthy people; very often PGE inhalation causes coughing and retrosternal soreness; occasionally, a paradoxical constrictor effect of PGE2 has been reported. By I.V. route, PGE2 elicits a moderate (30%) increase in SGaw, while in pregnant women a paradoxical bronchoconstriction (60% increase of Raw) has been observed.

Pharmacological actions of other PG's on the airways have been less extensively investigated. PGA1, 2 and PGB1, 2 are usually mild bronchoconstrictors in vivo and in vitro. PGD2 is about 5 times more active as a bronchoconstrictor than PGF2 on several preparation. PGG2 and PGH2 endoperoxides and TXA2 appear to be very potent bronchoconstrictor but their instability is a limiting factor for a precise evaluation of their relative potency. Only few data are available concerning the activity of prostacyclin; PGI2 has been reported to be a mild dilator on human smooth muscle in vitro but has not effect in vivo in normal subjects; however it could prevent bronchoconstriction induced in asthmatic patients by several stimuli.

PG metabolites (15-keto- and 13, 14 dihydro-15-keto-PG), TXB2 and 6-keto-PGF1α have only weak biological activities on respiratory smooth muscle.

Studies with various blocking agents demonstrate that the pharmacological actions of PG's upon the respiratory system are mainly mediated through direct actions on smooth muscle cells and accessorily through indirect actions, modifying the activity and/ or release of hormones or neurotransmitters. Structural requirements for a bronchoconstrictor effect include a C9- α -OH function and a   Δ 5-6 double bond while bronchodilation requires a ketone group in C9, substitution in C11 and 2 double bonds, mainly in   Δ 5-6; the   α-OH in C15 is necessary to both action.

Pharmacological actions on the pulmonary circulation

Pharmacological actions of PG's upon the pulmonary circulation have been extensively studied by many investigators. PGF2α is a potent pulmonary vasoconstriction in all species studies, constricting both pulmonary veins and arteries; PGF1α and PGF2 have similar although less prounced actions. Ath the opposite PGE1 decrease pulmonary vascular resistances (PVR) in lungs of several species including man. PGA 1 is a vasodilator compound upon pulmonary circulation while PGA2 is a modest pulmonary pressor substance. In the intact dog, PGBs increase PVR, PGB2 being 10 times more potent than PGB1. PGD2 has a pronounced pulmonary vasoconstrictor effect similar in potency to that of PGF2 α   . Stable PG endoperoxide analogues have marked vasoconstrictor effects, being approximately 10 times more potent than PGF2 α   ; this is contrasting with a weak activity of natural endoperoxides (PGH2) in vivo probably explained by their short half-life and spontaneous transforming in other compounds with opposite activities (TXs, PGF2α , PGD2α , PGD2 and PGE2 being vasoconstrictors while PG12 is a vasodilator). Although TXA2 has been shown to be one of the most powerful vasoconstrictor agent on several smooth muscle preparations in vitro, no information is presently available about its action upon pulmonary vessels in vitro and in vivo. However, it is likely that TXA2 is an effective pressor substance on the pulmonary vascular bed. At the opposite, PG12 and stable PG12 analogues have marked pulmonary vasodilating effects on several preparations.

PGs exert their vascular action mainly by direct effects upon vascular smooth muscle cells and not through the intervention of neurotransmitters or by an action on blood cells, particularly platelets.

POSSIBLE ROLES OF THE PROSTAGLANDINS IN LUNG DISEASES

Prostaglandins and anaphylaxis

The antigen-antibody interaction elicits from basophils and/or mastocytes of blood and lung tissue of allergic or sensitised subjects the release of several chemical mediators including preformed stored mediators (histamine, ECF-A, kallikrein) and de novo synthesised mediators including platelet activating factor (PAF), PGs and leukotrienes (SRS-A); the release process is controlled by cyclic nucleotides, cAMP inhibiting and cGMP enhancing the secretion of mediators, via their protein-phosphorylating activities on the microtubular system.

The site and mechanism of anaphylactic PG release remain to be determined.

Since any perturbation of the cell membranes evokes a release of PGs, it has been hypothesized that anaphylactic PG release could be a mechanical consequence of other phenomenon such as; i) contraction of bronchial smooth muscle elicited by other mediators; ii) distorsion of other cells of the lung parenchyma; iii) membrane distorsion of mast cells, induced by the histamine granule exocytosis.

However other experiments are more in favour of a direct synthesis of PG's by mast cells and probably eosinophils under the action of other mediators. It is now generally accepted that anaphylactic PG release is a secondary phenomenon due to the effects of other mediators, mainly histamine, which act probably via the activation of the phospholipase A2.

Although most studies concerning the role of PG's in anaphylaxis have been focused on primary PG's of the E and F types, it appears now that the role of these PGs is probably much less important than that of other compounds such as TXA2, and lipoxygenase products (leukotrienes, HETE's). Therefore during antigen challenge of sensitised guinea-pig whole lung, AA metabolism is mainly directed towards TXA2 and hydroxy-fatty acid synthesis, only 1 and 10% of AA being converted to PGE2 and PGF2α metabolites respectively. Other observations underlining the role of the lipoxygenase pathway in the anaphylactic reaction is that NSAI usually enhace and prolonge the anaphylactic bronchoconstriction, while SRS-A antagonists and inhibitors of both cyclo-oxygenase and lipoxygenase suppress this enhanceing effect of NSAI. Contrarily to initial statements this enhancement of the anaphylactic response to NSAI is not due to the depression of PGE biosynthesis but rather to the diversion of AA metabolism via the lipoxygenase pathway. Furthermore, mast cell, eosinophils and neutrophils which are involved in the immune reaction wo antigen challenge are known to release HETE's and leukotrienes which reinforce the anaphylactic reaction in two days: i) per se; ii) indirectly because the HPETE's are probably powerful mediator-releasers. (see after).

Recent reports of the biological properties of leukotrienes and HETE's have enlightened their possible role in the anaphylactic reaction.

HETE's are very powerful chemotactic and chemokinetic agents upon neutrophils and eosinophils but have no action on mononuclear leukocytes; in addition, they enhance the expression of C3b receptors and stimulate the synthesis of cGMP in these cells, the overall effect being an accumulation of neutrophils and eosinophils at sites of immunological reactions. LTB4 (i.e. 5, 12, dihydroxy-AA) and 5-HETE, are by far the more potent compounds followed by 8-HETE, 9-HETE, 11-HETE, 12-HETE, and 15-HETE; neither LTC4 and LTD4 exert substantial effects on polymorphonuclears. LTB4 and 5-HETE seem therefore to be able to recruit and regulate the polymorphonuclear leukocytic component of the hypersensitivity response.

Biological activities of LTC4 and LTD4 are qualitatively the same as those of native SRS-A. It is now known that SRS-A is a variable mixture of these 2 substances depending on the cell system and the experimental conditions. Both compounds are potent constrictors of bronchial smooth muscle, LTD4 being equipotent to LTC4 on human bronchial muscle and 3 orders of magnitude more potent than histamine; in addition, LTC4 and LTD4 seem to exhibit a regional specificity for small airways. In the guinea-pig, the bronchoconstriction induced by leukotrienes could be potentiated by the subsequent release of TXA2 induced by these substances. Other known biological actions of leukotrienes include an increase of vascular permeability and a probable vasodilator effect. These substances appear therefore to be able to contract smooth muscle and alter the permeability and tone of the microvasculature, all phenomenon which play a major role in the anaphylactic reaction.

Prostglandins and asthma

In asthma the inhaled PGs exert powerful, delayed and long-lasting pharmacological effects. Several studies have shown that asthmatic patients are 160 to 8000 times more sensitive to $PGF2\alpha$ inhalation than normal subjects with, however, important interindividual variations. The mechanisms which could account for this hyper-reactivity are not fully elucidated. Trials with aerosolised bronchodilating PGE in asthma have been disappointing. A bronchodilation, 5 to 10 times more potent than isoprenaline, has been observed in most patients but with frequent unresponsiveness and some mild bronchoconstrictive responses.

Very early a role of PG's in bronchial asthma was considered because of: i) the powerful bronchoconstrictor effects of some PG's (PGF2α , PG endoperoxides, TXA2 and more recently leukotrienes); ii) the hyperreactivity of asthmatics to PGF2α ; iii) the similarities between allergic asthma and the anaphylactic reaction specially in terms of chemical mediator release; iv) the increase in the excretion of PGF2α metabolites during asthmatic attacks.

However the precise role of PG's in bronchial asthma remains unclear. NSAI such as indomethacin and aspirin elicit very variable effects according to the patient; in most asthmatic subjects indomethacin preatreatment does not prevent, worsen or improve asthmatic attacks but NSAI are able to improve the disease in a small number of patients. Recent evidence suggests that the induction of broncho-constriction in aspirin-sensitive patients could be due to the diversion of AA metabolism towards the lipoxygenase pathway.

Role of PGs in other pathological conditions

Although PG's are released during pulmonary embolism their role in the haemodynamic response is not elucidated; however, since NSAI attenuate the hypocapnic airway constriction, an intervention of PG's has been suggested in the airway response to embolism.

Most reports have not been able to demonstrate any significant role of PGs in producing the pressor response to alveolar hypoxia; however, the observation that this response is potentiated by NSAI suggests that a circulating or lung produced vasodilator such as PGE or PG12 normally oppose the hypoxic vasoconstriction and maintain the pulmonary circulation in a dilated state.

REFERENCES

Dahlen, S.E., Hedqvist, P., Hammarstom, S., and Samuelsson, B. 1980. Prostaglandins, Arachidonic Acid and Inflammation. Science, 210 : 978.
Even, P., and Sors, H. 1981. Biological actions of prostaglandings and related substances on the respiratory system, in : "Scientific Foundation of Respiratory Medicine". J.G. Scadding and G. Cumming eds, W. Heinemann, London, 297-314.
Gardiner, P.J., and Collier, H.O.J. 1980. Specific receptors for prostaglandins in airways, Prostaglandins, 19 : 819.
Goetzl, E.J. 1980. Mediators of immediate hypersensitivity derived from arachidonic acid, New Engl. J. Med, 303 : 822.
Hyman, A.L., Spannahake, E.W. and Kadowitz, P.J. 1978. Prostaglandins and the lung. Amer. Rev. Resp. Dis., 117 : 111.
Kuehl, F.A., and Egan, R.W. 1980. Prostaglandins, Arachidonic Acid and Inflammation, Science, 210 : 978.

Mathe, A.A., Hedqvist, P., Standberg, K., and Leslie, C.A., 1977. Aspects of prostaglandin function in the lung (two parts). New Engl. J. Med. 296 : 850 and 910.

Samuelsson, B., Hammarstrom, S., Murphy, R.C., and Borgeat, P., 1980. Leukotrienes and Slow Reacting Substance of Anaphylaxis (SRS-A), Allergy, 35 : 375.

Smith, A.P. Prostaglandins and the respiratory system, 1976. in: "Prostaglandins : physiological, pharmacological and pathological aspects", S.M.M. Karim ed, MPT Press, Lancaster, p. 83.

Sors, H., and Even, P. 1981. Biosynthesis and metabolism of protaglandins and related compounds by the lung, in: "Scientific Foundations of Respiratory Medicine". J.G. Scadding and G. Cumming eds, W. Heinemann, London, 297-314.

DISCUSSION

LECTURER: Sors   CHAIRMAN: Cumming

DENISON: Two questions. Firstly you have pointed out the
                importance of the oxygenases, in particular
                lipoxygenase, have you any idea of its oxygen
                dependence, in particular its Michaelis constant,
                if you know it.

SORS: There have been quite a lot of discussion about the
                inter-relation between the oxygenation process and
                the lipoxygenase in some other systems such as
                plants.   They are very dependent on oxygen
                reactions because it begins with hydroxic
                peroxidation with some free radical mechanism.   I
                couldn't tell more about it because it is quite a
                complex subject involving the intervention of free
                radical mechanisms and some other oxygen species,
                but it does probably have very important relations
                with hydroxylation compounds.

DENISON: The other question is that if you look at the lungs
                of all terrestrial mammals they all breathe in the
                same way and they use their lungs in the same way.
                Why do you think there are then species variations
                in the way they respond to these compounds?

SORS: The most recent explanation concerning the
                pharmacology of prostaglandins in the airways and
                pulmonary vasculature concerns the initial state of
                the preparation, and this could explain quite a lot
                of differences in tone in different species.   I
                have no other explanation, but sometimes apparent
                discordances are not due to species differences but
                to the type of the preparation.   It is probable
                that the same response cannot be obtained on
                bronchial isolated smooth muscle and in vivo
                airways.

HEATH: At the moment many members of the academic staff of
                the University of Oxford are taking acetyl
                salicylic acid because it is supposed to protect
                them from coronary thrombosis.   Every time these
                gentlemen have a cup of tea they take an aspirin
                tablet with it and of course the basis for this is
                that it inhibits the action of thromboxane which as

you pointed out is an aggregative agent and a vaso constrictor; but you also pointed out that aspirin has a bad effect on the action of prostacycline which is a handy aggregatory vasodilator. Could you please tell me what should the dons of Oxford go on doing in the future?

SORS: Yes, concerning the action of aspirin in this situation someone could probably respond to your question. It seems that the inhibition of thromboxane synthesis by aspirin in platelets is much more potent than the inhibition of cycloxygenases in the vessel wall and probably more long-lasting. Therefore some workers have said and have shown that the inhibition of thromboxane synthesis in platelets was much more delayed than the inhibition of prostacyclin synthesis in the vessel wall and this could explain the beneficial action of small doses of aspirin in the prevention of thrombotic diseases, but of course it has been said that the best thing will be to take very small doses at quite large intervals so that prostacyclin synthesis by the vessel wall, which is inhibited for a small time, could be restored and then the two compounds were not together inhibited.

CUMMING: Can you tell us what you mean by a small dose and a long interval?

SORS: Yes, it has been said that the best way of preventing platelet aggregation in this condition could be to take about 200 to 300 mg. of aspirin two times a week.

SPENCER: I hope you forgive me, I may have misunderstood you. I know nothing about collagen but earlier on you mentioned, talking about the production of arachidonic acid, that it arose from the cell membrane and you mentioned that there was no circulating prostaglandins. Later, talking about the lung you talked about the lung removing circulating prostaglandin, now this seems to me a contradictory statement. Are they locally active or do they circulate.

SORS: They do not circulate and several techniques which have measured circulating prostaglandin, particularly radio-immunological techniques and mass spectrometry techniques have shown that there is no circulating prostaglandin, but of course the removal of prostaglandin by the lung is an unphysiological process and when you inject prostaglandins they are effeciently removed by the lung. Also in any pathological state which are accompanied by a large increase in prostaglandin production and its release into circulation, then the lung could catabolise these prostaglandins. It is probably not a physiological process and most of the prostaglandins which are synthesised in the cells are immediately catabolised in the cells without any release in the circulation.

BAUM: I understand that certain anti-malerial agents such as mepacrine are potent inhibitors of phospholipase activity. Do you know whether they have ever been tried in the treatment of asthma?

SORS: No, I do not have any information on this point, but a great deal of research has to be done on the use of specific lipoxygenase inhibitors which are not toxic, to see what might be the effect of these drugs in asthma. I don't think that there are at present any such inhibitors but I have no information on phospholipase A2 in asthma. The most important phospholipase A2 inhibitor is at present cortico-steroids.

THE PULMONARY ENDOTHELIAL CELL AND THE MYOFIBROBLAST

Donald Heath

Department of Pathology
University of Liverpool

The endothelial cell and the myofibroblast of pulmonary blood vessels differ from one another in many respects. The first has a fixed ultrastructure consistent with precise biochemical functions and appears to be easily damaged by a raised intravascular pressure. The second seems to be a much tougher reserve cell, able to adapt both its form and situation to prevailing conditions in the blood vessel.

In a transverse section of the pulmonary trunk, the endothelial cells are hardly perceptible, their presence being detectable only by their nuclei which bulge towards the lumen (Heath and Smith, 1979). However, on staining their margins with silver salts and looking at them en face, they assume a new interest. In rats endothelial cells from different vessels differ both in size and shape (Kibria et al, 1980). Those of the aorta are small and fusiform with the long axis in the direction of blood flow and are probably best suited to a high intravascular pressure. In contrast those of the inferior vena cava are much larger and rectangular, a shape likely to be compatible with a low pressure. The endothelial cells of the pulmonary trunk have their own distinctive size and shape, somewhere between those of the aorta and inferior vena cava. Those of the pulmonary veins like those of their systemic counterparts are large and round. These general impressions are confirmed by measurement of the diameters of the cells (Kibria et al, 1980, Kombe et al, 1980).

In young rats at least, the pulmonary endothelial cells have a certain limited capacity to change their form to conform to the haemodynamic environment. They change in response to pulmonary hypertension irrespective of its cause. The elevation of pulmonary

355

arterial pressure may be induced by hypobaric hypoxia in a decompression chamber or by ingestion of <u>Crotalaria spectabilis</u> seeds which contain a pyrrolizidine alkaloid, monocrotaline, which is capable of inducing pulmonary vasoconstriction. In both instances the pulmonary endothelal cell assumes a small, fusiform shape akin to that of the aorta (Kibria et al. 1980, Kombe et al. 1980). We have not yet confirmed these findings in young humans. Older subjects show a much more heterogeneous appearance than occurs in the animals and in fact commonly show multinucleated giant endothelial cells (Helliwell et al. 1981).

On scanning electron microscopy two new features of the endothelial cell appear. These are cytoplasmic tags and pits opening onto the luminal surface (Heath and Smith, 1979). On transmission electron microscopy the cytoplasmic tags are seen clearly and the pits prove to have a definite ultrastructure (Smith et al. 1971, Smith and Ryan, 1972, 1973). These caveolae intracellulares, some 50–90 nm in diameter, open onto the luminal surface by a stoma covered by a diaphragm composed of a single lamella. At the meeting point of the plasma and caveolar membranes and the caveolar diaphragm are dense osmiophilic bodies which may represent a circular skeletal structure that maintains the patency of the stoma and the integrity of the diaphragm. Sometimes several caveolae fuse together to form a complex structure but in doing so the shape and identity of the individual caveolae are maintained so that the whole comes to resemble a bunch of grapes. The wall of the caveolae is of unit membrane type with two lamellae. It is not uniform in appearance but is granular with the formation of small nodules. These appear to represent enzyme clusters or binding sites where the inactive decapeptide angiotensin I is converted into the active octapeptide angiotension II particularly in the pulmonary capillaries. Angiotensin–converting enzyme can also be demonstrated in the undifferentiated plasma membrane of the surface of the pulmonary endothelial cell (Ryan and Ryan, 1975).

Normal alveolar capillary endothelial cells are not fenestrated but abut bluntly or overlap so that a narrow cleft of variable size exists between adjacent cells. The adjacent cell membranes fuse (maculae occludentes) except for small slits 4 nm wide which allow communication between the capillary lumen and extravascular space. Hence these endothelial cells are much less resistant to the passage of oedema fluid and large molecules than the cells of the alveolar epithelium which are tightly fused by zonulae occludentes. In the lung both toxic substances and haemodynamic disturbances may lead to oedema fluid collecting in the ground substance of the fused basal lamina and then projecting into the lumen with the thinned endothelial cells being stretched over the vesicle (Heath et al. 1973).

In contrast to the fixed endothelial cell carrying out its

specialised functions at the interface between the blood and the wall of the pulmonary vessel the myofibroblast is a tougher reserve cell characteristically appearing in different sites in the wall of the pulmonary artery (Smith and Heath, 1980). Unlike the endothelial cell, it appears to have the capacity for altering its form, and one suspects its function, according to the place and circumstances in which it is situated. The myofibroblast is a type of connective tissue cell found as an integral part of granulation tissue and it has ultrastructural features of both the smooth muscle cell and the fibroblast. Its muscular nature is revealed by the presence in its cytoplasm of small bundles of myofibrils which however, do not permeate most of the structure of the cell as in the mature, definitive smooth muscle cell. There are also attachment points on the plasmalemma for these myofibrils and focal condensations within their course in the cytoplasm. At the same time the copious dilated endoplasmic reticulum throughout most of the cytoplasm indicates that it is also a fibroblast. These cells originate in the media and migrate into the intima.

The myofibroblast is responsible for the various types of intimal proliferation in the pulmonary blood vessels in age change and in different types of heart and lung disease. The behaviour of myofibroblasts and the form they assume in the intima depends on the functional conditions which exist there. Thus so fas as age change is concerned, the myofibroblasts in the intima of pulmonary veins lose their myofibrils and assume progressively the ultrastructural features of fibroblasts with associated collagen and elastin, the so-called 'non-specific intimal fibrosis' of classic histopathology. In contrast, the myofibroblast in the intima of the pulmonary artery is subjected to the stimulus of pulsation and this encourages the cells to maintain their myofibrils and retain their muscular features. Hence age change intimal proliferation is different in pulmonary arteries and veins (Smith and Heath, 1980).

In the presence of severe pulmonary hypertension, such as may occur with a large congenital cardiac septal defect, there is a greater tendency for these myofibroblasts to show their muscular origin. Typical muscle cells appear in the intima, their cytoplasm filled with myofibrils, and form fasciculi of longitudinally-orientated smooth muscle cells in the intima which are readily apparent on histological examination. Myofibroblasts appear to form elastin so that the typical intimal proliferation changes into one of prominent elastosis.

When pulmonary arterial hypertension is unusually servere, or its rise is unusually rapid, the myofibroblasts in the media are stimulated, often in a striking manner, to lay down ground tissue in a manner producing an appearance of concentric basement membranes. As a result the definitive smooth muscle cells of the media become widely separated by mucopolysaccharide (Heath and Smith, 1978). With such severe or abrupt elevation of pulmonary arterial pressure some of the muscle cells die and fibrin is forced from the blood stream into the walls of the pulmonary arteries. The fibrin is commonly held up at the inner and outer elastic laminae to give the appearance of fibrinous vasculosis and the classical histopathological appearance of the 'fibrinoid necrosis' of the so-called malignant phase of hypertension.

In the pulmonary circulation there is one further stage of reaction to hypertension, namely the formation of plexiform lesions. These comprise dilated branches of muscular pulmonary arteries which become the site of cellular proliferation composed of myofibroblasts. Part of the response of these cells appears to be fibrin washed into the dilated branches from their parent muscular pulmonary arteries showing fibrinoid necrosis. There is a second type of cell involved in this proliferation, namely the fibrillary cell (Smith and Heath, 1979) which from its ultrastructure is seen to be related to the 'vasoformative reserve cells' of American authors (Stein et al. 1969). This is closely related to the type cell of the cardiac myxoma and the so-called 'papilliferous tumour of the heart valves' (Stovin et al. 1973).

REFERENCES

Heath, D., Moosavi, H., and Smith, P. (1973). Ultrastructure of high altitude pulmonary oedema. Thorax, 28, 694.
Heath, D., and Smith, P. (1978). The electron microscopy of 'fibrinoid necrosis' in pulmonary arteries. Thorax, 33, 579.
Heath, D., and Smith, P. (1979). The pulmonary endothelial cell. Thorax, 34, 200.
Helliwell, T., Smith, P., and Heath, D. (1981). Endothelial pavement patterns in human arteries. J. Path. In press.
Kibria, G., Heath, D., Smith, P., and Biggar, R. (1980). Pulmonary endothelial pavement patterns. Thorax, 35, 186.
Kombe, A.H., Smith, P., Heath, D., and Biggar, R. (1980). Endothelial cell pavement pattern in the pulmonary trunk in rats in chronic hypoxia. Br. J. Dis. Chest 74, 362.
Ryan, J.W. and Ryan, U.S. (1975). Metabolic activities of plasma membrane and caveolae of pulmonary endothelial cells, with a note on pulmonary prostaglandin synthetase. In Lung Metabolism, edited by A.F. Junod and R. de Haller p. 399, Academic Press, New York.
Smith, P., and Heath, D. (1979). Electron microscopy of the plexiform lesion. Thorax, 34, 177.
Smith, P., and Heath, D. (1980). The ultrastructure of age-associated intimal fibrosis in pulmonary blood vessels. J. Path. 130, 247.

Smith, U., and Ryan, J.W. (1972).   Substructural features of
    pulmonary endothelial caveolae cellulares.   Tissue and Cell,
    4, 49.
Smith, U., and Ryan, J.W. (1973).   Electron microscopy of
    endothelial and epithelial components of the lungs:
    correlations of structure and function.   Fed. Proc. 32, 157.
Smith, U., Ryan, J.W., Michie, D.D., and Smith, D.S. (1971).
    Endothelial projections as revealed by scanning electron
    microscopy.   Science, 173, 925.
Stein, A.A., Mauro, J., Thibodeau, L., and Altey, R. (1969).   The
    histogenesis of cardiac myxomas: relation to other
    proliferative diseases of subendothelial vasoformative reserve
    cells.   In Pathology Annual edited by S.C. Sommers, vol. 4,
    p.293.   Butterworths, London.
Stovin, P.G.I., Heath, D., and Khaliq, S.U. (1973).   Ultrastructure
    of the cardiac myxoma and the papillary tumour of heart
    valves.   Thorax, 28, 273.

DISCUSSION

LECTURER: Heath   CHAIRMAN: Cumming

CUMMING: Regarding the first cell you discussed, the
                epithelial cell, and its changed ratio of length to
                breadth. If the vessel has the same diameter, then
                its circumference is the product of their breadth
                and the number of cells around it.   If these cells
                change their shapes so that their breadth becomes
                less, either there must be a diminishing calibre or
                else a proliferation of cells.   Have you any
                information about the turnover of epithelial cells
                and whether in your rats the diameter of these
                vessels have actually changed?

HEATH: We do have indications of this and you are quite
                right, considerations of this type depend on the
                speed with which the hypertension becomes
                established.   If the hypertension is slow in
                developing the cells could expand to cope with the
                pressure. This is the sort of thing one finds
                interestingly enough with pulmonary hypertension
                induced by crotellaria, because as many people in
                the audience will know, there is a latent period
                while crotellaria is being converted by the liver
                into a metabolite. We found that during this
                period there is a slow increase in the size of the
                cells; however, with hypoxia, where the pulmonary
                hypertension is very rapid in development, there
                you find that the division of cells into smaller
                cells is much quicker, so I think that your point
                is well taken and is valid.

SOLIMAN: Thank you very much Professor Heath for your nice
                lecture, I have two points to ask about the fact
                that in your experimental model of hypoxia there
                were induced endothelial changes. I'd like to ask
                about the reversibility of these changes after
                correction of hypoxia as well as if there is any
                measurement of the pulmonary artery pressure.

HEATH: No, we do not have measurements of the pulmonary
                artery blood pressure in these experiments and the
                reason for this is that we have made measurements
                of pressure using the same method in previous
                years. That is why we have not done it in these.

In the literature we are talking of systolic
pressures of something like 80 mm. of mercury.

SOLIMAN: Is there any evidence of reversibility of these
             changes after correction of hypoxia.

HEATH: The answer is almost certainly that the cells
             would revert back to normal, because it's very
             characteristic of every feature of hypoxic
             pulmonary hypertension, be it ventricular
             hypertrophy, an increase in the thickness of
             pulmonary trunk or muscularisation of pulmonary
             arteries, that it is reversible. I think the
             changes in the endothelial cells of the pulmonary
             trunk would be the same; we haven't done it but it
             is an excellent idea.

SOLIMAN: Just another point, Professor Heath; is there any
             described endothelial changes in high altitude.

HEATH: This is again an exceedingly interesting idea. I
             am hoping to go to Bolivia again in November and I
             think, as you suggest, that it would be a very
             interesting thing to examine the pulmonary arteries
             of the Indians who live permanently in altitudes.
             As you know, there are many features, when we sit
             here talking about the human lung, and David
             Denison has referred to it in several of his
             questions about the Michaelis constant in the past
             few days, we must never lose sight of the fact that
             we are talking about the lung with a certain PO_2,
             Now, if you go to another situation where the PO_2
             is different I imagine that both the structure and
             the cell biology and the cell biochemistry is
             different. I would think it is quite possible if
             you look at the endothelial pavement of the
             pulmonary arteries of high altitude Indians, they
             would be different from yours and mine.

SIMPSON: I am quite fascinated by your endothelial studies
             and I was wondering if you have at any time exposed
             these rats not just to hypertension but to
             hyperoxygenation, and whether the high oxygen
             tension makes a difference. The second question I
             have is related to this; in the changes in the
             pulmonary artery that you saw in the shunts was
             there any difference in the changes prior to the
             reversal of their shunt such as occurs in the

Eisenmenger complex to what occurs afterwards? Is there any change as the oxygenation of the pulmonary artery blood increases?

HEATH: I am afraid I have answers for neither of these questions but I agree it would be interesting to know.

SPENCER: I am intrigued by the myofibroblasts. I'd like to know things about myofilaments in the endothelial cells, because this was described by Charles Benjon many years ago and as he pointed out endothelial cells change their shape, sometimes they are columnar and sometimes cuboidal, and he attributed this to the contraction of myofilaments within those cells. Other myofibroblasts are anyway related to endothelial cells, but where do these myofibroblasts come from? Do they come from the media, do they come from endothelial cells or are they already present in the intima?

HEATH: I must say my present feeling is that the myofibroblasts I see are derived from the media and in many of the conditions I have described one can see a direct continuity between the medial and what is going on in the intima. So I would say from that sort of evidence that the endothelial cell is not involved. I wonder whether Professor Corrin has any ideas on that. My only view is that the endothelium was not concerned.

DENISON: Donald, you spoke earlier on in your pictures of the endothelial cell as if there were some desirable virtue in having long thin cells in the axis of the vessel when there is high pressure inside and I want to question that view because I am not sure that there is any mechanical advantage to having the tails of a vessel a particular shape and I would like to know what you think about that. That is not a direct consequence of Gordon's question, that you were looking at the recent change in diameter and it is just betraying whether it has recently constricted.

HEATH: I must be very careful to point out there, David, that I did not say it was desirable at all, this is something you just put into the question; what I did was to use that splendid word which pathologists all use, it was "associated". I believe that this shape of endothelial cell is associated with high pressure. Now whether that is advantageous or not I don't know but surely one might perhaps make the assumption that since the Almighty arranged the aorta to be lined by endothelial cells of that shape it is probable they don't find too much of a disadvantage, but of course there are people in this room like yourself and the Chairman who are much better able than I am to determine whether that particular shape is best suited in that situation.

CUMMING: If the Chairman could be forgiven a comment, it depends very much on whether the stress applied to the single cell is sustained more or less readily than the stress applied to the tight junction. If the tight junction had a very high yield stress, the more tight justions per unit length there were, the greater will be the resistance to radial stress; that would be advantageous, but as to which of the two is correct I have no idea.

               Other questions?

CORRIN: From your photographs I was unable to pick out any intracellular detail at all and when you showed with the aged endothelium that there were large areas with no intracellular junctions and went on to tell us that these represented multinuclear giant endothelial cells, I am afraid that my mind arrived at the conclusion, anticipating what you are going to say, that these perhaps represented areas where the endothelium was lost.

HEATH: You are of course quite right but what I was illustrating was with a simple silver nitrate stain. At the time we were not using a nuclear stain as well, and Timothy Honeywell in our department is using haemotoxylin with silver stain and is presenting some of this work at the Pathological Society in January.

TRICOMI: Thank you Professor Heath. In atrial septal defect in long-surviving subjects, one can exceptionally observe not only pulmonary artery dilatation but also wall calcification. Then is it correct for a radiologist to say that there occurs a sort of "aortisation" of the pulmonary artery in case of long-standing hypertension. That's my first question. This seems to me to be relevant to what you taught us from a histological standpoint. My second question is: is it correct for a radiologist to assume that the signs of precapillary hypertension result from a disproportion between the big artery of the elastic hilus and the small pulmonary arteries of the fundamentally peripheral muscle-type branching? To end with, and I apologise to the Chairman if I, as a radiologist take advantage from this opportunity to be taught by an anatomopathologist, could you please tell us something more on the pathogenesis of pulmonary oedema in subjects with precapillary hypertension at high altitude?

HEATH: Question number three - the pathogenesis of oedema in subjects at high altitude. Firstly, no-one knows the answer, but there are several clues. I think the most important clue is the fact that the lung apparently does not seem to have dramatic changes in the partial pressure oxygen to which it is exposed. Every time high altitude pulmonary oedema has been described people have pointed out that subjects who are really aggressed are the people who normally live at high altitude in a state of hypoxia and who descend to sea level. From the early literature from Peru and India this was made very clear and there was an interesting paper by Dr. Lafabo from Colorado and recently by Stoggings and his associates who pointed out that people who go from Lebville in Colorado down to Denver for just six hours , with exposure to normal oxygen pressure and then go back to hypoxia, this is enough to induce high altitude pulmonary oedema. So I think this change from low pressure to high pressure and then back to low, is important. Another feature is age because it is children and young adults who get it. There is also a very strong genetic predisposition. In the medical literature there were many reports of fathers who had altitude pulmonary oedema and then 20 years later when their sons have been exposed to similar conditions they also got it. With regard to your first question, I myself wouldn't regard calcification of the walls of pulmonary arteries as being a very significant part of a pulmonary vascular pathology at all. I know one gets calcification in the large elastic pulmonary arteries in sustained pulmonary hypertension, but I wouldn't have thought this was a very important striking feature of pulmonary hypertension.

CORRIN: I just wonder whether the so called calcification in branches of the pulmonary vessels may have possibly resulted from pulmonary venous occlusion, because this in itself is a cause of pulmonary hypertension and in the damaged vessels you get elastic fibres impregnated both with iron and calcium and this could very readily be mistaken for a calcification in pulmonary vessels.

CUMMING: Tricomi's second question Donald, if I can remind you.

HEATH: But that was a third question.

CUMMING: It was question number two and that referred to the ratio between the diameter of the large central vessels in the mediastinum and the small peripheral vessels and is this an adequate diagnostic criterion for hypertension.

HEATH: I don't know, I have no data.

CUMMING: A paper by Laws a few years ago has described this. I think I should at that point terminate the proceedings and have a coffee and reassemble at 11.30.a.m.

METABOLIC ASPECTS OF THE PULMONARY ENDOTHELIUM

Alain F. Junod

Respiratory Division, Department of Medicine
Hôpital cantonal universitaire
CH - 1211 Geneva 4, Switzerland

The title of this presentation would have been unthinkable thirteen years ago. At that time, there were only a few isolated reports in the medical literature mentioning the possible role of the pulmonary circulation in the handling of circulating substances. The pioneering work done by Vane and his collaborators in the late sixties revealed that the lung was capable of metabolizing a number of circulating vasoactive agents and was the start of numerous investigative studies to analyze and define this new function of the lung. This research area has become wide enough to give the opportunity to cover a number of metabolic processes, among which are lipid metabolism, coagulation, and intermediary metabolism, fields that several groups are actively investigating. But instead of giving a superficial overview of a number of metabolic processes, which would take the form of a catalogue, I prefer to discuss more specifically how the lung is able to deal with a number of circulating substances and considerably alter their plasma concentration. I will also discuss the meaning of these metabolic properties and how we may take advantage of them in order to try testing the functional state of the pulmonary endothelium, in clinical situations.

Several categories of vasoactive substances are affected in various ways during their passage through the lung; biogenic amines, such as 5-OH tryptamine (5-HT) and noradrenaline (1-NE); lipids (prostaglandins); peptides (bradykinin, angiotensin I); nucleotides and nucleosides. A well defined process has been shown to take place for each of these types of compounds.

Removal of the biogenic amine, 5-HT, by the pulmonary circulation is extensive (up to 95% of 5-HT infused into the lung

367

can be extracted within one passage) and is determined by its
uptake by vascular endothelial cells. Experiments done in the
presence and absence of inhibitors of monoamine oxidase (MAO), the
main enzyme responsible to the oxidative deamination of 5-HT, have
shown that the uptake of 5-HT by the lung was limited by the
transport step, and not by the subsequent intracellular metabolism
of the amine[12]. Radioautographic experiments carried out in
presence of the MAO inhibitor, iproniazid, have also clearly
demonstrated that endothelial cells were involved in this uptake of
5-HT since more than 95% of the silver grains were found over
endothelial cells[20]. The transport of 5-HT by endothelial cells
has been rather well characterized [10, 11, 12] as being Na^+-
dependent and carrier - mediated, similar to amine transport in the
central and peripheral nervous systems. The accumulation of amine
inside the cell is thought to result from the utilization of the
Na^+-K^+ activated ATPase or Na^+ pump. Hence the transport of 5-HT
is reduced by inhibition of the Na^+ pump: directly by metabolic
inhibitors such as 2-deoxyglucose or lack of energy substrate.

This type of uptake has been further pharmacologically
characterized. Various drugs like cocaine, tricyclic
antidepressants, chlorpromazine, phenoxybenzamine have been shown
the inhibit transport of 5-HT, whereas other investigators have
measured the sensitivity of the intracellular degradative process
to agents such as clorgyline, deprenyl and semicarbazide. The
studies on intracellular degradation show that, beside the A and B
forms of mitochondrial monoamine oxidase, there is a third form of
MAO involved in the pulmonary catabolism of 5-HT[16].

l-NE, another biogenic amine, is also removed from the
pulmonary circulation by a mechanism very similar to that of 5-HT.
The rate of extraction of NE is also determined by the transport
step, a temperature-sensitive Na^+-dependent and saturable process.
The magnitude of the lung effect, however, is not as spectacular as
for 5-HT : only 20-40% of the l-NE infused into the pulmonary
artery is lost during its passage through the lung.
Radioautographic and fluorescent studies have also established the
endothelial localization of this phenomenon[10,14], but, instead of
the even labeling of the endothelium along the pulmonary vascular
bed found with 5-HT, there were preferential sites of uptake for l-
NE in pre- and post- capillary vessels and veins.
Pharmacologically, the uptake process for l-NE is nearly identical
to that of 5-HT, with only a few exceptions among which is the
effect of phenoxybenzamine. This finding, together with the fact
that no real competition seems to exist between 5-HT and l-NE for
their intracellular transport, suggests that different carriers are
used by these two amines[11, 14].

Other catecholamines and sympathomimetic drugs have been
tested and, interestingly enough, are handled in many different

ways. Thus, epinephrine is hardly affected during its passage through the lungs. Neither is dopamine, although some authors have reported that dopamine was partially metabolized during its pulmonary transit, but without being taken up by the lung. Metaraminol is probably taken up by the same mechanism as that operating for l-NE. Phenylethylamine is markedly degraded and this process appears to be limited by the availability and activity of intracellular MAO, and not by the transport process itself[1].

The removal of circulating prostaglandins (PG) of the E_1, E_2 and $F_{2\alpha}$ series is not so different from that of the biogenic amines. Here again, according to Bito, removal (up to 95%) of PGs from the circulation seems to be determined by a carrier-mediated type of transport showing evidence of saturability and substrate specificity[2]. Once inside the cells, PGs are metabolized, mainly by 15-OH PG dehydrogenase and by 13-14 PG reductase. PGAs are apparently not a good substrate for the carrier operating for PGEs and $PGF_2\alpha$, although some authors think that PGA binding to albumin prevents its uptake by the lung[9]. Organic substrates, like bromcresol green are inhibitors of PG uptake by the lung, whereas ethacrynic acid, frusemide and polyphloretin phosphate are among the drugs known to interfere with the metabolism of PGs, after the PGs have been taken up.

Completely different from the preceding processes are the mechanisms of hydrolysis of angiotensin I and bradykinin, which result in the generation of both a biologically active peptide, angiotensin II, and inactive metabolites of bradykinin[5]. In fact, angiotensin converting enzyme (ACE) or bradykininase is one single enzyme acting on these two substrates. This enzyme, as shown by Ryan et al, is located on the membrane of the endothelial cell[17]. Hydrolysis takes place at the surface of the cell, and does not require intracellular transport of the substrate. There is no doubt that the lung is the major site of activation of angiotensin II, since plasma converting enzyme activity can account only for a small fraction of angiotensin II generated after infusion of angiotensin I. Other sites of generation of angiotensin II exist however, since, during cardiopulmonary bypass, angiotensin II generation proceeds in the absence of pulmonary circulation[7]. For bradykinin, the role of the lung does not appear as critical, the half-life of bradykinin in blood being rather short. Various substances like EDTA, 2-mercaptoethanol, and N-ethyl maleimide reduce ACE activity, but most of the interest has recently focused on peptides first isolated from the venom of Bothrops Jararaca, known as BPF (bradykinin potentiating factor), which act as specific inhibitors of ACE. Development of research has led to the synthesis of a new anti-hypertensive agent, D-3-mercapto-2-methyl propranoyl-L-proline, capable of inhibiting ACE activity even when given orally. ACE has been isolated and purified by several groups of investigators and shown to be a glycoprotein with a M.W. of 130 000 - 140 000[19].

The last type of substrate to be discussed is ATP which undergoes first hydrolysis and then uptake[4]. ATP is first transformed
into adenosine by adenosinetriphosphatase and 5-nucleotidase, ectoenzymes found in endothelial cells, and adenosine is
rephosphorylated into nucleotides, but, depending on adenosine
concentration, some of it can also be deaminated under the action
of adenosine deaminase, with the subsequent formation of inosine
and hypoxanthine. The uptake step is blocked by dipyridamol, a
powerful inhibitor of adenosine transport. Unlike the other
processes described above, ATP hydrolysis and adenosine uptake are
not specific properties of endothelial cells, but are present in
most cellular populations.

Now that I have generally reviewed various phenomena occurring
in the pulmonary circulation, it is appropriate to discuss their
physiological importance by asking the following questions.

1   Do all endothelial cells share the same properties?

The answer is yes and no. Yes, as far as angiotensin
coverting enzyme is concerned since immunofluorescent studies have
confirmed its localization in the vessels of every organ tested.
My collaborators and I have measured identical enzyme activities in
endothelial cells isolated from the pulmonary artery and the aorta.
This generally is apparently true, too, for 5-HT, whose uptake is
evenly spread along the pulmonary vascular bed and is also found in
endothelial cells freshly isolated from both the pulmonary artery
and the aorta. From 1-NE data, on the other hand, there is
evidence of heterogeneity of function of the endothelium since its
uptake is limited to certain segments of the pulmonary vascular
tree and cannot be found in the endothelium of large vessels.
Finally, if the site of uptake of PGs in the lung is likely to be
the endothelial cell, this has not yet been established; the only
experimental fact reported so far is that freshly isolated and
cultured endothelial cells from large pulmonary and systemic
vessels do not exhibit this phenomenon[15]. The specific features of
pulmonary endothelium are its strategic location and its density.

2   Are these processes involved in the regulation of the plasma
       concentration of vasoactive substances?

It is difficult to give a precise answer to this question.
The activity of ACE in the lung appears to be controlled by the
oxygen partial pressure, as ACE activity decreases with lower
PO_2[13]. But whether this decrease represents a process designed to
regulate angiotensin II and bradykinin concentrations cannot yet be
determined with certainty. It is probably fair to say that plasma
concentrations of these substances are considerable altered during

their passage through the lung, but that does not mean that the lung exerts a true regulatory action, since the lung cannot control the conditions known to affect these extraction or activation processes: no physiological action, except for the effect of low PO_2 on ACE, is known to modulate the role of the lung in the disposal of these vasoactive substances. The only variable that has been shown to affect the magnitude of the lung effect is the amount of available pulmonary vascular surface[6], but, again, it is difficult to attribute to this factor an active regulatory role. In cases when pulmonary hypertension or emphysema, there has been no report on marked disturbances in the control of blood pressure or coagulation.

3   Can these properties of pulmonary vascular endothelial cells
     be used to assess endothelial function?

This question, I believe, is the expression of the most interesting, although indirect, application of this research field. Data already suggest that the magnitude of l-NE or 5-HT extraction by the lung or angiotensin II generation is modulated as a function of the pulmonary vascular area and of the metabolic state of the endothelium. Treatment of rats with monocrotaline results in pulmonary hypertension, with a concomitant decrease in 5-HT and l-NE uptakes[8]. Exposure of rats to hyperoxia leads, in 18 hours, to a significant reduction of 5-HT uptake by the lung, before any histologic change can be detected[3]. It appears therefore possible, by the use of appropriate markers, to test the functional integrity of the pulmonary vascular endothelium. The main problem to be solved is the distinction between a decrease in the pulmonary vascular surface, with normal endothelial cells, and a dysfunction (or qualitative defect) of the endothelium. The use of other substrates such as basic lipophilic amines (imipramine, propranolol) which are accumulated by the lung without the need for the existence of an energy requiring transport step and could therefore be considered as markers of endothelial surface, is of great potential interest. The complexity of this accumulation process, made of at least two main components (partitioning process and saturable binding) requires, however, additional experiments before this approach can be validated. In the meantime, thanks to the combined efforts made by different groups of investigators working in various directions, we see the emergence of new concepts emphasizing the involvement of endothelial cells in several pathological conditions. The nearly concomitant reports on the mechanisms of lung adaptation to hyperoxia and on the vulnerability of endothelial cells to the toxic effect of oxygen, have stimulated studies on the antioxidant defenses of endothelial cells. This experimental situation, in which oxygen radicals are thought to play a major role, is thus closely related to another protocol testing the cytotoxic effect of activated neutrophils on endothelial cells[18] through the release of the same O_2 radicals.

The road from the discovery of extraction of 5-HT by the lung to the study of endothelial function in general inflammation may appear somewhat tortuous, but is in fact the result of a logical sequence of assumptions and experimental confirmations. The example of a research effort going from the study of an apparent organ specific property to that of a general biological phenomenon is rare enough to deserve mention.

References

1. R.R. Ben-Harari and Y.S. Bakhle. Uptake of — phenylethylamine in rat isolated lung. Biochem. Pharmacol. 29:489 (1980).

2. L.Z. Bito, R.A. Baroody and M.E. Reitz. Dependence of pulmonary prostaglandin metabolism on carsier-mediated transport process. Am. J. Physiol. 232:382 (1977).

3. E.R. Block and A.B. Fisher. Depression of serotonin clearance by rat lungs during oxygen exposure. J. Appl. Physiol. 42:33 (1977).

4. Y. Dieterle, C. Ody, A. Ehrensberger, H. Stalder, and A.F. Junod. Metabolism and uptake of adenosine triphosphate and adenosine by porcine aortic and pulmonary endothelial cells and fibroblasts in culture. Circ. Res. 42:869 (1978).

5. E.G. Erdos. Angiotensin I converting enzymes. Circ. Res. 36:247 (1975).

6. B.L. Fanburg and J.B. Glazier. Coversion of angiotensin I to angiotensin II in the isolated perfused dog lung. J. Appl. Physiol. 35:325 (1973).

7. L. Favre, M.B. Vallotton, and A.F. Muller. Relationship between plasma concentration of angiotensin I, angiotensin II and plasma renin activity during cardiopulmonary bypass. Eur. J. Cl;in. Inv. 4:135 (1974).

8. C.N. Gillis, R.J. Huxtable, and R.A. Roth. Effects of monocrotaline pretreatment of rats on removal of 5-hydroxy-tryptamine and noradrenaline by perfused lung. Br. J. Pharmac. 63:435 (1978).

9. H.J. Hawkins, A.G.E. Wilson, M.W. Anderson, and T.E. Eling. Uptake and metabolism of prostaglandins by isolated perfused lung: species comparisons and the role of plasma protein binding. Prostaglandins 14:251 (1977).

10. Y. Iwasawa, C.N. Gillis, and G. Aghajanian. Hypothermic inhibition of 5-hydroxytryptamine and norepinephrine uptake by lung: cellular location of amines after uptake. J. Pharmacol. Exp. Ther. 186:498 (1973).

11. Y. Iwasawa, C.N. Gillis. Pharmacological analysis of norepinephrine and 5-hydroxytryptamine removal from the pulmonary circulation: differentiation of uptake sites for each amine. J. Pharmacol. Exp. Ther. 188:386 (1974).

12. A.F. Junod.   Uptake, metabolism and efflux of ^{14}C-5-hydroxytryptamine in isolated perfused rat lungs. J. Pharmacol. Exp. Ther. 183:341 (1972).

13. P.J. Leuenberger, S.G. Stalcup, R.B. Mellins, L.M. Greenbaum and G.M. Turino.   Decrease in angiotensin I conversion by acute hypoxia in dogs. Proc. Soc. Exp. Biol. Med. 158:586 (1978).

14. T.E. Nicholas, J.M. Strum, L.S. Angelo, and A.F. Junod.   Site and mechanism of uptake of ^{3}H- -norepinephrine by isolated perfused rat lungs. Circ. Res. 35:670 (1974).

15. C. Ody, Y. Dieterle, I. Wand, H. Stalder, and A.F. Junod. PGA_1 and PGF_2 metabolism by pig pulmonary endothelium, smooth muscle and fibroblasts. J. Appl. Physiol. 46:211 (1979).

16. J.A. Roth and C.N. Gillis.   Multiple forms of amine oxidase in perfused rabbit lung. J. Pharmacol. Exp. Ther. 194:537 (1975).

17. U.S. Ryan, J.W. Ryan, C. Whitaker, and A. Chiu.   Localization of angiotensin converting enzyme (Kinase MI).   II Mmmunocytochemistryand immunofluosescence. Tmssue and Cell 8:125 (1976).

18. T. Slacks, C.F. Moldow P.R. Craddock, T.K. Bowers, and H.S. Jacob.   Oxygen radicals mediate endothelial cell damage by complement-stimulated granulocytes.   An in vitro model of immune vascular damage. J. Clin. Invest. 61:1161 (1978).

19. R.L. soffer, R. Reza, and P.R.B. Caldwell.   Angiotensin converting enzyme from rabbit pulmonary particles. Proc. Natl. Acad. Sci. USA 71:1720 (1974).

20. J.M. Strum and A.F. Junod.   Radioautographic demonstration of 5-hydroxytryptamine - ^{3}H uptake by pulmonary endothelial cells. J. Cell. Biol. 54:456 (1972).

DISCUSSION

LECTURER: Junod   CHAIRMAN: Cumming

ROSSI: Is there any evidence that the cytotoxic agents
                which are particularly toxic for the lung, are
                picked up by endothelial cells in the lung or not?
                I mean bleomycin, bisolvon and maybe also
                methotrexate.   Also is it a different kind of
                reaction because it seems to be a hypersensitivity
                instead of a toxic mediator through oxygen
                radicals.

JUNOD: It is possible that there is uptake of these drugs
                by endothelial cells both in a perfused lung system
                and in an in vitro system with cultured endothelial
                cells. This is a possibility but I think there is
                no evidence that confirms this possibility.

CUMMING: I think Herve Sors may be able to help us on this
                question. No? Sorry, then ask your question.

SORS: I would like to ask two questions. You have shown
                that in vitro endothelial cells do not take up
                prostaglandins and we know that this phenomenon can
                probably occur also in vivo. Do you have any
                explanation for this discrepancy?   The second
                point: do you have any notion whether low partial
                pressures of oxygen can change anything in the
                uptake of substances in the lung? I have in mind a
                work of Guillos showing that prostaglandin uptake
                in vivo could be lowered in some broncho pulmonary
                diseases with a low PO_2.

JUNOD: The first question about the uptake of
                prostaglandins by endothelial cells, there is a
                basic discrepancy between what you said and what I
                have said.   That could be explained by the
                assumption that capillary endothelial cells do take
                up prostaglandins, but that endothelial cells taken
                from large vessels, and that's what we do, do not
                take up prostaglandin.   That would be indirect
                evidence of the specificity of function of
                capillary endothelial cells toward prostaglandins,
                but so far there has been no experimental
                confirmation that capillary endothelial cells do
                take up prostaglandin, although is very likely.
                The second question related to the effect of a high

PO$_2$ or a low PO$_2$?   I am not aware of the work by Guillos on prostaglandin inactivation by the lung at low PO$_2$;   I think if you do that again, there would be a question as to whether there is a decrease in the pulmonary vascular surface, or whether the reduction in prostaglandin clearance stems from a dysfunction of a normal number of endothelial cells;   so far nobody has been able to offer any satisfactory answer. It has been shown that in rat lung exposed to hypoxia there is decreased inactivation of prostaglandin, as it was shown to be the case for tryptamin.

KAY:

I have a comment and a question please. Fairly recently we did some experiments in which we were able to demonstrate that histamine receptors were particularly well expressed on alveolar macrophages in the guinea-pig and that following the proper stimulus one could demonstrate formation of partial reductive products of oxygen including the superoxide radicals using familiar methods. We speculated at the time that this may have a deleterious effect on pulmonary tissue. Unfortunately the young lady who was doing these experiments had to go back to Chile and we never got around to doing them. But what we intended to do was to take hydrogen peroxide generators such as glucose-glucose oxidase or xanthine-xanthine oxidase to see whether any particular pulmonary cell was susceptible to partial reductive products prepared in that way. I wonder if you have done that type of experiment.

JUNOD:

Yes, that's what we intend to do now but with endothelial cells it is not so easy a task. One problem is that under most conditions oxygen radicals are generated inside the cell and when we add exogenous oxygen radicals or hydrogen perioxides it is not exactly the same situation. A second is that what we see if there is release of oxygen radicals by cells, it means that this is the normal flow of what has been produced inside the cells and we see only the net outflow of these oxygen radicals. You can approach this problem by using various indicators but when you add them to the external media it doesn't mean that they have access to the inside of the cell and so you can only use an indirect approach. We intend to do this with endothelial cells,   and it would be

interesting to do it also with other pulmonary
cells if they can be isolated. Your question
raises the important feature of associating
different cells in the lung and I believe many
people have tried very hard to isolate cells but it
is remarkable in the lung that there are so many
relationships between other cells.

CORRIN: I have two questions. When you quote figures for
the uptake of circulating substances, is it
customary to quote figures for a single passage
through the pulmonary circulation or is this the
figure which might be ultimately achieved when the
blood circulates many times through the lung?

JUNOD: This is a multiple passage phenomenon.

CORRIN: The second question: you have compared certain
aspects of the pulmonary artery with the aorta and
at least in one respect you can detect no
differences. Can either you or Professor Heath
tell us whether the pits or caveolae that were
shown in the pulmonary arteries are also present in
systemic vessels or are they unique to the lung?

JUNOD: We have seen the same caveolae in isolated
endothelial cells whether they came from pulmonary
artery or the aorta. I think, as far as converting
enxyme is concerned, this activity of diapeptidase
is found all over the endothelial surface, but is
limited to the caveolae at least in endothelial
cells, but what is not true for other cells is the
presence of five prime nucleotides.

CUMMING: If the caveolae are seen both in the pulmonary
circulation and in the systemic circulation, can
you attribute the specific metabolic activity to
the caveolae?

JUNOD: No, they are not, and in fact for converting enzyme
there is no difference between pulmonary
circulation and systemic circulation.

CUMMING: Can you hazard any guess about specific activity in
the pulmonary circulation, and where that might be
sited?

JUNOD: I think there is very little specific in the
pulmonary endothelium as far as I am aware. PO_2

for example is also responded to endothelial cells originating from systemic vessels. In everything that we have tested so far we have got similar results from aortic and pulmonary endothelial cells; what might be different are the end links of some specific substrates depending on the localisation of the endothelium, namely there is a difference but it may be different between capillaries and large vessels, and that the function of capillaries may be much more important than the function of the endothelium in large vessels. But qualitatively speaking I think we have found no evidence of a specific function of the pulmonary endothelium.

BAUM: As you know, Glutathione perioxidase is a selenium dependent enzyme. I wonder whether you know if there is any evidence at all in those parts of the world where selenium deficiency has been shown in man, for example in regions of China, whether there is any evidence there of any kind of pulmonary pathology.

JUNOD: Firstly, I'd like to make a comment on the slide that I showed to you where there was no difference in Glutathione perioxidase between control and hypoxic cells, but both these cells showed a marked decrease in Glutathione perioxidase and this is due to the fact that 1% of the complement will switch in a 2% medium and at that time you have determined the concentration of selenium dependent Glutothione perioxidase decrease. Since this decrease was observed both in normal and hypoxic cells the toxicity of oxygen is unlikely to be related to this drop of glutathione perioxidase. My answer therefore would be that it is unlikely to be the case, or maybe there is a compensation by an increase in the non selenium-dependent form of glutathione perioxidase. It has been shown that in experimental animals with a diet depleted of selenium that they are more susceptible to the effect of hypoxia but again I think it depends very much on the possibility of compensation by the other enzymes.

SORS: Do you have any explanation for the vasoconstrictive reaction to hypoxia? Do you think it may be related to the prostaglandin metabolism or to the angiotensin converting enzyme?

JUNOD: A decrease in diapeptide converting peptidase
                resulting from hypoxia would inhibit the formation
                of a vasoconstrictive substance and further the
                maintenance of vasodilator substance, so hypoxia
                (if you believe that it is affected through
                inhibition of converting enzyme) would have the
                opposite effect and would lead to a further
                vasodilation. As far as I know, there has been
                very little study on the conditions known to
                decrease the production of prostaglandins by the
                lung and I don't know if there is any good evidence
                that hypoxia per se leads to the generation of
                prostaglandins which would have a vasoconstrictor
                effect or would lead to a decrease in vasodilator
                prostaglandins that would then secondarily increase
                the pulmonary vascularity. What has been shown is
                that indomethacin can reduce the hypoxic
                vasoconstriction. it's difficult to know whether
                this is the direct effect of production of
                prostaglandins or whether it is mediated by another
                pharmacological action of indomethacin.

SORS: Maybe I could add a comment on this question.
                There is no evidence currently in the literature
                that prostaglandins mediate the vasoconstrictive
                effect of hypoxia. As Junod has told us
                indomethacin in some conditions and in some
                species, in very limited conditions, could induce
                prevention of hypoxic vasoconstriction response.
                We have ourselves performed some experiments in man
                submitted to hypoxic conditions and we couldn't
                detect any production of prostaglandin in various
                effluents in these conditions, nor can we prevent
                hypoxic vasoconstriction by cyclo oxygenase
                inhibitors. This was at a time when we have only
                measured primary prostaglandins and we would say
                that some other systems could be implicated.

KAY: I'd like to ask you another question about this
                intriguing relationship between superoxide and
                related compounds and possible damage to the
                pulmonary endothelium. You mentioned the
                respiratory burst of phagocytic cells and, as is
                well known, some sort of stimulus is required.
                With phagocytic cells the stimuli are varied
                through immunological receptors and the electron
                donor is generally thought to be an NADPH. In the
                endothelial cells what is the electron donor? Its

it related to NADP or NADPH, and is there any pathological defect in pulmonary endothelial cells in chronic granulomatous disease?

JUNOD: I cannot answer any of these questions.

KAY: When animals were exposed to 95% oxygen or 80% oxygen, I believe they showed an adaptation and protection of toxicity, which is suggested was due to increase in superoxide dysmutase. How much of that increase in superoxide dysmutase is located or localised in the endothelial cells and how much might be in other cell types in the lung?

JUNOD: Nobody knows. The only thing that I can say is that endothelial cells could account for this increase. I cannot make any prediction as to whether endothelial cells participate in that increase; what is known is that alveolar macrophages when exposed to hypoxia do not increase superoxide dysmutase activity. The interesting feature about this is that the lungs showed marked damage and mathematic measurements showed that the lungs had become adapted to hypoxia and there was a marked reduction in the number of endothelial cells. The question was also raised as to whether there are two different types of cells, one that has a low enzymic activity and dies and only the cell with higher enzymic activity survives, and that will be a process of selection rather than one of induction. So far the question has not been solved.

KAY: Could I ask you to expand on that, and try and relate it to paraquat toxicity where, if you expose the animals to oxygen you get a protective effect and there the damage is more to type I and type 2 cells rather than to the pulmonary endothelium.

JUNOD: The problem with respiratory pathology is that once the endothelial cells have been damaged there are associated changes in different cells. It is difficult to know, for example with hypoxia whether the first cell to be damaged is the endothelium and then after a few days, depending on the species, there might be marked changes affecting pneumocytes, type 2 cells. The lung responds in a fairly non-specific fashion in general and I would be very reluctant to assume tht there has been no damage to endothelial cells by these drugs. There is however some interesting statement to make, for example when you exposed the rats to hypoxia and they receive superoxide dysmutase prior to the hypoxia, they do not show any decrease in clearance, there is a protective effect of superoxide dysmutase. I don't know whether superoxide dysmutase has access to endothelial cells; if it does not, that could mean that in an in-vivo situation, the damage due to hypoxia would result from the exogenous supply of these oxygen radicals of superoxides; but in rats exposed to paraquat, superoxide dysmutase has no protective effect and it's very likely that the production of superoxide radicals is an endogenous process when superoxide dysmutase given intravenously has no protective effect.

CUMMING: I think that at this stage I should bring the proceedings to a close, reminding you that lunch is now available in San Rocco and the coaches will leave from the square at 1.15 p.m.

FEYRTER CELLS

W. Taylor

Department of Pathology
University of Liverpool
P.O. Box 147, Liverpool L69 3BX

The lung endocrine or Feyrter cells (Feyrter, 1954) may be identified with the light microscope by means of their argyrophilia. Argyrophil cells are capable of reducing silver salts to metallic silver in the presence of an additional weak reducing agent. The most commonly employed technique is that of Grimelius (1968). The black silver granules which indicate a positive result are easily seen in the sections (Fig. 1). The electron microscope demonstrates dense core secretory vesicles measuring about 120nm in diameter in the cytoplasm of these cells (Fig. 2). These vesicles are characteristic of cells which produce polypeptides. By demonstrating the silver granules deposited over the secretory vesicles, electron microscopy of silver-stained material has confirmed that the reducing power of such cells is localised in the vesicles (Creutzfeld et al., 1975). Substances identified in these cells include dopamine (Hage, 1972) and 5-hydroxytryptamine (Lauweryns et al., 1974). Bombesin-like activity has been detected in human foetal and neonatal lung by Wharton et al., (1978), and probably originates in these cells. Argyrophilic endocrine cells are thus present in the lung in large numbers but we know little of their function.

This paper is concerned with two areas. Firstly the possibly physiological role of the lung argyrophil cells as illustrated by the effects of high altitude hypoxia, and secondly a discussion of one of the pathological lesions to which they give rise, the pulmonary tumourlet.

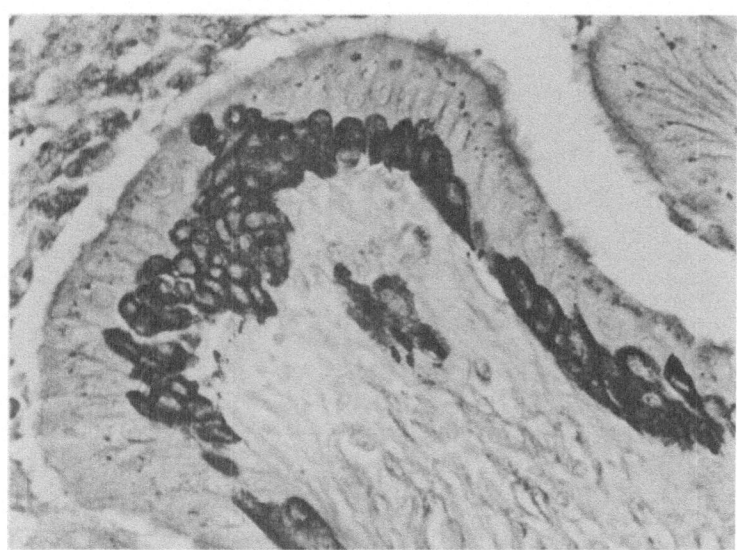

Fig. 1.   Small bronchus in human lung. Large numbers of darkly
          stained argyrophil cells are present. Grimelius stain.
          By courtesy of the Journal of Pathology.

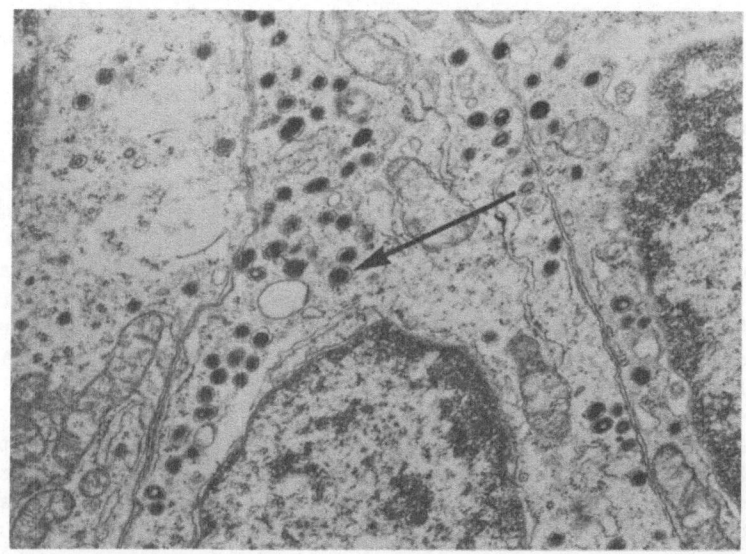

Fig. 2.   Rabbit lung. Dense core secretory vesicles (arrow) in the
          cytoplasm of a group of argyrophil cells in a bronchiole.
          Electron micrograph. By courtesy of the Journal of
          Pathology.

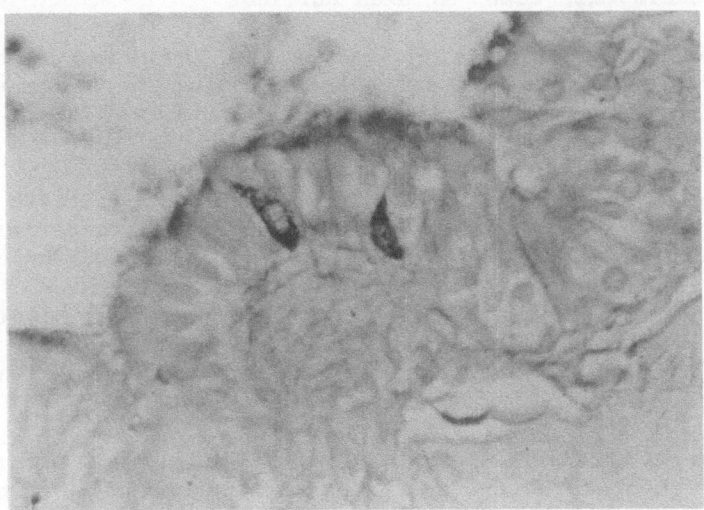

Fig. 3.   Sea level rabbit lung.   Two individual argyrophil cells in
          bronchial epithelium.   A thin, tapering process extends
          towards the free surface of the epithelium.   Grimelius
          stain.

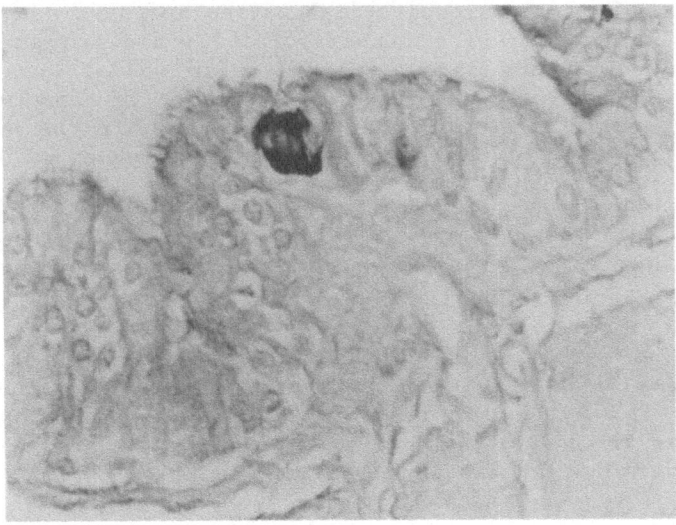

Fig. 4.   Sea level rabbit lung.   A group of argyrophil cells in a
          small bronchus.   Groups of cells have a corpuscular
          structure.   Grimelius stain.

PULMONARY ARGYROPHIL CELLS AT HIGH ALTITUDE

At high altitudes such as in the Peruvian Andes, the
barometric pressure is low and animals living there are subjected
to chronic hypoxia. The native high altitude dweller copes with
this chronic hypoxia remarkably well and is said to exhibit
"natural acclimatisation" (Heath and Williams, 1977). The basis of
this "natural acclimatisation" is not understood but it is possible
that it involves inherited as well as acquired characteristics. If
lung argyrophil cells have any function in relation to hypoxia then
they might be expected to be more common in conditions of chronic
hypoxia. The following experiment compared the frequency of
argyrophil cells in naturally acclimatised high altitude rabbits
living 4300 m above sea level in the Peruvian Andes, with the
frequency in control rabbits living at sea level.

Six high altitude rabbits were killed by intraperitoneal
injection of pentobarbitone sodium, their lungs rapidly removed and
cut into blocks 3-4 mm thick approximately at right angles to the
major airways. The blocks were fixed in Bouin's solution for 48
hours and then transferred to 70 per cent alcohol for storage until
they could be processed into paraffin wax. Six New Zealand white
rabbits of corresponding ages were used as sea level controls.
Sections were stained by the Grimelius argyrophil technique. In
the rabbit, argyrophil cells occur as individual cells (Fig. 3) and
as groups of cells (Fig. 4) in both the bronchial tree and the
alveolar walls. Lauweryns and Cockelaere (1973) have referred to
the groups of cells as neuroepithelial bodies and have demonstrated
a nerve supply to them. The numbers of cells and groups of cells
in the airways and lung parenchyma (alveolar walls) were counted
and the area of each irregularly-shaped section of lung was found
using a point-counting technique. For each rabbit the number of
cells and groups of cells were expressed per square cm of tissue.

In the lung parenchyma (Fig. 5) there were more individual
cells per square cm of section in the hypoxic test animals than
in the sea level controls. Using a ranking test of significance
the difference between the test and control groups was significant
at the 1 per cent level. In the airways (Fig. 6) the test animals
contained more groups of cells than the controls and this
difference was significant at the 5 per cent level. Individual
cells in the airways and groups of cells in the parenchyma were
also commoner in the test animals than in the controls, but the
differences were not statistically significant.

The presence of more argyrophil cells in the high altitude
rabbits than in the sea level controls could be inherited or
acquired. If inherited, it could be a genetic characteristic of
the native high altitude rabbit, i.e. a result of natural selection
occurring over the many years that Peruvian rabbits have lived at

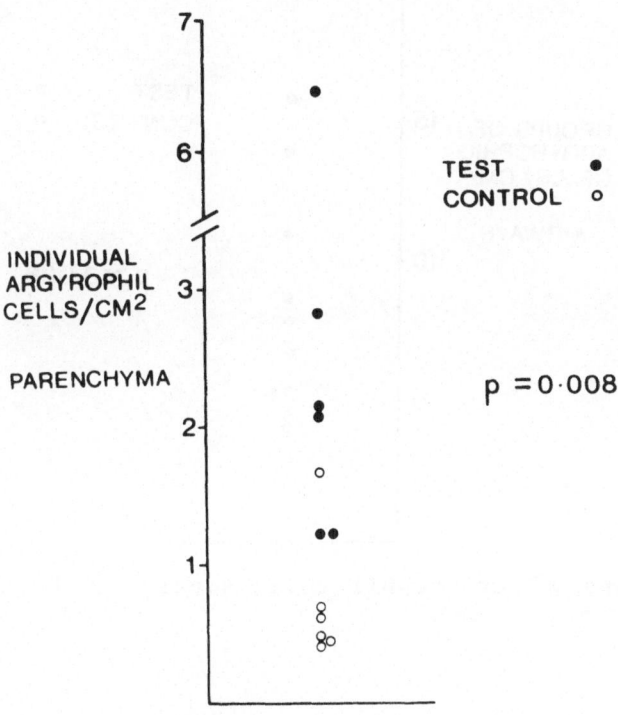

Fig. 5. Individual argyrophil cells/square cm in the lung
         parenchyma (alveolar walls).

high altitude. If acquired, it is probably the result of
hyperplasia in early life. The outstanding environmental
difference between the test and control animals was the hypoxia of
high altitude. It is suggested that the difference in incidence of
argyrophil cells in test and control animals, whether it is
inherited or acquired, is an effect of hypoxia.

     If this difference was acquired, it would probably occur
during the neonatal period, rather than in utero. The intrauterine
environment is probably no more hypoxaemic at high altitude than it
is at sea level. This is because the effects of the hypoxic high
altitude environment are compensated for by an abnormally heavy
placenta which probably has a greater density of histological
components (Heath and Williams). Lauweryns and Cokelaere have put
forward the hypothesis that pulmonary argyrophil cells act as
hypoxia-sensitive chemo-receptors. They propose that groups of
argyrophil cells, to which they have demonstrated a nerve supply,

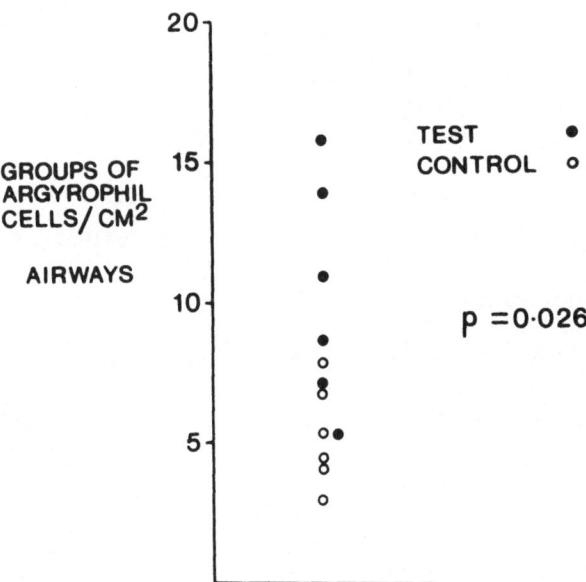

Fig. 6. Groups of argyrophil cells/square cm in the airways (bronchial tree).

detect hypoxia in the bronchial tree and alveoli and release a polypeptide in response to it. This polypeptide could then mediate some local response, eg. the vasoconstrictor response to hypoxia. In this context Lauweryns and Cokelaere (1973) have demonstrated that in acute experimental hypoxia the dense core vesicles of the cells undergo exocytosis across the bronchial basement membrane. The results obtained on hypoxic high altitude animals support this hypothesis that the lung argyrophil cells act as hypoxia-sensitive chemoreceptors.

PULMONARY TUMOURLETS

Although little is known of the physiology of the argyrophil cell in the lung, this cell is known to give rise to at least three different pathological lesions - the small cell or oat cell carcinoma, the carcinoid tumour and the tumourlet. The term tumourlet was first used by Whitwell to describe lesions which he found adjacent to bronchietatic airways in lobectomy specimens (Whitwell, 1955), a site which might be expected to be hypoxic. Tumourlets also occur in the periphery of lungs without other major

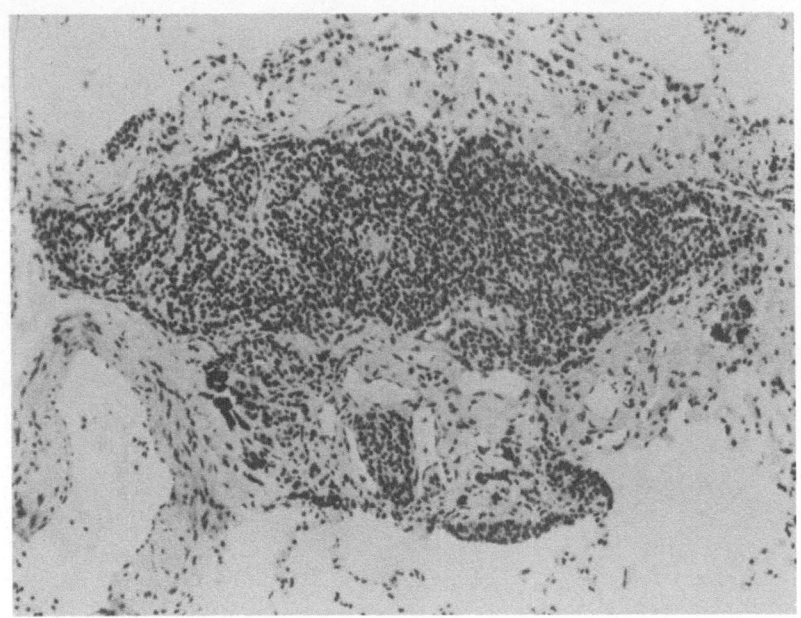

Fig. 7.   Tumourlet in the lung of a 75 year old woman.   It consists
          of nodules of cells, with uniform rounded nuclei, embedded
          in connective tissue.   HE.

pathology.   They consist of nodules of cells embedded in connective
tissue (Fig. 7).   The cells have rounded, oval or spindle-shaped
nuclei and a moderate amount of eosinophilic cytoplasm.   They often
form a layer on the external surface of the connective tissue.   The
Grimelius stain (Fig. 8) demonstrates that the tumourlet cells are
argyrophilic.   Examination of the same tumourlet at a lower
magnification illustrates that it is closely related to a pulmonary
artery and to a terminal bronchiole (Fig. 9).   An elastic stain on
an adjacent section demonstrates that much of the tumourlet is
enclosed within the elastic lamina of the adjacent terminal
bronchiole (Fig. 10).   The tongue of cells which extends out from
the terminal bronchiole in Fig. 10 is the branch of the bronchiole
which is giving rise to the tumourlet.   The cells which are lining
the parent bronchiole where it is close to the tumourlet are
argyrophilic (Fig. 11).   They extend around much of the lumen of
the parent bronchiole and are clearly demarcated from the ciliated
cells.   It is likely that at least some of the changes referred to
in the past as basal cell hyperplasia are really hyperplasia of
argyrophil cells (Auerbach et al., 1956).

    Many of the nodules which constitute the tumourlet are still
within the elastic tissue of the old bronchiolar wall.   Some of
the nodules, however, are outside its limits.   This must be so in
the case of groups of tumourlet cells on the external surface of

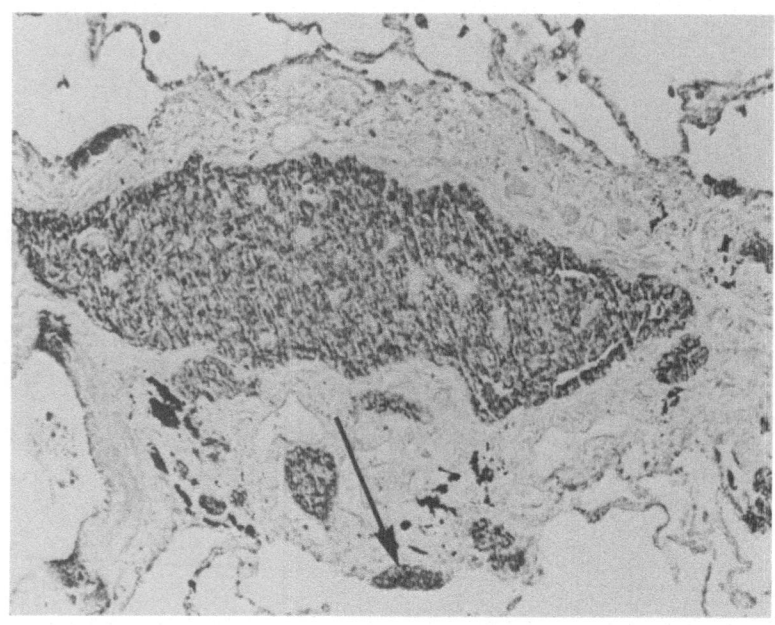

Fig. 8.   Tumourlet.   The same field as Fig. 8.   The arrow indicates
a layer of cells on the external surface of the connective
tissue.   Grimelius stain.

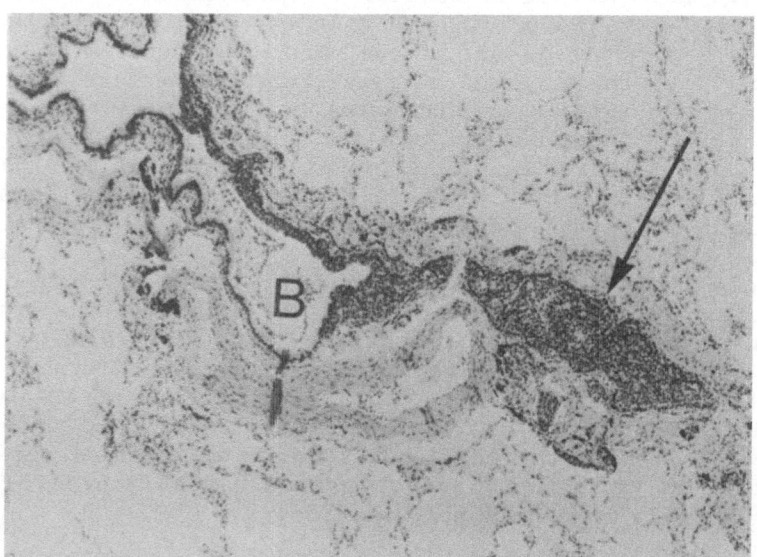

Fig. 9.   Tumourlet (arrow) with adjacent pulmonary artery and
terminal bronchiole (B) HE.

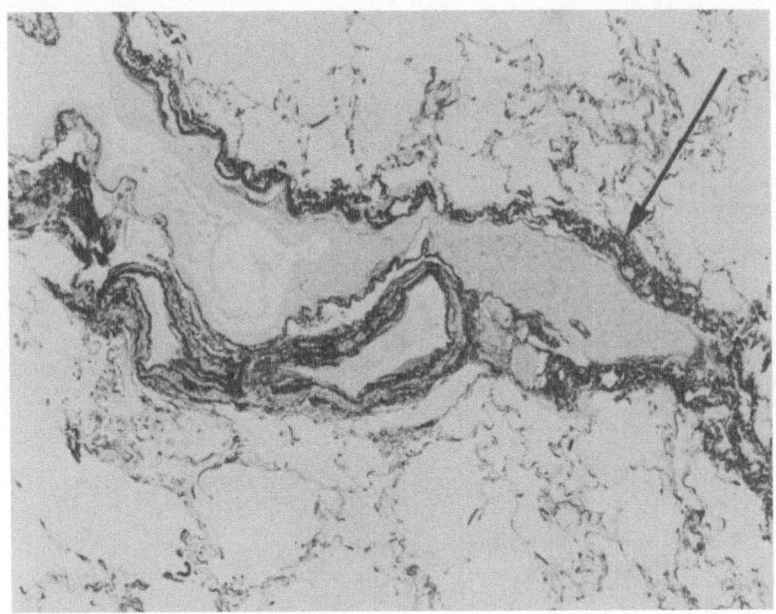

Fig. 10. The same field as Fig. 9.   Much of the tumourlet is
enclosed within the elastic tissue (arrow) of the adjacent
terminal bronchiole.   Elastic van Gieson.

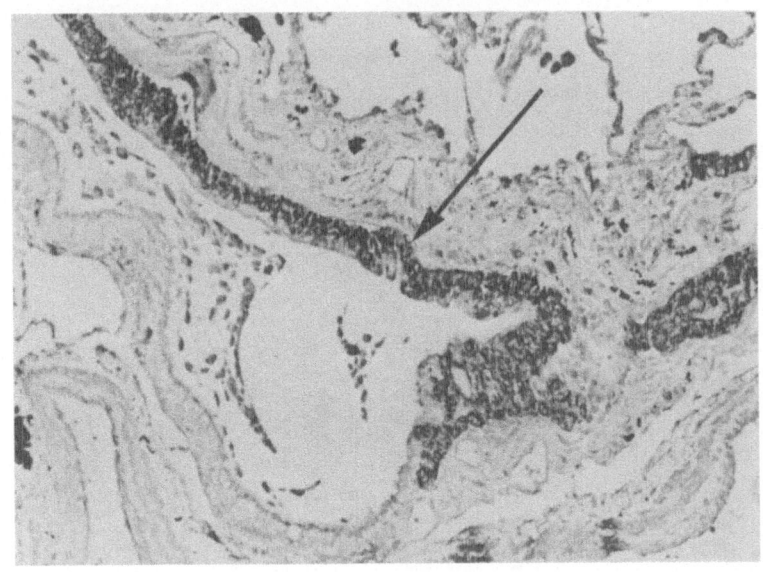

Fig. 11. Terminal bronchiole seen in Fig. 9, showing argyrophil
cells (arrow) extending around much of its lume.
Grimelius stain.

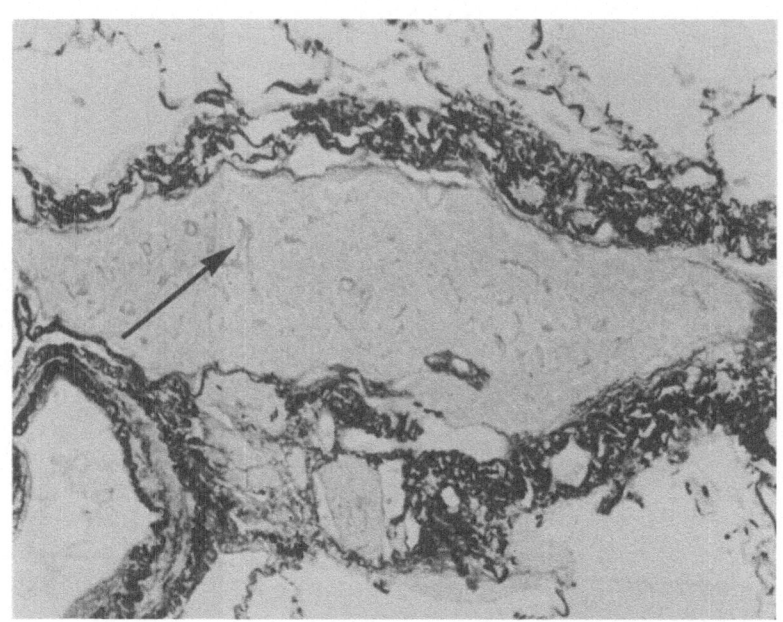

Fig. 12. Capillaries (arrow) are present among the hyperplastic
cells of this large tumourlet nodule which lies within the
old bronchiolar wall.   Elastic van Gieson.

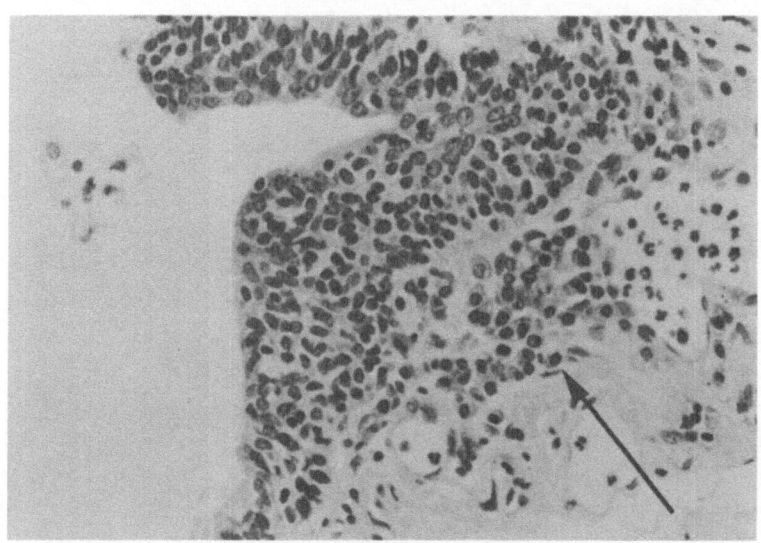

Fig. 13. Hyperplastic argyrophilic bronchiolar cells.   In the
submucosal tissue there are a dilated vessel and an
infiltrate of plasma cells (arrow). HE.

the bronchovascular connective tissue (Fig. 7).   Serial sections
demonstrate continuity of these cells with the intrabronchiolar
cell mass (Ranchod, 1977).   Capillaries are present in the
tumourlet nodules which lie within the old bronchiolar wall (Fig.
12) and migrate in among the proliferating bronchiolar cells from
the dilated submucosal vessels (Fig. 13). Some of the connective
tissue in which the tumourlet nodules are embedded is within the
old bronchiolar wall so presumably fibroblasts have migrated in
also.   There is often a substantial inflammatory cell infiltrate,
mainly of plasma cells, in the connective tissue beneath the
bronchiolar basement membrane (Fig. 13), and some of these
inflammatory cells invade the hyperplastic mucosa. With vessels,
inflammatory cells and fibroblasts penetrating the bronchiolar
basement membrane it is not difficult to envisage groups of
hyperplastic argyrophil cells extending outside the normal limits
of the basement membrane. The elastic tissue of the bronchiolar
wall does not form a continuous covering and hyperplastic
argyrophil cells can clearly extend through gaps in it to form
nodules in the adjacent peribronchial connective tissue.
Eventually a lumen reappears in the largest tumourlet nodule and
becomes progressively wider as the bronchiole returns towards a
normal structure (Fig. 14).

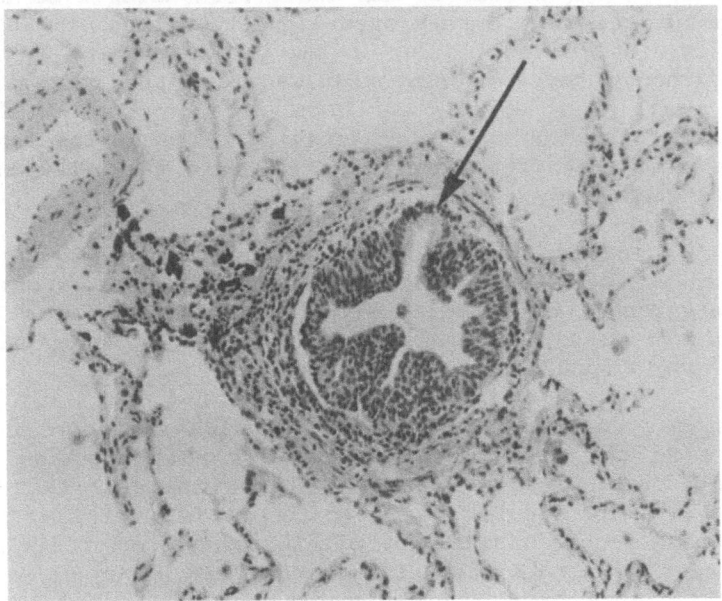

Fig. 14.   The lumen of the bronchiole is reestablished and although
            some hyperplastic basal cells remain, ciliated cells have
            reappeared (arrow) HE.

At least some tumourlets are thus derived from basal cell or argyrophil cell hyperplasia in a terminal bronchiole, but it is not entirely clear from examination of the morphology of the tumourlet whether it is a hyperplastic lesion or neoplastic one. Churg and Warnock (1976) have described tumourlets as peripheral pulmonary carcinoid tumours, and a few cases of metastases to regional lymph nodes have been recorded (Hausman and Weimann, 1967). These cases must be regarded as carcinoid tumours. Such cases are uncommon however, and, at least in the case of tumourlets arising in the periphery of relatively normal lung, tumourlets should be regarded as hyperplastic rather than neoplastic basal argyrophil cells in bronchioles. Such hyperplasia is a common and probably non-specific response to a wide variety of injuries. In the series which follows, tumourlets were found in about a quarter of the elderly population of Liverpool coming to post mortem examination. It is difficult to believe that such a high proportion of elderley persons have pulmonary carcinoid tumours. This series is small – 25 cases of which 6 had tumourlets, but even if the proportion of tumourlets were less than a quarter, many more cases than that have areas of argyrophil cell hyperplasia not amounting to tumourlets, but still extensive. Most pulmonary carcinoid tumours are situated in the proximal part of the bronchial tree and are within reach even of the rigid bronchoscope. Tumourlets are situated in the periphery of the lung close to the terminal bronchioles. Taking the frequency of these lesions and their peripheral site together, even if tumourlets were merely premalignant lesions, it would mean that most carcinoids would occur in the lung periphery, which they do not. Although the tumourlet nodules are intimately associated with many small blood vessels and lymphatics, in none of the cases examined has there been vascular invasion. The mediastinal lymph nodes were also free from metastases, as were the other organs of the body in each case.

If this kind of tumourlet is predominantly a hyperplastic rather than a neoplastic lesion, its presence may be used as an index of argyrophil cell hyperplasia. The following small series examines the frequency of tumourlets in various pathological states in which hypoxia would be expected to occur.

The series consists of 25 cases of which six cases contained tumourlets in the lungs. They were all patients who died in hospital and on whom post mortem examinations were carried out. The only criterion for inclusion in the series was that the post mortem examination be carried out within about five or six hours of death, so that autolysis was not too extensive. Even so, autolysis was found in some cases and this was usually associated with aspiration of gastric contents. The mean age of the subjects was 69 years with a range of 35-94 years. Cases with tumourlets a mean age of 69 years, with a range of 67-78 years, and cases without tumourlets a mean age of 67 years, with a range of 35-94 years.

The series included 13 males and 12 females. Four females and two males had tumourlets. The series is clearly too small for comment on the sex incidence of tumourlets, but Whitwell found that 70 per cent of his tumourlet cases were female. In the 25 cases of this series the percentage of emphysema in the lungs, the ratio of the cardiac ventricular weights and the percentage medial thickness of the muscular pulmonary arteries were measured. One lung from each case was distended with Bouin's solution and cut into 1 cm thick sagittal slices after 48 hours. Although Bouin's solution is unpleasant to work with, it is a more satisfactory fixative than formal saline for this purpose. It hardens the tissue, enabling the lung to be sliced sooner than would provide satisfactory results with formol saline. Simple distension provides adequate material and is less cumbersome than inflation with formaldehyde vapour. After cutting, the slices were left to fix for a further twelve hours before they were examined. The lung slices were then placed one at a time under water in a shallow container. Point counting of the emphysema was carried out using a perspex grid with holes drilled in it at the angles of equilateral triangles with sides 1 cm long (Dunnill, 1962). Twelve blocks of tissue were then cut randomly from the slices of distended lung. About 30 blocks were taken from the other, non-distended lung at the time of the post mortem examination and fixed in Bouin's solution. The blocks from each lung were processed into paraffin wax and 5µm sections cut and stained with haematoxylin and eosin, the elastic van Gieson stain and Grimelius stain.

The slides from the distended lung were used to measure the percentage medial thickness of twenty muscular pulmonary arteries. Muscular pulmonary arteries were taken to be vessels of external diameter between 100 and 1000um with two definite elastic lamine (Harris and Heath, 1977). Only vessels of approximate transverse section were measured. Two measurements of the diameter and four measurements of the thickness of the muscular media between the elastic laminae were taken in each artery using an eyepiece micrometer. The means of these measurements were used to calculate, for each artery, the percentage of the overall external diameter which was represented by the width of the muscular media. This is referred to as the percentage medial thickness of the muscular pulmonary arteries (MT per cent). From the values for the twenty arteries measured the mean MT per cent for the case was found. At the time of the necropsy, the left cardiac ventricle togehter with the intaventricular septum (LV + S), and the right cardiac ventricle (RV) were weighed and their ratio (LV + S : RV) calculated.

The blocks from both lungs were examined for tumourlets and their prescence or absence noted in each case. The argyrophil stain was much less effective on the distended lung. This has been noted previously in rabbit lungs. Most tumourlets were, however,

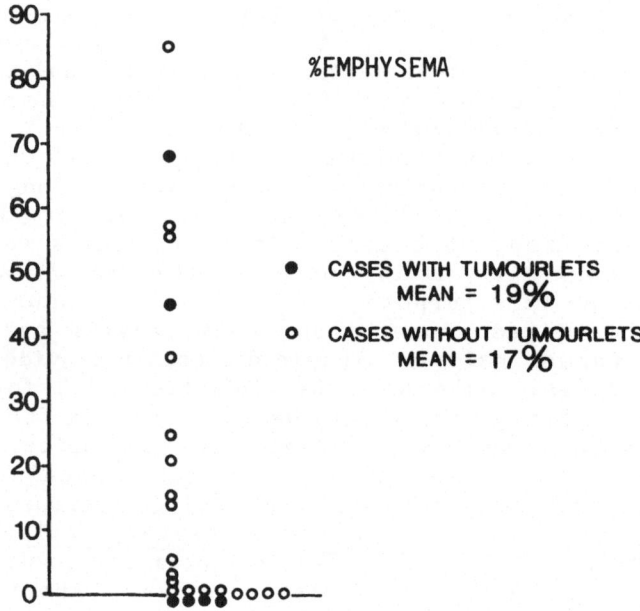

Fig. 15. Percentage of emphysema in cases with and without
tumourlets.

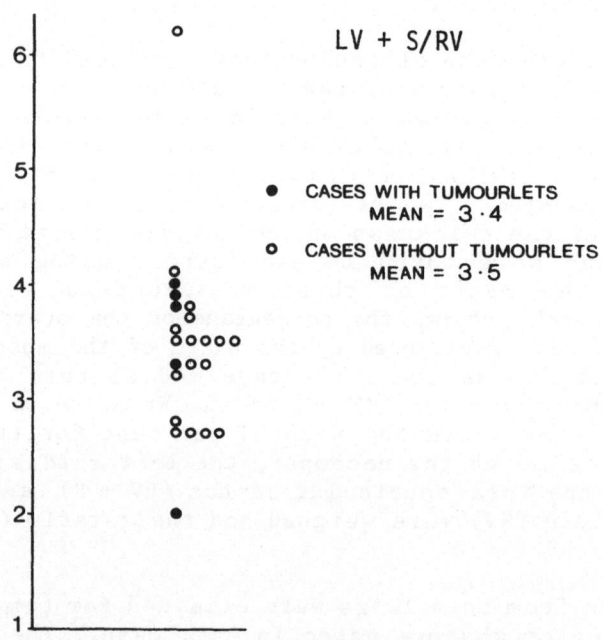

Fig. 16. Ratio (LV + S) : (RV) in cases with and without
tumourlets.

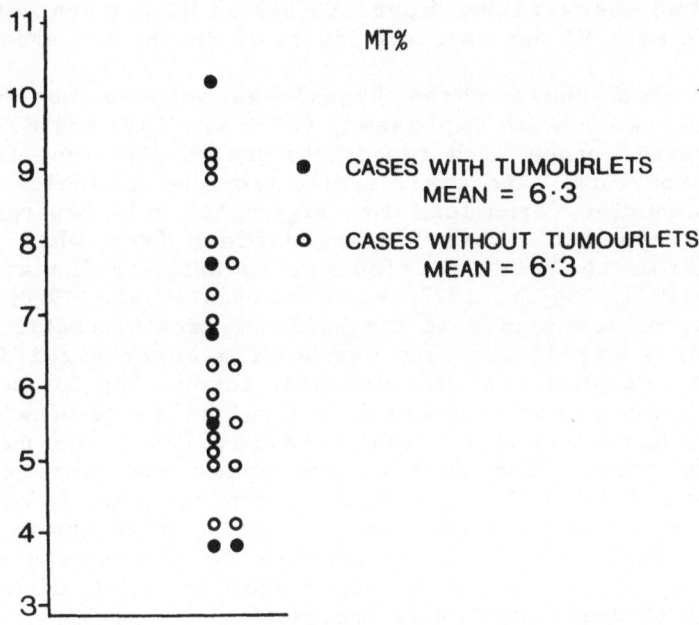

Fig. 17. MT per cent in cases with and without tumourlets.

found in the non-distended lung since more blocks were taken from that side.

Fig. 15. shows the percentage of emphysema in cases with and without tumourlets. Two cases with tumourlets had significant emphysema, the other four had non. The mean percentage of emphysema measured in the distended lung by point counting was almost the same in cases with tumourlets as without. Emphysema is known to be associated with hypoxia, or at least with intermitten hypoxic episodes which occur with recurrent bouts of infection and fluid retention (Whitaker, 1954).

The ratio (LV + S) : (RV) in cases with and without tumourlets is shown in Fig. 16. The normal ratio lies between 2.3:1 and 3.3:1 (Fulton et al., 1952). Values less than 2.3:1 indicate isolated right ventricular hypertrophy and are likely to be associated with hypoxia. Only one case fell into that category. Although that case contained numerous tumourlets, the mean ratios (LV + S) : (RV) for all cases with and without tumourlets were almost the same.

Fig. 17. shows the MT per cent for cases with and without tumourlets. Values greater than 7 per cent may be taken to indicate hypertrophy of the medial smooth muscle (Heath and Best, 1958) such as occurs in pulmonary hypertension and inhypoxia.

Although two cases with tumourlets had an MT per cent of greater than 7, the mean MT per cent was identical in the two groups.

Thus when these three hypoxia-associated morphological indices, percentage of emphysema, (LV + S) : (RV) and MT per cent are considered, cases with tumourlets are no different from those without tumourlets. The small series provides no indication that, in man, tumourlet formation, i.e. argyrophil cell hyperplasia, is induced by hypoxia. This finding differs from what might be expected from the work carried out on animals (Lauweryns and Cokelaere, 1973, Taylor, 1977, and Lauweryns et al., 1978) and from the finding of tumourlets in the walls of bronchiectatic cavities. Local hypoxia within the lung may be of greater significance in inducing argyrophil cell hyperplasia, accounting for cases with tumourlets who do not have morphological evidence of widespread hypoxia. More detailed analysis involving the numbers of tumourlets found, the type of emphysema and more extensive examination of the pulmonary vasculature may provide additional evidence for or against any association between tumourlets and hypoxia. The fact that tumourlets do sometimes occur in physiological degranulation is not a problem which might prevent recognition of small argyrophil lesions.

REFERENCES

Auerbach, O., Petrick, T.G., Stout, A.P., Statsinger, A.L., Muehsam, G.E., Forman, J.B., and Gere, J.B., 1956. The anatomical approach to the study of smoking and bronchogenic carcinoma. Cancer, 9:76.

Churg, A., and Warnock, Martha, 1976. Pulmonary tumourlet. A form of peripheral carcinoid, Cancer. 37:1469.

Creutzfeldt, W., Arnold, R., Creutzfeldt, Cora, and Track, N.S. 1975. Pathomorphologic, biochemical and diagnostic aspects of gastrinomas. (Zollinger-Ellison syndrome), Human Path., 6:47.

Dunhill, M.S., 1962. Quantitative methods in the study of pulmonary pathology. Thorax, 17:320.

Feyrter, F., 1954. Zur pathologie des argyrophilen helle-zellen-organes in bronchialbaum des menschen, Virchows Arch. Path. Anat. Physiol. 325:723.

Fulton, R.M., Hutchinson, E.C., and Morgan Jones, A. 1952. Ventricular weight in cardiac hypertrophy. Br. Heart J. 14:413.

Grimelius, L. 1968. A silver nitrate stain for a2 cells in human pancreatic islets, Acta. Soc. Med. Upsal. 73:243.

Hage, Esther, 1972. Endocrine cells in the bronchial mucosa of human foetuses. Acta. Path. Microbiol. Scand. Section A, 80:225.

Harris, P., and Heath, D., 1977. "The Human Pulmonary Circulation", 2nd ed., Churchill Livingstone, Edinburgh, p27.

Hausman, D.H., and Weimann, R.B. 1967. Pulmonary tumourlet with hilar lymph node metastasis. Cancer, 20:1515.

Heath, D. and Best, P.V., 1958. The tunica media of the arteries of the lung in pulmonary hypertensions. J. Path. Bact. 76:165.

Heath, D., and Williams, D.R., 1977, "Man at High Altitude", 1st ed., Churchill Livingstone, Edinburgh, pp 220–221, 227–228.

Lauweryns, J.M. and Cokelaere, M. 1973. Hypoxia-sensitive neuro-epithelial bodies: intrapulmonary secretory neuroreceptors modulated by the CNS. Z. Zellforsch. Mikrosk. Anat., 154:521.

Lauweryns, J.M., Cokelaere, M., Theunynck, P., 1978. Cross-circulation studies on the influence of hypoxia and hypoxaemia on neuroepithelial bodies in young rabbits. Cell. Tiss. Res., 193:373.

Lauweryns, J.M., Cokelaere, M., Theunynck, P., and Deleersnyder, M., 1974. Neuroepithelial bodies in mammalian respiratory mucosa: light optical, histochemical and ultrastructural studies, Chest, 65:22. (supplement).

Ranchod, M., 1977. The histogenesis and development of pulmonary tumourlets. Cancer, 39:1135.

Taylor, W., 1977. Pulmonary argyrophil cells at high altitude. J. Path. 122:137.

Wharton, J., Polak, Julia M., Bloom, S.R., Ghatei, M.A., Solcia, E., Brown, M.R., and Pearse, A.G.E., 1978. Bombesin-like immuno-reactivity in the lung, Nature, 273:769.

Whitaker, W., 1954. Pulmonary hypertension in congestive heart failure complicating chronic lung disease. Quart. J. Med. (New Series), 23:57.

Whitwell, F., 1955. Tumourlets of the lung. J. Path. Bact., 70:529.

                                    DISCUSSION

LECTURER: Taylor   CHAIRMAN: Cumming

CUMMING: Thank you, the paper is now open for discussion.
                Could I begin by asking a question about the
                tumourlets? They occlude the terminal bronchiole
                and apparently distal to the occlusion the
                structure is roughly normal. Is it normally
                distended, do you think it is ventilated and what
                relationship to function do you think might exist?

TAYLOR: This is a very localised lesion and any relation to
                pulmonary function in general probably can't be
                predicted. If you have a patient with half a dozen
                or ten tumourlets in the lung, the amount of lung
                tissue which is in danger of not being aereated
                properly is relatively small. The fact that you
                don't get collapse distal to the tumourlets or even
                infection is, I think, in keeping with what is
                known about obstruction of terminal bronchioles.
                It is possible for ventilation to take place from
                other sources.

JEFFERY: Dr. Taylor thank you for your beautifully
                illustrated talk. Can I just ask two points: one,
                in recent studies by Dr. Wolcan and J. Pollack they
                have been using immunocytic chemical methods to
                demonstrate bombasine like immune activity at sites
                which they imply would correspond with the Feyrter
                cells you have just described. Have you done any
                such studies on these tumourlets to demonstrate
                that sort of immunoreactivity? It needs doing.
                Various substances have been described in relation
                to argyrophilic cells in the lung and I thought the
                biochemists who are around may take you up on this.
                I think you have got to make a distinction between
                substances which actually have been demonstrated
                within agryrophilic cells and· catecholamines,
                hydroytryptamine and so on and other substances
                like bombasine which have been demonstrated in lung
                tissue. It is supposed and generally assumed that
                they come from D cells but it is not yet proven. I
                think bombasine seems to be a very interesting and
                promising substance in this line. These cells are
                provisional members of the POC series and to be
                regarded as final bona fide members of this series

you have got to demonstrate a polypeptide and I think bombasine may well be the one we are looking for, but it isn't quite tied up yet.

HUTCHISON: It seems to me that this is a very interesting study but the comparison of the patients with the animal work is rather difficult. The animals have been subjected to a general level of quite severe hypoxia, whatever we mean by that, but in the patients you don't really have any definite evidence of what level of hypoxia they had and it's too much to expect that you had any lung function tests or measurements of light gases or anything like that in these patients who subsequently underwent autopsy. So we are left with rather indirect evidence. As to the question of hypoxia, animals may have general hypoxia, and there may be variations within the lung itself in the general level of oxygenation. In the emphysematous case you showed many cases of emphysema do not have much in the way of hypoxemia in the sense of a low arterial oxygen tension and in the case you showed, I'll throw caution to the winds and suggest that it was a case of centrilobular emphysema limited largely to the upper zones. The rest of the lung is pretty normal in which case you wouldn't expect very severe abnormality of lung function in such a patient. If these tumourlets are really related to hypoxia, are they related to local hypoxia, in which case the question really is: what is the distribution? Have you noticed any relationship of the tumourlets to the distribution of abnormalities of the lung, in cases of emphysema? A second question, have you studied patients with severe hypoxia or hypoxemia, such as severe cardiac shunts or severe lung disease?

TAYLOR: Yes, I have three cases. Firstly the suggestion that this is indirect evidence, I can't disagree with that; it is indirect evidence. We never feel justified in looking round the ward to say: he looks as if he might die tomorrow so let's have his pulmonary function test done; nor would it be very relevant at that stage of someone's life to do so. So indirect morphological evidence is what we are left with. Secondly you said that in the slide of an emphysematous lung I showed, you thought that it was centrilobular emphysema involving the upper part of the lung and the remainder of the lung was

normal. Well, from our own understanding most of
the lower lobe in fact was solid pneumonia. I
can't be sure that there wasn't alveolar duct
emphysema there, but it was far from being a normal
lung. Whether it was centrilobular emphysema or
not is another interesting question. I think it's
quite difficult to diagnose panacinar emphysema
destroying a large area of the lung. Certainly
there were some centrilobular foci in the upper
part of the lower lobe in that case. You may well
be right that it was centrilobular emphysema
because this is the type of emphysema which is
particularly related to hypoxia and there need not
be enormous areas of lung tissue involved. One of
the things we don't really understand about
emphysema is that sometimes patients with
relatively small amounts of emphysema seem to be in
really serious trouble over it with big right
ventricles, cor pulmonale and so on, whereas others
with very extensive panacinar emphysema seem to
progress fairly normally. Although the radiologist
has spotted it because it is gross it doesn't seem
to have any very drastic effect on the patients.
The third point, you said: are there any ties
between local lesions in the actual site in which
tumourlets are found? Well, these tumourlets are
not particularly situated in seriously damaged
areas of the lung. They are not necessarily in the
emphysematous areas, if the emphysema is in the
upper lobe they may be there but they may also be
in the middle lobe or in the lower lobe and so on.
I can't answer that quite as accurately as I'd like
to do because most of the search for tumourlets has
been done in lungs which have not been distended
and point counted for emphysema. That's because of
the effect of distension on the argyrophyl stain.
You did also raise a fourth point: have we looked
at any people who are known over a long period to
be hypoxic, with pulmonary function tests and so
on? I don't have any pulmonary function tests
taken before death when a patient might be regarded
as being in his usual state; unfortunately the
patients who have the worst lungs, those with
severe bronchiectasis, honey-comb lung and so on,
and with cor pulmonale often have the most
tremendous infection by the time they die and it's
very difficult to get round the problems of
autolysis. An ideal group to look at, as you
suggested, would be people with cardiac shunts and

I haven't done that because we don't have access to such patients very often.

SIMSON: I was very interested in Dr. Hutchinson's comment because I feel that one of the significant points of bronchiectasis is the increase in the bronchial artery supply, and I wonder whether this may play a part in the development of bronchoalveolar tumourlets. My question is more related to the clinical situation where we have had difficulties in switching from the oat cell carcinoma to the small cell carcinoma because it is subdivided now into the small cell, lymphocyte-like type, the polygonal cells and the spindle cell. Do you think that these are derivatives from the same Feyrter cells that you described in those three various situations? The second question is: is there any difference in the nature of each of these cells. I particularly say this in relation to the chemotherapy, we have had some hesitation in using small cell type chemotherapy in the spindle cell lesion.

TAYLOR: I haven't any very strong views on this. Certainly when you look for morphological evidence these lesions are derived from these endocrine cells. I think most people would agree that you don't find dense-core vesicles in all cases and you may be dealing with several different cell types, we don't in fact have a chest unit for which we do the surgical pathology. Professor Spencer may well have a view on this and Brian Corrin has more experience on this than I do. I believe Professor Spencer also has a view about the bronchial circulation in asthma. Is that correct? Sorry, in bronchiectasis in relation to the development of tumourlets.

SPENCER: One question and one answer; the question is: have you looked at the lung as a whole to see what is the normal distribution of these cells? Are they more in the upper part of the lung, in the lower part of the lung or are they evenly distributed through the lung? In answer to the question about oat cells, small round cells, spindle cells and large cells, we had exactly the same problem when we tried to classify these for WHO. The answer is: yes, they are all derived from Kulchitsky cells. One can get a spindle cell tumour, a small cell tumour and a large cell tumour which is almost indistinguishable from an anaplastic squamous cell, but they are all derived from Kulchitsky cells.

ANON: I'd like to ask about multiple lesions in pulmonary tissue at different sites, in different lobes.

TAYLOR: This is what Professor Spencer has just asked and I haven't answered him yet. We have done some work on the distribution of argyrophilic cells in the human lung. So far as I know they are not more common or less common in the better aerated areas. I think that the upper lobes are better ventilated than the lower lobes, the cells do not tend to be more common in lower lobes as far as I know.

DENISON: Would you mind if I came back to the physiological role of these cells and your thoughts about hypoxia? They are placed very close to the surface of the airway and they would be exposed to a PO_2 that would be very little different from that which exists in the gas in the airspaces.   Now even if you underventilate very markedly sufficient for example to double alveolar P_{CO2} all that you can do then is to drop the PO_2 by something like 40 mm. of mercury. This would be the equivalent of going up about 1500 to 2000 m. altitude. This may not be sufficient to stimulate the development of these cells at all, I think it is misguided to suppose that there is any point in looking at the indexes that you have chosen of hypoxic disease. If, in the case of shunt, the shunts are going to by-pass the blood from the regions of interest and so the hypoxic blood won't be able to influence the PO_2 detector cells. I think you should stick to the concept that these cells are there to sense alveolar PO_2 or airway PO_2, but anticipating that the thing that is most likely to change it and give it sufficient signal would be a certain altitude. That's a comment and the question is: supposing that they really are oxygen sensors, then histologically how closely do they compare with the other oxygen sensitive tissue in the carotid body and elsewhere?

TAYLOR: I'll answer the second point first. Under the light microscope they don't look like carotid body tissue. Under the electron microscope the type 1 cells show the dense-core secretory vesicles, so there is that similarity. As far as the other point goes, you are presenting me with data which I don't know about but which I am interested to hear. Apart from the pathological lesions, the function of these cells is mainly concerned perhaps with the transition from intrauterine to extrauterine life and the very early part of any individual's life. It is hypoxia and not hypoxaemia that they are sensitive to. I don't know enough about respiratory physiology or about lung function test in neonates to know whether there could be a role for them at the perinatal period. Perhaps you can help me here but I think this is much more likely to be the time at which, physiologically at any rate, they are doing something useful to the people who have them. I think the lesions which we see later on in life are not physiological but pathological and may be a response to a variety of stimuli, hypoxia being only one of the possibilities. We may be dealing with something which is the result of cigarette smoke whether or not it acts through emphysema or acts in some other way.

FONZI: I'd like to know your view on the tremendous discrepancies among various authors regarding the site and the number of Feyrter cells both in different animal species and at different levels of the bronchial tree. In a recent paper in "Thorax" Kleinerman ascribed these discrepancies to the use of different techniques. I find this observation an extremely important one albeit alarming, particularly if referred to a comparative Feyrter cell test in treated and non-treated animals. It might also be difficult to distinguish them from neuroepithelial bodies.

TAYLOR: Yes, I remember Kleinerman's paper in "Thorax". I think this is a very important point but in the Kleinerman series, he was talking about a number of different techniques as well. For instance one of the techniques which should be in use was the Van Giesens strain rather than the chromium stain; in fact Van Giesen was Professor Laurent's predecessor at Liege in Belgium, he was a very distinguished gentleman but I haven't found his stains to be very useful or very effective at all for these cells and I would be rather suspicious of any counting done using that stain. Similarly, I can't quote

individual bits of this paper from memory, but I am
pretty sure that some of the techniques which were
used in the distention of the lung exerted a fairly
serious effect on whether or not the stain works
effectively. I have looked at rabbits' lungs to
examine these effects of distention as compared
with non-distention. If you give the cells a
staining score and allot them +, ++, or +++, there
is a tremendous difference between the staining on
a distended lung and on a non-distended lung and
this results in counting a smaller number of cells
from the distended lung. Apart from these factors
which may explain some of the features that you are
talking about, I can only say that in my rabbits'
lung there was consistency of observation since I
looked at them all myself. One becomes fairly
consistent although one must have fairly rigid
criteria of recognition. There are other factors
like how you feel that day or what you have had for
breakfast, or whether you have had a hang-over and
so on, but even so, one tries to maintain a fairly
consistent set of criteria. I think it is possible
to obtain consistent results when there is just one
individual involved, if there is a half dozen then
I think it would be difficult.

CUMMING: We have gone over time as usual and I terminate the
                discussion at that point and invite Herbert Spencer
                to give his second paper.

LEGIONNAIRES' AND SOME VIRAL PNEUMONIAS

Herbert Spencer

Emeritus Professor of Morbid Anatomy
University of London

In 1976 a hitherto unrecognised form of bacterial pneumonia with a 15 percent mortality rate struck many of the participants at an American Legionnaires' Convention held in Philadelphia. Subsequently this new form of pneumonia became known as Legionnaires' disease. Fatal infections are almost entirely restricted to persons already suffering from other disease such as cancer, leukaemia or a malignant lymphoma. Since the original outbreak many sporadic and several small epidemics have occurred in many countries throughout the world.

Many serotypes of the causative organism Legionella pneumophila exist, and are greatly varying pathogenicity. L. pneumophila is classified as a bacterium in the family Legionellaceae. It has been isolated in nature from stream water and mud. Several epidemics have now been traced to the inhalation of water droplets derived from the warmed cooling water used in heat exchange and evaporative condensers employed in large scale air conditioning plants and electricity generating stations. The organism is inhaled within aqueous droplets and gives rise to a primary alveolitis (lobar type of pneumonia). Initially the alveoli fill with inflammatory oedema fluid but fibrin rapidly appears in the intra-alveolar exudate together with a few acute and some mononuclear inflammatory cells. The resulting pneumonic changes spread rapidly throughout the lobe or lobes of the lungs and fill the small air passages. At this stage the alveolar wall structure becomes difficult to identify but can still be demonstrated by a reticulin silver stain. Organisation of the pneumonic exudate soon occurs both by interstitial cell reaction and by accretion of the intra-alveolar exudate and often results in

405

extensive and very severe lung fibrosis resulting in permanent damage to lung function.

The causative organism L. pneumophila is difficult to demonstrate in the pneumonic lesions and it was not until Dieterle's silver staining method was used that the minute cocco-bacilli 0.5 um wide by up to 2.0 um long were demonstrated both lying free and within macrophage cells within the alveoli. The stain is however very capricious like many silver stains. The author has also successfully employed Machiavello's stain and toluidine blue to show the organisms. The L. pneumophila are however, best demonstrated in tissue sections by immunofluorescence using fluorescein labelled polyvalent antiserum.

Macroscopically, the affected lobes of the lungs rapidly become consolidated and closely resemble pneumococcal lobar pneumonia in the stage of grey hepatisation. Terminal changes due to endotoxic shock may occur. L. pneumophila is a difficult organism to culture but has been successfully grown on Mueller-Hinton agar containing traces of cysteine and ferric salts.

Although influenza viruses A and B remain the best known and the most important respiratory viral pathogen, other viruses have been increasingly recognised as not infrequent causes of serious and fatal lung disease and of these viruses only four will be described.

Measles virus still remains the most important world wide cause of fatal lung infections in young children. Because of widespread infant malnutrition in the Third World and also in some places in Western nations young newly weaned children may develop impaired immunity and consequently can develop very severe measles infections. Instead of developing an immunity quickly and overcoming the infection, both such children and occasionally immunodepressed adults may develop persistent measles lung infection which is known as giant celled pneumonia. The pathology of both classical severe measles and giant celled pneumonia do not differ, and the damage to both bronchial and alveolar epithelium may be very severe, predisposing the victim to secondary bacterial invasion. The bronchial epithelium undergoes metaplasia to an undifferentiated multilayer epithelium with loss of both ciliated and mucous goblet cells. It is however, in the alveoli that the most striking changes are seen. The alveolar epithelium is replaced by syncytial masses of type 2 pneumocytes in the cytoplasm of which viral inclusions are present formed from aggregates of the virus. Gradually the syncytial cells are replaced by a multilayered epithelium and if recovery occurs normal alveolar epithelium is reconstituted.

Adenovirus lung infections usually affect young children or

adults with impaired immunity. The lining epithelium in many of the smaller air passages is either totally destroyed or replaced by a mutilayered undifferentiated epithelium. Also hyaline membrane disease is frequently seen in the alveoli indicating necrosis of the lining alveolar epithelium. Some less damaged alveolar epithelial cells contain intranuclear inclusion bodies which are aggregates of the virus and such cells are enlarged and are called smudge cells. Adenovirus infections in childhood may complicate measles, and in adult life may occur in transplant patients.

Respiratory syncytial virus mainly attacks infants resulting in severe proliferative changes in the lining epithelium of the small air passages with loss of the normal ciliated cells. As a consequence secondary bacterial invasion commonly occurs. Cytoplasmic eosinophilic inclusion bodies may be found in occasional respiratory bronchiolar cells. The alveolar epithelium may also undergo metaplasis to a type 2 cell.

Cytomegalovirus pneumonia is much more frequent than in former years and usually accompanies a generalised infection. It is seen both in immunodepressed adults and children. In adults it is one of the more common 'opportunist' viral infections complicating therapeutic immunodepression for organ transplantation.   It frequently coexists with other opportunist infections especially pneumocystis pneumonia. Although the pneumonic changes cause minimal damage to the lungs the infection can be readily diagnosed by enlarged alveolar epithelial cells which contain the characteristic intra-nuclear inclusion bodies giving to the nuclei the 'owl's eye' appearance.

Although many viral pneumonias are disorders of limited duration it is becoming increasingly recognised that some, notably influenza and adenovirus infections, may persist and cause increasing damage to the affected lungs after the acute illness has subsided. Such lungs are characterised by severe fibrosis and secondary bronchiectasis.

DISCUSSION

LECTURER: Spencer   CHAIRMAN: Cumming

SOLIMAN: Why does Legionnaires' pneumonia heal by fibrosis
                in .contra-distinction to the pneumococcal type
                which does not normally resolve by fibrosis.

SPENCER: The answer is I don't know, the fatal cases are
                very severe and they occur in people who are
                already suffering from some other disease and this
                may interfere with the whole healing process. I
                don't know whether this is the answer but I suggest
                it as a possibility.

ROUSSOUW: How does one differentiate between a giant cell
                derived from fusion of intra-alveolar macrophages
                and type 2 giant cells? That's my first question.
                The second one: you have stated previously in your
                first paper, that type 2 alveolar epithelial cells
                may be derived from type 1 and in your summary you
                also suggest that the alveolar epithelium may
                undergo metaplasia to a type 2 cell. Would you
                please comment on the direction of the
                differentiation?

SPENCER: Giant cells first, how do you differentiate between
                tuberculous giant cells, the giant cell macrophage
                derivation and the giant cell of pneumocyte
                derivation. The answer is by electron microscopy.
                The macrophage giant cells will have the
                characteristics of a macrophage, there may be
                microvilli, there will be the typical intracellular
                organelles whereas in type 2 giant cells you have
                the characteristics of type 2 cells. The second
                question I think was about type I and type 2
                transformation. Type 1 cells are, so to speak, a
                fully differentiated cell, once a type I cell is
                damaged and dies it is only replaced from a type 2
                cell. The type 2 cell is a progenitor cell and
                type 2 cell can regenerate first of all producing
                more type 2 cells which, if the stimulus is
                removed, will then gradually transform into type I
                cells. So the type 2 cell is a progenitor of the
                alveolar epithelium, it's a progenitor cell and
                type I cell is a cell which is incapable of
                replication, as far as we know. Does this answer
                your question?

HEATH: I'd like to make two comments if I may Chairman. The first is to just relate our experience in Liverpool where we opened a brand new large teaching hospital not so long ago with air conditioning. Within a short time we had cases of Legionnaires' disease, one of which occurred in one of my colleagues from which I am glad to say he recovered successfully, but this again underlines the point that Professor Spencer was making that air conditioning seems to have a lot to do with this. The other point which lends support to Professor Spencer's statement is that my colleague did in fact have bronchial asthma and it is interesting that it was he who acquired this infection. What I'd like to ask Professor Spencer on this point is this: when these cases occurred, the hospital water supply was immediately heavily chlorinated and the authorities apparently seemed quite determined on this course of action. What does he regard as the type of chlorination which is necessary to keep a large hospital complex or medical school free of possibility of re-infection by Legionnaires'. That is question number one. Question number two is, I am glad to say, a point of total disagreement with Professor Spencer and it brings me back to earlier on in the week. Professor Spencer has just answered this question from Dr. Roussouw and he rather implied that it was easy to distinguish between these giant cells derived on the one hand from macrophages and on the other from granular pneumocytes and he quoted the typical macrophage inclusions on the one hand and typical inclusions like lamellar bodies on the other. This brings me back to the problem I have, which I mentioned during the week, that unfortunately so often macrophages will ingest surfactant derived from the lining granular pneumocytes and it is virtually impossible to distinguish them in many cases and I remember that Professor Corrin had the utmost difficulty in papers which he published in the Journal of Pathology in distinguishing between macrophages and cells either single or giant cells derived from the granular type 2 pneumocytes. You realise, of course, I am raising the whole subject once again gentlemen, of BAL. Having disgraced myself I'll say no more.

SPENCER: Thank you Donald. I don't know how you are going
                to improve the quality of your water for making tea
                or coffee. I am not a public health expert and I
                can't answer, I don't know what concentration of
                chlorine is required to destroy these organisms, so
                I don't feel capable of advising you on that
                question. Turning to your second question: how do
                you differentiate a type 2 cell from an alveolar
                macrophage cell; you touch a subject which is much
                more difficult. I hesitate to say this but one of
                the questions asked on Monday was:- how does a lung
                macrophage get into the alveolus? It has always
                seemed to me that the type 2 cells are the
                progenitor cells of alveolar macrophages. This is
                absolutely unorthodox but I sometimes wondered
                whether the type 2 cells were the progenitor cells
                of macrophages and when they were dropped onto the
                alveolar surface they assumed phagocytic potential.
                This is a long argument and i won't go into it
                further; it is quite unorthodox but it is an idea
                which I just keep in the back of my mind.

CUMMING: Perhaps I can be forgiven a comment in the public
                health aspect of destroying the Legionnaires'
                bacillus. I think the answer lies in separating
                the water supply that does the heat exchanging from
                the normal water supply of the hospital. You can
                then treat this as a separate system and put in as
                much chlorine or other antibacterial drugs as you
                choose. The problem then is how to keep the two
                water supplies separate and supply the cooling from
                a refrigeration system, but the present practise of
                air conditioning has not yet assimilated this
                problem.

WILLIAMS: Professor Heath was sitting in front of me so he
                got his comment in faster, but I should try to
                demonstrate in my talk and then in the table which
                I showed from a recent publication that the
                distinction between these cells is very difficult
                and you need to do a whole host of different tests,
                it is not acceptable to do it by one criterion
                whether it is light microscopy or electron
                microscopy, pure morphology does not give the
                answer.

SIMSON: Professor Spencer, you have outlined several
                predisposing causes to the viral infections in the
                lung. One of the interesting features we have had

is the incidence of adenovirus in Swire-James or McCloud's syndrome. We are wondering if there is any basis on which this can occur because it does certainly seem a very common and recurrent problem in these patients.

SPENCER: No, I have no information. I am interested to hear this because the Swire James people think it is due to a bronchiolitis and one can't help wondering whether the bronchiolitis was caused by the adenovirus infection. Perhaps people are not sufficiently aware of the fact that a lot of these virus infections in the lung are persistent. Adenovirus infections do persist and I can't help wondering whether the Swire James syndrome which is centrally due to bronchiolitis, mightn't in fact be caused by adenovirus and then later you come back and still find the adenovirus present. Maybe that's the explanation.

TAYLOR: I wish to ask a question about plumbing. It is not only the water used in ventilation in which these organisms may be present, but the shower heads as well which are on the same system as the ordinary tap water.

CUMMING: Yes, they are indeed and that' the problem of separating the fittings that you use for potable water from those you use for cooling, showering, washing and so on. This is a plumbing problem which I think has not even been conceived of yet. Well, it's now 11 o'clock and we will break at this point. Thank you Herbert for that most entertaining presentation. We will reconvene at 11.30.

NATURAL HISTORY OF BRONCHIAL CARCINOMA

Peter Mitchell-Heggs

Consultant Physician Epsom
Hon. Sen. Lecturer, Charing Cross Hospital
Medical School

Bronchial carcinoma is a common disease killing over 30,000 people in Great Britain every year. It affects subjects from the age of 40 to over 70 with an average incidence around the age of 50-60. Men are more commonly affected than women. Smoking is strongly associated with squamous and anaplastic small cell carcinoma of the bronchus.

There are no <u>characteristic</u> presenting symptoms or signs of bronchial carcinoma, although some presenting features are common. The outlook for a patient with a lung cancer is appalling. Of one hundred patients presenting, perhaps six or seven will be alive five years later - most dying within the first three years after diagnosis. Treatment influences the prognosis marginally, but improving methods of screening prevents unnecessary surgery and other treatments in many patients. Radiotherapy is a useful palliative and symptomatic agent but has little effect on the natural history of the disease.

Chemotherapy, in 'oat cell' carcinoma has exciting prospects for real improvement in prognosis, the overall picture is, however, depressing.

Presenting features of bronchial carcinoma

Malignant disease of the lower respiratory tract presents usually insidiously. There are no specific diagnostic features although certain groups of symptoms and signs may be highly suggestive. Subjects with bronchial carcinoma often have chronic bronchitis; both bronchial carcinoma and chronic bronchitis being closely related to tobacco smoking. However, there may be no symptoms or signs associated with a bronchial carcinoma until comparitively late in the natural history.

The classically described signs such as clubbing, hypertrophic pulmonary osteo-arthropathy (HPOA), muscle wasting and hoarseness of voice are usually absent. If symptoms are present, the commonest are persistant cough, weight loss, haemoptysis and deteriorating breathlessness. If signs are present, the commonest are those of a slowly resolving consolidation or collapse or marked focal chest signs in the absence of fever or other symptoms.

Rate of progression of bronchial carcinoma

The rate of progression of a bronchial carcinoma depends on several features. In the first place the type of tumour must be considered. There are many different pathological sub-groups of bronchial carcinoma, - squamous cell carcinoma, 'adenocarcinoma' of the bronchus and anaplastic (oat cell) and undifferentiated carcinoma. In general in malignant disease and certainly with bronchial carcinoma, prognosis related to cell mass depends on the rate of cell turnover; this, in turn, depends on the degree of differentiation of the tumour. In general with bronchial carcinoma the greater the degree of differentiation of the tumour the better the prognosis.

The rate of progression of disease due to bronchial carcinoma depends on the degree of spread of the lesion. The more scattered the lesions and the more organs affected, the worse the prognosis.

The site of the primary (or other lesions) is important in determining the rate of progression. A lesion next to a vital structure may invade this structure with dramatic effect on prognosis, even when the tumour itself is quite small. For example, some bronchial carcinomas may invade the pericardium and myocardium producing fatal tamponade or dysrhythmias as early presenting symptoms.

The age of the subject and his coexisting 'general health' (especially including here any condition which tends towards an immune paresis) are determinants of prognosis. The prognosis of tumours in older subjects is often relatively better than in younger subjects; this is not due to an undue excess of poorly differentiated tumours in younger subjects and there is at present no satisfactory explanation.

Even without treatment the best prognosis is in adenocarcinoma - the prognosis usually quoted being between nine and twenty-four months from diagnosis to death. An intermediate prognosis between six and eighteen months is found in subjects with squamous cell carcinoma of the bronchus. The worst prognosis is in patients with 'anaplastic' or 'oat cell', poorly differentiated bronchial carcinoma; here prognosis is from days to nine months.

Effect of treatment on the natural history

Fortunately in some subjects the bad prognosis provided by the 'untreated natural history' can be mitigated. Treatment depends on the cell type of the tumour and the degree of spread of the lesion at the time that any treatment is planned.

The tumour cell type can be determined by biopsy or brushing of the lesion at exploratory bronchoscopy or analysis of sputum cytology. The staging of the lesion is initially clinical and later uses scanning techniques and radiology. The more thorough and painstaking the means of staging, the smaller the percentage of really localised lesions is found to be. Since most patients with bronchial carcinoma die of bronchial carcinoma even when surgery has been carried out, it must be that the growth had already spread at the time of surgery rather than that a second growth had occurred to produce the local or distant secondary growths which determine prognosis.

For squamous cell carcinomas of the bronchus whatever their degree of differentiation, and for adenocarcinoma, surgery provides the best prognosis, if technically feasible and where the lesion is localised to the lung

For poorly differentiated carcinomas of the bronchus surgery in general offers a slim chance of cure. All surgeons have cases of very long survival after surgery for 'oat cell' carcinoma; these individual cases do not, however, unfortunately dispel the worldwide overall data suggesting the inability of surgery to stem the advance of these unpleasant growths.

For generalised, or non resectable tumour, radiotherapy is an excellent palliative agent which does not overall increase prognosis; radiotherapy usually improves quality but not quantity of life. Well differentiated squamous and adeno carcinomas are often slow to respond to radiotherapy. Anaplastic cell, poorly differentiated carcinomas often respond well to radiotherapy. Here, however, chemotherapy excels especially with adjuvent radiotherapy. Chemotherapy for adeno and squamous carcinoma is at present of no proven value.

Overall prognosis with bronchial carcinoma

There is less than 10% five year survival for any group of subjects presenting with a variety of bronchial carcinoma.

Of 100 subjects presenting, 80 will be inoperable due either

to spread of disease or general disability. Most of these patients
will be dead by 3 years, but a few, 1 or 2, may survive, with
treatment to 5 years. Of the remaining 20 subjects with apparently
localised disease only 6 or 7 will be alive at 5 years. Improved
methods of staging do not increase this final number.

Estimation of prognosis from tumour growth rates

Making assumptions about the means and rate of growth of
tumours usually by extrapolating from animal models, various
authors have attempted to calculate prognosis and to assess the
beneficial effects of any therapy by assessing the growth rate of
tumours.

Assuming that a tumour grows from a single cell, that its
growth is exponential rather than linear, its volume and mass at
any time may be calculated if the rate of division is known. The
rate of growth is expressed as the 'volume doubling time'. In this
way it is easy to see why tumours which appear to be small,
suddenly grow and cause apparently dramatic physical effects. An
example of this phenomenon, which appears in several publications
on this theme, compares the principle of doubling of cell number
per division to putting one grain of rice on the first square of a
chessboard, two on the second, four on the third, etc. By the
other end of the board more grains of rice than have ever appeared
anywhere would have to be found.

However, a number of malignant cells die before dividing and
cells develop clones which behave differently from the parent
cells. One of the differences found is a difference in the rate of
division and thus the doubling rate of the tumour. The natural
history and the prognosis depend on the ability of the metastatic
cell to colonise its site of impaction and to divide effectively at
that site. In this way prognosis may depend on the efficiency of
the process of metastasis and the rate of growth of metastases.

Age and 'aggressiveness' of bronchial carcinoma

It is thought that some cancers, including lung cancer, may
lie dormant for many years. The 'dormant' period is not however
one in which no action occurs, the tumour is dividing but at a very
slow rate. Assuming that a certain critical size is the smallest
detectable because of its appearance on 'Mass X-ray' or its causing
a specific symptom, this size will be reached after a definite
number of divisions.

If a thirty year old subject develops (i.e. starts to 'hatch')
a carcinoma with a doubling time of 30 days, the lesion will be
presenting three years later as a rapidly growing tumour. If the
tumour doubling time is three hundred days, he will present 30

years later. The first subject will have an early presentation of an aggressive cancer, the second a late presentation of a slow growing lesion. The concept that a tumour may have been present, growing very slowly for many years, may explain how a carcinoma may present twenty years after a subject has last smoked, or been exposed to asbestos.

Conclusion

The natural history of bronchial carcinoma is that of a lesion, usually attributable to smoking, which occurs in both sexes from the midthirties onwards with a peak incidence in the fifties and early sixties age groups.

It is inevitably progressive without ablative treatment which is technically possible and effective in less than 10 per cent of cases. Other therapies delay the natural history for between one and twenty-four months and very occasionally longer.

Earlier diagnosis or even better prevention of this killer disease, responsible for 30,000 deaths annually in Great Britain, is urgently needed.

Suggested reading:

Geddes, D.M. (1979) Brit. J. Dis. Chest 73, 11.
Straus, M.J. Lung Cancer. Grune and Stratton
    New York, San Francisco and London, 1978.

DISCUSSION

LECTURER: Mitchell-Heggs   CHAIRMAN: Cumming

CUMMING: The paper is now open for discussion. Can you
                    come down Dr. Tricomi?

TRICOMI: A brief comment from a radiologist.
                    Radiologically the best thing a radiologist can
                    do is to prevent the occurrence of a pulmonary
                    tumour since as we have heard, its therapeutic
                    outlook is extremely poor. For a radiologist it
                    is easier to spot the tumours stemming from the
                    large bronchi, the main bronchi and the lobar
                    bronchi down to a certain level. By means of
                    tomography we are able to detect a tumour's
                    development in the bronchus prior to the
                    occurrence of a possible shadow or obstructive
                    emphysema. All radiologists in Italy, and I am
                    sure also abroad, have had cases of a total
                    neoplastic occlusion of a main bronchus without
                    considerable changes or parenchymal signs. As
                    regards doubling time, our experience showed
                    that in only 15% of all tumours we see and
                    follow up, is it possible to measure the
                    doubling time. It is clear that the most likely
                    tumours to be assessed by means of doubling time
                    are the peripheral ones which are not so easily
                    detectable with doubling time technique because
                    inflammatory phenomena frequently interfere,
                    which prevent us from relying on the doubling
                    time technique.

MITCHELL-HEGGS: I agree with you that it is extremely difficult
                    to make a judgement in relation to volume
                    doubling time. I was not intending to suggest
                    that it was likely to be a clinical means of
                    assessing a particular tumour, I was discussing
                    it only as a means of looking at the natural
                    history of the disease. Radiologically I
                    absolutely agree with you, I think that there
                    are an enormous number of variations of
                    radiological presentation of central and
                    peripheral lesions and often central lesions
                    present with no radiological change but only
                    with a vague awareness of increased
                    breathlessness. A patient described difficulty

in breathing which he said he never had before –
he just couldn't breathe properly.   His
respiratory function testing and also his
bronchoscopy showed a very small tumour which
was narrowing one of his major bronchi with no
radiological change, so here we have something
which can even antedate radiology.

SCANO: I'd like to know what sort of criteria you
establish before deciding to operate in cases of
pulmonary tumours and the importance you attach
to the functional assessment of the patient.

MITCHELL-HEGGS: When one is not particularly bothered about
abnormal respiratory function after surgery, one
has less need to carry out a more difficult
respiratory function testing.   However, in
patients who have quite severe chronic
obstructive airway disease together with their
bronchial carcinoma, there is a place for
assessing very carefully whether or not it is
the lesion which is producing some of the
disturbed respiratory function and whether
surgery is indicated. One might have to go as
far as attempting by means of an occlusive
catheter to block off one of the major bronchi,
or even the pulmonary artery, to the area that
one was considering for section.   This is
apparently quite an invasive procedure but it
would be well worth while doing in this very
difficult case, where there is no spread outside
the primary site and it is technically feasible,
for one does not want to leave a patient as a
respiratory cripple, substituting one form of
death for another.   Does that answer your
question.

CUMMING: I wonder if I might make a comment at that
point. It's certainly true that assessing the
function of the lung by laboratory techniques is
of value in eliminating those people who you
would kill by doing a lobectomy. Your second
comment related to regional assessment to see
what were the regional problems and you
suggested occlusion. I am now going to ask
David Denison to make a point where regional
assessment is far less invasive than that and it
might be helpful in the assessment of pre-
operative function.

DENISON: It is practical to assess the function of each part of the lung during fibroptic bronchoscopy by mass spectrometry using single breath tests, so that it is possible to study each lobe of the lung and turn and even each segment within half an hour. You need to have a very clear idea of what you would like to achieve by this. As the Chairman has said there are two decisions: firstly what is the quality of the bit of the lung that you intend to take away even though it has a tumour in it, and the other thing you want to know is how good is the function of the rest of the lung; is it capable of supporting life if that bit were taken away? I agree with Gordon Cumming that it is a minority of cases where it really does influence your decision; its worth doing but regrettably it doesn't often change the decision.

CUMMING: I think the problem, the major decision, is really whether the tumour is invading the carina, since that's an absolute contraindication to surgery, all other things become more or less unimportant.

KAY: I have three questions. Perhaps the first one is the ignorance of the non-cognoscenti, but it seems to me that your doubling time must be a gross underestimate of the time to reach that particular tumour mass because you have a large number of cells which are necrosing and therefore are not doubling. You also have a large number of cells which have been sloughed off, so those estimates must be actually far less than you are presenting.

MITCHELL-HEGGS: That would be a very nice explanation which would fit in with the clinical associations. If one is extrapolating purely from doubling times, then one must know that they are precise. These are the figures that we were left with but I agree with you that it would be a very nice explanation.

KAY: It seems a rational one. Its the case when patients are seen that they have a large number of necrotic cells in their tumour.

MITCHELL-HEGGS: Yes, but one has to suggest that there is almost an equal mass of necrotic tissue which has developed during tumour growth. The rate of necrosis of cancer cells might vary with both tumour type and site.

KAY: The second point I wanted to make was in terms of chemotherapy. Chemotherapy effect seems to be proportional to the doubling times of the tumours, in other words the more rapidly the tumour divides, like the oat cell, the better it responds to chemotherapy. That's because all the therapeutic agents rely on interfering with DNA synthesis and in rapidly dividing cells this is most active and so the chemotherapeutic agents act best on the oat cell with the shortest doubling time.

MITCHELL-HEGGS: That's certainly so. I think one can take it a little further than that because it's surprising how other cancers with doubling times rather like adeno and squamous cell cancers, can be more susceptible to some chemotherapeutic agents than squamous cells and adenocarcinoma cells. So although I would take that as one means of differentiation I don't think it explains the whole story.

KAY: Could you comment finally on the stamp cell concept in terms of chemotherapy?

MITCHELL-HEGGS: In relation to normal stamp cells or cancer stamp cells?

KAY: Cancer stamp cells.

MITCHELL-HEGGS: We are looking at one of two possibilities, either they are specific cells which are the active germ centres of the carcinoma, and which we need to destroy, or we are looking at a whole series of cells which are randomly dividing and we can just pick off the largest possible number.

SIMSON: I agree very much with the previous comment that the more people investigated pre-operatively the fewer patients come to thoracotomy, but I think by the same token, the more patients you investigate, the fewer patients have an open-

and-close thoracotomy, which I think is the most disastrous situation of all. I think that carcinoma of the lung is an eminently treatable disease, 25% of patients die of infection, 15% die of other conditions. These patients require supportive treatment, they require symptomatic treatment, not necessarily a curative treatment but the majority of patients with carcinoma of the lung are amendable to our help, not in cure but certainly in symptomatic treatment in improving the quality of their lives. So often we have tended to make a diagnosis, say that its incurable and assume that it was untreatable. I take issue with your statement.

MITCHELL-HEGGS: I don't think we are really at issue. I take your point but within the context of our discussion I don't think they are treatable for what I was describing as treatment. Of course we look after them and care for them and there are all sorts of things that we can do to make them better. Sleeve resections are absolutely fascinating, what happens to our much discussed mucociliary escalator when this is done? Is there a build up of mucus on the distal side of the sleeve which can't get over? Have you done rebronchoscopy to see whether mucus transport is normal across that area?

KAY: When we do a sleeve resection it consists of approximating the two portions. There was no intervening prosthesis and so there is only a suture line and there was no build up apparent at that line. We had experience in regard to this in sleeve resection of the right upper lobe where the tumour is actually taken out together with a part of the main bronchus. Such patients have no difficulty at all in ventilating, perfusing or mucus clearance.

MITCHELL-HEGGS: This procedure disturbs the integrity of the ciliary escalator and yet somehow the mucous manages to pass over the junction.

TAYLOR: My question is rather peripheral to what you are talking about, I am afraid, and is also rather academic but as a man who has to look for finger clubbing every day of his working life have you any views of the mechanisms by which finger clubbing occurs?

MITCHELL-HEGGS: I have little to add to the already turgid literature on the subject. The only thing that I could add is that I do not like the overuse of repeated reports of query early clubbing and query partial atypical clubbing, because of its associations with bronchocarcinoma. I think there is a clear definition of what clubbing is in answer to your question in terms of its mechanism I have nothing to add.

CUMMING: I think it's true to say that the person who first clearly defines the aetiology of clubbing of the fingers will probably get the Nobel Prize.

HUTCHISON: You made the interesting suggestion that you thought that metastases may grow faster than the primary growth. I wonder whether you have clinical or experimental evidence to support that idea; clinically it might seem so because the metastates often end up in a key site. If you have such evidence one wonders how it is brought about because many of these tumours are very vascular and therefore the question of the protection from local or general immune mechanisms would then not be relevant. Can you give us any information on that?

MITCHELL-HEGGS: There has been some information in the literature about this; my personal experience relates in fact to two patients both of whom had widely disseminated oat cell carcinoma of the bronchus and both were receiving chemotherapy which had no effect on the tumour, and was stopped. One was then studying them from the point of view of an experimental preparation whilst developing their metastases. One had a primary in the lung and metastases in bone and liver. The rate of increase in size of the liver mass was greater than the rate of increase of size of thoracic mass as defined by tomography. I don't know if there is any experimental evidence to back this result and perhaps other members of the audience could help on that.

KAY I'd just like to make a comment on that. Does

the mass of the tumour necessarily correlate with the actual number of tumour cells. I know it is the only thing one can use, but I think it's a poor correlation.

MITCHELL-HEGGS: I think you are right, it's the best we can do, a sort of repeated biopsy. The rate of necrosis of a tumour may vary from site to site as well; all that one can observe is that the growth rate of the mass can vary.

CUMMING: It seems to me that someone should go home and work out what is the message which tell the cell when to divide and when to stop dividing and then we might understand this process a bit more clearly. Does anyone have the answer to that question? No, I have no more questions on my list, it would be convenient since we have a few minutes before the end of the session if we went up and had a photograph taken. Would the members of the British party stay behind for a couple of moments after the photograph for some administrative discussion? Thank you very much.

THE INDUCTION OF CANCER BY CHEMICALS

Gerald M. Cohen

Toxicology Unit, Department of Pharmacology
School of Pharmacy, University of London
29/39 Brunswick Square, LONDON WC1N 1AX
England

INTRODUCTION

Cancer is one of the major causes of death in our society. It has been estimated that approximately 80% of all human cancers are of environmental origin.[1] This estimate is based primarily on epidemiological studies which show an enormous variation in the incidence of various forms of cancer is seen in different geographical areas of the world. Studies of migrants, who readily acquire a cancer incidence similar to that of their new country of residence, also imply that environmental rather than genetic factors are of major importance.

With the exception of skin cancer, for which solar ultraviolet light is an important causative agent, much emphasis is now placed on the possible role of environmental chemicals as major causative factors in the aetiology of human cancer.[2] However, such environmental chemicals must also include ill-defined life style factors such as diet, social and cultural habits.[3] This stress on chemicals is due partly to the recognition that certain chemicals have been identified as the causes of human cancer (Table 1) but also that they may induce both cancers in experimental animals and mutations and malignant cell transformation in various short-term tests for the detection of potential mutagens and carcinogens. The identification and removal of such carcinogenic chemicals from our environment should therefore lead to a marked decrease in cancer incidence. This simple logic was first applied over 200 years ago when the Danish Chimney Sweepers' Guild urged its members to take daily baths after the astute observations of Percival Potts who, in 1775, noted the high incidence of scrotal cancer in this profession and correctly ascribed "its origin to a lodgement of soot in the rugae of the scrotum".

425

TABLE 1: CHEMICALS GENERALLY RECOGNISED AS CARCINOGENS IN HUMANS

CHEMICAL	SITE OF CANCER
Industrial Chemicals	
2-Naphthylamine	Urinary bladder
Benzidine	Urinary bladder
4-Aminobiphenyl	Urinary bladder
Bis(chloromethyl)ether	Lungs
Nickel Compounds	Lungs, nasal sinuses
Chromium Compounds	Lungs
Arsenic Compounds	Lungs, skin
Asbestos	Lungs, pleura
Benzene	Bone marrow
Vinyl chloride	Liver
Bis(β-chloroethyl)sulphide (Mustard gas)	Lungs
Drugs	
Chlornapthazine	Urinary bladder
Diethylstilboestrol	Vagina
Chemical Mixtures	
Cigarette smoke	Lungs, urinary tract, pancreas
Soots, tars, oils	Skin, lungs
Betel nut	Buccal mucosa

Modified from E. Miller: Cancer Res., 38, 1479-1496, 1978.

CHEMICAL CARCINOGENS

A large number of chemicals, with a wide diversity of chemical structures and occurences, have been shown to induce cancer in animals.   These include polycyclic aromatic hydrocarbons, nitrosamines, nitro-aromatic amines and mycotoxins.   These may be subdivided into two major clases, ie, direct acting carcinogens and indirect carcinogens.[2] Most of the chemicals fall into the latter category and require metabolic activation by one or more steps, to a more reactive form(s), invariably an electrophile, which then combines with critical cellular macromolecule(s) leading in some, as yet undefined, way to tumour formation (Figures 1).   Many factors, including age, sex, diet, hormonal status, inducing agents and inhibitors of microsomal mixed function oxidase enzymes, DNA repair, immune competence and tumour-promoting agents may profoundly influence the ultimate tumour response.   These may act at any one or more of the many steps outlined in Figure 1. One of the more important and better understood of these factors is the ability of the host to metabolise the carcinogen.

Figure 1: SIMPLIFIED SCHEME OF THE RELATIONSHIP BETWEEN METABOLIC
ACTIVATION AND TUMOUR FORMATION

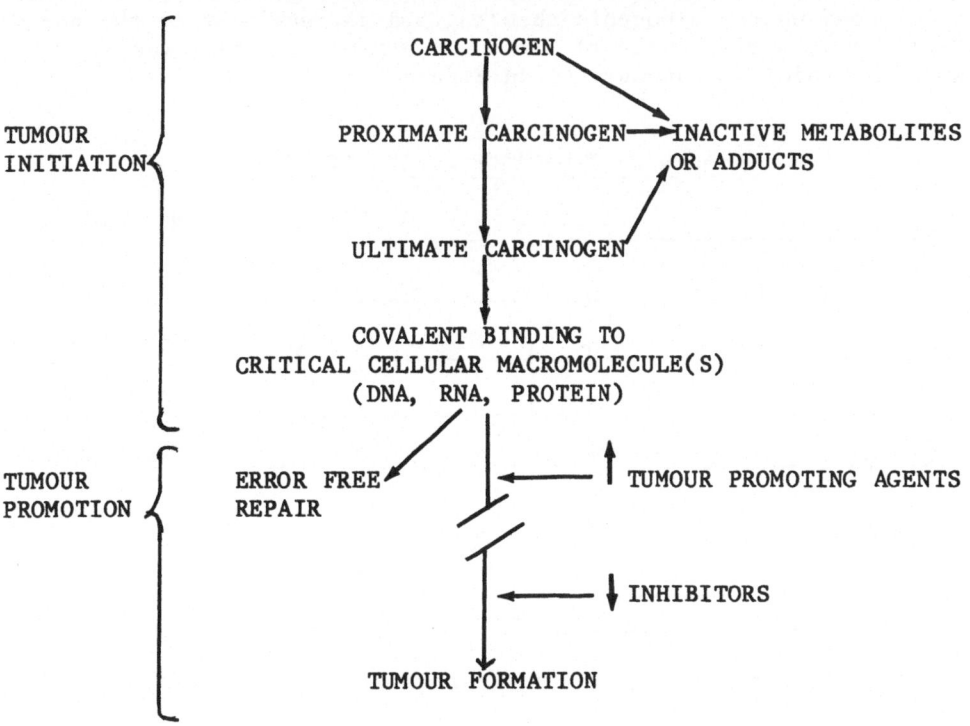

It should be readily apparent (Figure 1) that it is the balance of the different metabolic pathways, involving oxidation and conjugation reactions, as well as reactions with different cellular macromolecules including DNA, RNA and protein, which finally determines the amount of an ultimate carcinogen available for covalent binding with critical cellular nucleophiles. Although the key macromolecule(s) with which carcinogens covalently bind are not unequivocally known, a good correlation has often been observed between carcinogenicity and binding to DNA[4], although in some cases the qualitative as well as the quantitative nature of the carcinogen-DNA binding appears to be of major importance for both carcinogenesis and mutagenesis.

TWO-STAGE OR MULTISTAGE CARCINOGENESIS

Most chemicals have been tested for potential carcinogenicity after single or more often repeated administration to experimental animals, ie, they have been tested to see if they are complete carcinogens. However, since the work of Berenblum and others in the late 1940's, it has been evident that carcinogenesis, in mouse skin, is at least a two-stage process.[5] The two stages are referred to as initiation and promotion. Tumour initiation requires only a single application of a sub-threshold dose of a chemical carcinogen and is generally considered to be irreversible. Tumour promotion on the other hand, requires multiple applications of a second non-carcinogenic chemical and is, at least in the early stages, reversible. Some of the most interesting features of this two-stage model are summarised in Figure 2.

Figure 2: TWO-STAGE CARCINOGENESIS

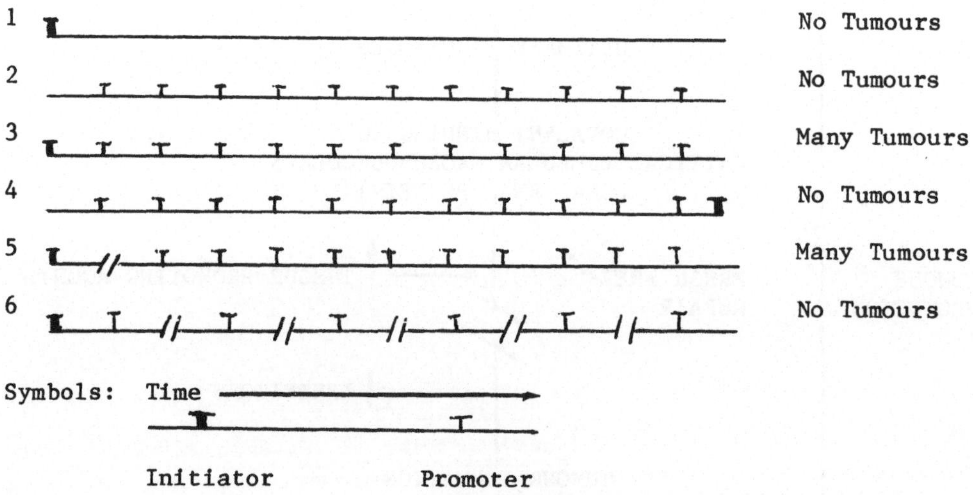

Symbols: Time

Initiator   Promoter

If the skin is treated with a single dose of an initiating agent, such as benzo(a)pyrene or ethyl carbamate (urethane), then no tumours are formed (1 - Figure 2). Similarly if the skin is treated with multiple applications of a tumour promoter, eg, 12-0-tetradecanoyl-phorbol 13-acetate (TPA) the active principle from the seeds of Croton tiglium, then no tumours are formed (2 - Figure 2). However, if TPA is given repeatedly after the initiating agent then a large number of papillomas are formed and followed after several months by carcinomas (3 - Figure 2). The sequence of these events is critical, if the initiator is applied after the promoter then no tumours are formed (4 - Figure 2). However, provided the promoter is applied after the initiator, its application may be delayed for several months and a large number of tumours still result (5 - Figure 2). However the interval between sequential applications of the promoter must not be too great (6 - Figure 2). This model is very valuable experimentally in that it enables one to determine those properties specific either to tumour initiators or promoters. However, one must be extremely cautious in the extrapolation of such animal data to the human situation where one is most frequently exposed simultaneously to both initiators and promoters. This is a situation known as cocarcinogenesis.

Recent work has suggested that two-stage or multistage carcinogenesis may be of importance for a number of systems besides the skin. Whilst the best evidence has been obtained for rat liver, data in support of multistage carcinogenesis, although not as yet conclusive, has also been obtained for lung, bladder, mammary gland, stomach, oesophagus, thyroid, and in various cell culture systems. In the human situation the importance of tumour promotion or cocarcinogenesis is far from clear but may be of far greater importance than currently realised. If such two or multistage carcinogenesis is important for humans then there are many important implications in terms of cancer prevention. For example the reversibility of tumour promotion, particularly in its early stages, should lend itself to the design of inhibitors of carcinogenesis. In this regard it should be noted that recently, in an elegant series of experiments, Slaga and co-workers[9] have recognised at least two different stages in tumour promotion and have identified inhibitors specific to both stages. The possible involvement of promoters of cocarcinogens in humans has been implicated in the causation of a wide number of different human cancers. (Table 2).

APPLICATIONS OF THE ABOVE CONCEPTS TO LUNG CANCER IN HUMANS

Lung cancer was responsible for more than 37,000 deaths in the United Kingdom in 1975. A large number of both prospective and retrospective epidemiological studies have shown a clear dose dependent relationship between the number of cigarettes smoked and the incidence of lung cancer.[7] The risk of lung cancer is also related to the age at which smoking started and the depth of

TABLE 2: SUSPECTED TUMOUR PROMOTERS OR COCARCINOGENS IN HUMAN CANCER

Possible Tumour Promoter or Cocarcinogen	Organ System
Cigarette Smoking	Lung
Asbestos	Lung
Alcohol (and Smoking)	Oral Cavity, Larynx, Oesophagus
High Fat Diet	Breast
Nutritional Deficiencies	Stomach, Cervix, Thyroid
High Fat and Protein Diet	Colon

inhalation. Whilst cigarette smoking is undoubtedly the major cause of lung cancer in our society, epidemiological studies in industry have identified other respiratory carcinogens, including arsenic, asbestos, chloromethyl i.e. methyl, ethers, chromium, coal- and petroleum- related carbon compounds, mustard gas, nickel and radiation.[8]

Tobacco smoke is a very comlex mixture of more than 3,000 chemicals.[9] Whilst the carcinogenicity of tobacco smoke cannot be attributed to any one class of compound, it is most probably due to the interaction of carcinogens, tumour initiators, promoters and cocarcinogens (Table 3).

The tumour initiators in tobacco smoke may be primarily poly-cyclic aromatic hydrocarbons as fractionation experiments resulting in the removal of this fraction of cigarette tar (0.6% of dry weight) led to a diminution of greater than 50% of the tumourigenic activity.[9] In this regard it is also of importance to know that cells withing the human respiratory tract may metabolise carcinogens such as polycyclic aromatic hydrocarbons[10] and certain nitrosamines to reactivate metabolites capable of binding to cellular DNA.[11]

In human lung cancer the role of tumour promotion and cocarcinogenesis is particularly difficult to discern. The relative risk of lung cancer in cigarette smokers declines after

they have given up smoking compared with those who continue.[7,9]
This decline is related to the length of time that the individual
has stopped smoking and reflects a tumour-promoting activity rather
than an initiating activity of tobacco smoke.   The possible
importance of cocarcinogenesis or tumour promotion is also very
well illustrated by the high incidence of lung cancer in cigarette
smokers exposed to asbestos. Asbestos workers who do not smoke
have only a modest increased risk of developing lung cancer,
whereas those who also smoke have a greater than 90-fold increased
risk.[12]

Table 3: CARCINOGENS, TUMOUR PROMOTERS AND COCARCINOGENS IN TOBACCO
SMOKE[*]

Carcinogens
Benzo(a)pyrene, 5-methylchrysene, dibenz(a,h)anthracene, dibenzo(a,h)pyrene, dibenzo(a,i)pyrene, dibenz(a,j)acridine, benzo(b)fluoranthene, benzo(j)fluoranthene, benz(a)anthracene, chrysene, methylchrysenes, methylfluoranthes, dibenzo(c,g)carbazole, benzo(c)phenanthrene, N-nitrosonornicotine, nitrosopiperidine, nitrosopyrrolidine, polonium-210, arsenic, nickel, cadmium, o-toluidine, β-naphthylamine.
Tumour Promoters
Volatile Phenols, benzo(e)pyrene.
Cocarcinogens
Catechol, 4-alkylcatechol, pyrene, fluoranthene, benzo(e)pyrene, naphthalenes, methylfluoranthenes, 1-methylindoles, 9-methylcarbazoles
[*]Most of the data obtained from activity on mouse skin

Modified from Wynder & Hecht (1976)

SUMMARY

     The evidence that chemicals are an important aetiological
factor in the causation of human cancer is reviewed. Chemical
carcinogens may generally be classified as either direct- or
indirect-acting, the majority falling into the latter category.
Such indirect-acting carcinogens require metabolic activation to
reactive electrophiles which, in order to exert their carcinogenic
action, combine covalently with critical cellular nucleophiles,
possibly DNA. Chemical carcinogenesis in many tissues, in
particular, skin, may be divided into at least two stages, ie,

tumour initiation and tumour promotion. Tumour initiation is considered to be irreversible whereas tumour promotion, at least in its early stages, is considered to be reversible. The recognition and removal of cocarcinogens, tumour initiators and promoters from our environment as well as the use of inhibitors of promotion offer possible approaches to reducing the high incidence of cancer in our society. These general comcepts of chemical carcinogenesis are related to lung cancer in man.

ACKNOWLEDGEMENTS

Much of the work by the author has been supported by a grant from the Cancer Research Campaign of Great Britain.

REFERENCES

1. R Doll, Strategy for detection of cancer hazards to man. Nature 265: 589 (1977).
2. E C Miller, Some current perspectives on chemical carcinogenesis in humans and experimenta animals: Presidential address. Cancer Res., 38: 1479 (1978).
3. J Higginson and C S Muir, Environmental carcinogenesis: misconceptions and limitations to cancer control. J. Natl. Cancer Inst. 63: 1291 (1979).
4. P Brookes, Covalent interactions of carcinogens with DNA. Life Sci., 16: 331 (1975).
5. T J Slaga, A Sivak and R K Boutwell (eds), Carcinogenesis Vol. 2. Mechanisms of tumour promotion and cocarcinogenesis. Raven Press New York (1978).
6. T J Slaga, S M Fischer, K Nelson and G L Gleason, Studies on the mechanism of skin tumour promotion: evidence for several stages in promotion. Proc. Natl. Acad. Sci. 77: 3659 (1980).
7. Royal College of Physicians, Smoking or Health (1977) Pitman, London.
8. J F Fraumeni, Respiratory carcinogenesis: an epidemiologic appraisal. J. Natl. Cancer Inst. 55: 1039 (1975).
9. E L Wynder and S Hecht (Eds), Lung Cancer, UICC Technical Report Series, Vol. 25. International Union Against Cancer, Geneva (1976).
10. G M Cohen, R Mehta and M Meredith-Brown, Large interidividual variations in metabolism of benzo(a)pyrene by peripheral lung from lung cancer patients. Int. J. Cancer, 24: 129 (1979).
11. H Autrup, F C Wefald, A M Jeffrey, H Tate, R D Schwartz, B F Trump, and C C harris, Metabolism of benzo(a)pyrene by cultured tracheo-bronchial tissues from mice, rats, hamsters, bovines and humans. Int. J. Cancer, 25: 293 (1980).
12. I J Selikoff and E C Hammond, Multiple risk factors in environmental cancer. Persons at high risk of cancer – an approach to cancer etiology and control, pp 467-483. J F Fraumeni (ed). Academic Press, New York (1975).

DISCUSSION

LECTURER: Cohen   CHAIRMAN: Cumming

CUMMING: Perhaps I can begin. The dream of Paul Ehrlich
               which was to create a magic bullet is very much
               your concept of gluconate conjugations. Would this
               happen to the other cells' detoxicating mechanisms
               in the body that used the same pathway, if you
               could get such a drug?

COHEN: It would be nice to think that we could design a
               drug which was acting in that way. The problems
               are still immense. We have to remember that the
               early concept of selective toxicity was one which
               met great success because we were dealing with
               specific bacteria. In some forms of cancer
               chemotherapy we have also been extremely successful
               but cancer is a multitude of different diseases and
               that has to be accepted first. Some of those,
               particularly faster growing tumours such as a colon
               carcinoma, have been treated very successfully,
               others have been far less successful. To go more
               specifically to your point we would need to direct
               the chemicals specifically to the targer tissue and
               the ways this might be done in terms of lysosome
               entrapment or immunological mechanisms. If we
               could design the right drug I think the delivery
               process would follow without too much trouble.

CUMMING: So you really need a magic bullet with a built-in
               guidance system.

LAURENT: I would like to comment on reversibility in the
               initiator. It seems to me that there are two
               mechanisms which may operate which would mean that
               this is not an irreversible step; one is that
               cells are continually turning over so that if you
               did have such a binding with an effect on DNA, that
               cell will be lost fairly soon so that's the first
               way of preventing irreversibility. The second is:
               there are varieties of enzymes that are capable of
               correcting defective sequences in DNA in vivo and
               it may be possible that such mechanisms may be
               operative. It seems to me that it's far too simple
               a scheme to have an initiator which is irreversible
               and I'd like to suggest that the decreased
               incidence of cancer that we have seen in smokers
               after a time they had given up smoking, may be

related in some way to these mechanisms as well as some of the promoting effects you were talking about.

COHEN: To take the point about DNA repair first. I didn't go into all the data which were available suggesting that promotion is now appearing to be important in many other models as well as in the mouse. Nor did I go into some of the experimental data suggesting two-stage carcinogenesis or multi-stage carcinogenesis in other organs. But I did briefly mention you can paint Benzpyrene in a dose which produces no tumours on the mouse skin and you can leave that for up to a year. If repair was important you should be able to repair that particular lesion. If you do not repair the lesion and you then apply the tumour producing agent and get the same cancer incidence, it doesn't seem to me that DNA repair in that model is operative.

LAURENT: I suppose we were both speaking in generalities; I would like to suggest in that particular case that you may be dealing with tissue which isn't disturbed by the rapid turnover in the epithelial cells.

COHEN: The skin turns over in a few days and all the epithelium comes off. Perhaps you can visualise it best by considering the basal cell which has been initiated – that would be on the basement membrane. It then divides into daughter cells, but for the initiator to remain all you have to visualise is that one daughter cell will stay on the basement membrane as a basal cell, the other cells will divide and then survive. In that model the cells are constantly turning over but one initiated cell still remains with any number of initiated cells present.

DENISON: You presented a model of cancer, as I understand it, in which cells are given a wrong instruction but eventually this wrong instruction has to appear on the outside of the cell as an abnormality of behaviour. Is there any hope of attacking cancer by that route rather than by preventing the cells being given the wrong instruction? Can you identify that form of tumours by some behavioural quirk rather than a defect of command?

COHEN: If we consider the effect of a tumour promoter and
                illustrate it by an example. A tumour promotor
                will cause many different effects in a cell,
                probably its major effect being to give the
                transformed phenotype and this involves cell
                surface changes, that then would enable the cell to
                be invaded. So we can think either of the tumour
                promotor or other phenomena causing similar effects
                to a tumour promotor. We think in terms of
                plasminogen activator or anchorage independent
                growth or effects on cell membrane. All these are
                later effects of tumour promotors and are what you
                are referring to in terms of the D-differentiated
                form of the initiated cell, when it begins to
                inhibit the tumour phenotype. What you want the
                cell to do is somehow we programme it so that it is
                no longer the tumour phenotype and you can perhaps
                do that by masking changes in the cell surface. A
                classical example is if you take malignant cells,
                let's say a malignant fibroblast, and grow them in
                culture. They are very unsociable in culture and
                show a typical picture of transform cells in that
                they pile up, one on top of the other and form
                criss-cross colonies. That's a characteristic of a
                malignant transformed fibroblast. If you have a
                normal fibroblast it will not pile up but will grow
                just to confluency, without transformed fibroblasts
                and you add a promotor which affects the cell
                surfact, then those malignant transformed
                fibroblasts in the presence of the promotor will
                just form a model, but they won't form criss-cross
                colonies nor will they pile up. Something has
                apparently got them temporarily to stop the
                expression of the tumour phenotype.

KAY: I enjoyed what you said very much, I enjoyed your
                talk and what laudable impartiality you have in
                your field, in the sense that it is obviously one
                of the many areas of carcinogenesis that have to be
                considered. I wonder if you could clarify one
                point. Am I right in saying that in one part of
                your talk you were explaining how there can be
                gross misunderstanding in the terms of substitution
                of bases which could lead to a neoplastic situation
                and you went on to give an example of this and you
                landed up with a group which I think had an
                hydroxyl group and it was linked to an amide group.
                Did you actually do the experiment of taking the
                metabolite from the carcinogen, labelling it,

introducing that into a system and showing the sort of effect that you would expect from the previous site.

COHEN: I have not myself done that work but it has been done by others in the literature. Workers in the United States exposed cells to that kind of situation I showed on the slide. If they allowed the cell to replicate before the lesion has been removed, then they obtain mutations in the cells. If on the other hand they allow the cells to repair and remove the lesion before they then sub-culture them and allow them to express themselves, then they obtain mutation, in other words the data they obtained were perfectly compatible with what I suggested, it was that the presence of those lesions in the DNA is mutagenic, if you remove that lesion as in that example it is not mutagenic. But let me say that's the principle for that particular carcinogenesis. If we go to other carcinogens it could be that the damage occurs only when you remove the lesion, in other cases it might be the presence which is important. The system is complicated, each carcinogen is different and it would depend on where the lesions are, some lesions in DNA are mutagenic, others are not mutagenic and that's probably best illustrated with nitrosamines.

ROUSSOUW: Mr. Chairman, just a short question on the technical procedures. I have noticed that you have done your centrifugation experiments on the 10,000 fraction. Have you done further differential centrifugation to see what the differences between other products are?

COHEN: If we start with naphthalene which doesn't have an hydyroxyl group, then you have to consider mixed function oxidation because the naphthalene would be converted to naphthol. We by-pass that pathway when we start with naphthol and when we use the intact tumour tissue, then we obtain only two major metabolites which are the sulphate and the glucuronide. So whilst mixed function oxidation is important and we were looking at that in several experiments, it is not important in that particular compound.

BELTRAMINI: I have got two questions. The first one is: what do you think is the role played by retinoids in the carcinogenesis process?

COHEN: Let me answer the first one first. Can we have the reserve slide 3 please? The question was: what is the role of the retinoids and it is an extremely important role in carcinogenesis. Slayden and co-workers in the States have recently described in a very elegant series of experiments that it is in 3 or 4 stages and they have been able to define it clearly. The first stage is initiation which is induced by the TPA, which according to them is the critical step – there is the induction of the primitive skin cell. The second stage can be performed both by QPA and by many other compounds, appears to involve polyamines and it is this second stage which is susceptible to retinoids. The polyamine step is probably susceptible to the retinoids and there is a lot of evidence implicating a relationship between vitamin A and lung cancer. I'd just like to mention two studies, one done by Basson and co-workers which showed that people with lung cancer had lower levels of vitamin A, but one does not know if that was a result of the disease or predisposition to the disease. Secondly the early work which in the lung cancer field implicated vitamin A was that of Saffrotti. They had an experimental model of respiratory tract cancer in a hamster where they gave thoric oxide to animals which induced tumours of the trachea histologically similar to that of man. If they gave thoric oxide first, which was by repeated intratracheal installations, they obtained tumours. However if after they finished the carcinogen administration then they give the animals vitamin A the obtained a decreased tumour incidence, in other words the vitamin A in that model was decreasing tumour incidence by acting most probably at a tumour promotion step or one of the latter stages in cancer formation rather at an initiating step.

THE MICROSOMAL NADPH OXIDATION SYSTEM IN LUNG CANCER

R. G. Butcher

The Midhurst Medical Research Institute
Midhurst, West Sussex

The lung is exposed to a wide variety of chemicals both from the environment and in the systemic circulation. These chemicals may be metabolised by the lung possibly increasing or decreasing their toxicity or pharmacological action. The first stage of this metabolism (Phase I reactions; see Cohen (1981), this volume) is carried out in the endoplasmic reticulum (microsomes) of the cell and is catalysed by a mixed function oxidase system (Fig. 1).

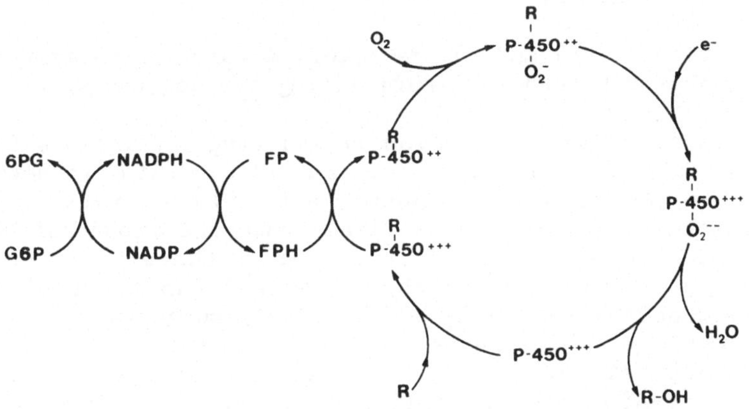

Fig. 1. Schematic representation of the electron flow from NADPH in the endoplasmic reticulum. Substrate is indicated as R and product as R-OH; FP is the flavoprotein, NADPH cytochrome reductase. Modified after Remmer (1975) and Estabrook et al (1978).

439

Cytochome P-450 is the terminal oxygen-activating enzyme catalysing the hydroxylation of diverse substrates in the presence of NADPH and molecular oxygen.   The flavaprotein NADPH cytochrome reductase transfers reducing equivalents from NADPH to cytochrome P-450, which is reoxidised by atmospheric oxygen.

However, NADPH serves not only as a source of electrons for the mixed function oxidase system.   It is also essential for a number of different biosynthetic reactions:

a)   In the biosynthesis of fatty acids both in the mitochondria and in the cytoplasm.

b)   In the biosynthesis of cholesterol.

c)   In purine biosynthesis to form inosine monophosphate (IMP).

d)   As the coenzyme for thioredoxin reductase in the conversion of ribonucleotides to deoxyribonucleotides.

e)   In the conversion of folic acid to tetrahydrofolate:-

dihydrofolate + NADPH   tetrahydrofolate + NADP$^+$.

Tetrahydrofolate serves as an intermediate carrier of methyl, hydroxymethyl and formyl groups in a large number of one carbon transfer reactions involved in the intermediary metabolism of amino acids, purines and pyrimidines.

Reactions c) d) and e) are likely to be particularly important for active growth such as occurs during carinogenesis.

In normal tissues the flow of reducing equivalents from NADPH in the endoplasmic reticulum is channeled according to need towards either the reduction of cytochrome P-450 and oxygen or towards biosynthetic reactions.   It is likely that the amount of NADPH used by cells at any given time for either of these two requirements will vary according to the physiological functions of the cell. Recent evidence reviewed by Apffel (1978) suggests that during the carcinogenic process, there is an impairment of transport in the endoplasmic reticulum from NADPH to atmospheric oxygen via the cytochrome P-450 pathway leading to a greater influx of reducing equivalents into the biosynthesis of proteins, purines and pyrimidines.

The major source of NADPH is the pentose phosphate pathway of glucose oxidation (alternatively known as the hexose monophosphate shunt and the pentose shunt).   The first two steps in the pathway involve dehydrogenation reactions which require the coenzyme NADP$^+$. Glucose-6-phosphate dehydrogenase catalyses the oxidation of glucose-6-phosphate to 6-phosphogluconate with the concomitant production of NADPH. The 6-phosphogluconate is oxidised in the presence of the enzyme 6-phosphogluconate dehydrogenase and NADP$^+$ to yield a pentose sugar and more reduced coenzyme. This pathway thus provides the ribose essential for the synthesis of RNA and DNA and therefore for protein synthesis and nucleic acid replication.

Evidence from Weber (1977 a and b) indicates that certain enzymes in cancer metabolism are key enzyme. Particularly important amongst these are the transformation-linked discriminants, enzymes that are increased or decreased in all cancers, the change occuring at an early stage in the development of a neoplastic cell from a normal one. Glucose-6-phosphate dehydrogenase is such a transformation-linked discriminant, and numerous studies have shown a markedly increased activity of this enzyme in carcinomata from many different sites in man (for review see Livni and Laufer, 1975).

Methodological Approach

From the above considerations it would seem important to be able to measure four separate activities in normal and cancer tissues:-

1. The optimal activity of NADPH cytochrome reductase.

2. The total activity of glucose-6-phosphate dehydrogenase.

3. That proportion of the total NADPH derived from the dehydrogenation of glucose-6-phosphate which is used in biosynthetic reactions.

4. The rate of use of NADPH, derived from glucose-6-phosphate oxidation, in mixed function oxidase activity.

How can these different biochemical parameters be studied in normal bronchial epithelium and in carcinoma of the bronchus? Many of the earlier chapters in this volume have highlighted the extremely heterogenous nature of the normal lung; a study of the biochemical characteristics of any individual cell type by conventional homogenisation procedures is almost certainly not possible. Moreover the cell population of the lung is going to change during carcinogenesis; any change in the biochemical properties of the lung may reflect only changes in the cell population.

It would be advantagous to relate the biochemical activity directly to the cell from which it is produced. To do this it is

necessary to precipitate the end product of the chemical reaction
in that cell and to measure the amount of reaction product at its
site of precipitation. This can be achieved by the use of thin
tissue sections; such sections allow free diffusion of reactants
from the surrounding incubation medium into the protoplasm even
while the enzymes remain in their natural matrix. Moreover, since
only a few sections are required for each biochemical measurement,
several assays can be obtained from a small piece of tissue such as
might be available from a human lung biopsy procedure. Of most
importance, each biochemical activity can be related directly to
the structure by reference to a serial section stained in the
conventional manner (for a review of this approach see Chayen et
al. 1974).

        Due consideration must be given to optimal biochemical
incubation conditions; conditions conducive to maximal
precipitation of the end product; and the stoichiometry of the
reaction to permit the chemical activity to be related to unit time
and unit volume of tissue. Recent advances in the development of
these techniques have included:- the incorporation into the
incubation medium of an inert colloid stabiliser to prevent loss of
proteins, including enzymes, from the undenatured tissue section
(reviewed by Altman, 1980); the quantitation of the end product by
scanning and integrating microdensitometry (Butcher, 1972); the
determination of the absorption characteristics and extinction
coefficients of the end products to enable the activity to be
calculated in absolute units (Butcher and Altman, 1973); and the
direct measurement of the enzyme kinetics in an individual cell or
a group of cells of the same type (Pette et al, 1979).

The Determination of Reducing Equivalents in Tissue Sections

        Oxidative metabolic activity can be demonstrated in tissue
sections by the incorporation into the incubation medium of a
tetrazolium salt. On reduction this is converted to a coloured
insoluble formazan which is precipitated at the site of the
biochemical activity. The activities of several oxidative enzymes
determined from serial sections of a human bronchogenic carcinoma
are shown in Table 1. Sections of bronchial epithelium from the
same lung but from a site distant from the tumour were used as
controls. As in all the studies discussed in this chapter, every
third and subsequent fifth serial section was stained with
haematoxylin and eosin, so that no section used in the biochemical
assays was more than 2 sections (20 um) distant from the
histological reference.

        Of the reactions studied and shown in Table 1, the activities
of the two dehydrogenases of the pentose-phosphate pathway were the
only ones to show a significant change. It is particularly
noteworthy that the activity of the mitochondrial isocitrate

dehydrogenase, an enzyme which generates NADPH, was only slightly elevated in the cancer cells.

TABLE I

The activities of various dehydrogenases in control bronchial epithelium and in carcinoma of the bronchus

DEHYDROGENASE	ADDED COENZYME	$nmolH_2 \cdot cm^{-2} \cdot 10 \ min^{-1}$	
		CONTROL	CARCINOMA
Succinate	-	25.9	23.4
Glycerophosphate	-	21.0	21.6
Malate	NAD^+	13.1	18.7
Glutamate	NAD^+	3.4	3.9
Lactate	NAD^+	32.8	39.2
Isocitrate	$NADP^+$	4.2	6.5
Glucose-6-phosphate	$NADP^+$	30.9	345.3
6-Phosphogluconate	$NADP^+$	20.6	267.8

It is evident from the above discussion that NADPH is used in the endoplasmic reticulum in two quite dissimilar functions. Reducing equivalents are produced at an electrode potential of about -320 mV and biosynthetic reaction which require NADPH probably take place at a potential close to this value. In contrast the reduction of cytochrome P-450 by reducing equivalents which are oxidised through the microsomal electron transpent chain almost certainly occurs at a considerably lower (i.e. more electropositive) potential. Chayen et al (1974) suggested that the amount of NADPH used in the two different pathways could be determined by measuring the amount of reduction which occurs at these two different potentials. The more electropositive tetrazolium salts, such as neotetrazolium chloride, cannot be reduced by NADPH when it and the tetrazole are mixed together at a physiological pH in a test tube (Table II; reaction 2). Yet formazan is precipitated in a tissue section incubated in such a mixture. The enzyme responsible for this reduction is almost certainly NADPH cytochrome reductase (Table II; reaction 3). It follows that, if a section is incubated with the substrate, glucose-6-phosphate, oxidised NADP$^+$ and neotetrazolium chloride, then the tetrazole will be reduced (reaction 4). The amount of formazan precipitated cannot, of course, exceed the optimal rate of activity of the NADPH cytochrome reductase system (reaction 3). The maximal rate of dehydrogenation of glucose-6-phosphate can be measured separately if an intermediate electron carrier, such as phenazine methosulphate (PMS) is incorporated into the incubation medium. This accepts electrons directly from NADPH and transfers them directly to the tetrazolium salt (reaction 5).

TABLE II

Reactions involving NADPH and the production
of neotetrazolium formazan

	SUBSTRATES	ADDITIONS	ENZYMES	PRODUCTS
1	G-6-P + NADP$^+$	-	G-6-P dehydrogenase	6-P-G + NADPH
2	NADPH	NT	-	NADPH + NT
3	NADPH	NT	NADPH reductase	Formazan
4	G-6-P + NADP$^+$	NT	G-6-P dehydrogenase NADPH reductase	Formazan
5	G-6-P + NADP$^+$	PMS NT	G-6-P dehydrogenase	Formazan

G-6-P = Glucose-6-phosphate
6-P-G = 6-Phosphogluconate
NT   = Neotetrazolium chloride
PMS = Phenazine methosulphate

Reducing equivalents from NADPH which are transfered through the microsomal transport system and can therefore reduce the neotetrazolium (reaction 4) have been defined by Altman and Chayen, (1970) as "Type I". The difference between this activity and the maximum rate of generation of NADPH (i.e reaction 5 - reaction 4)

is a measure of those reducing equivalents used at the negative
potential of about -320 mV in biosynthetic reactions.   This NADPH
had been designated "Type II" (Altman and Chayen, 1970).

Support for this concept has come from results using sections
of different tissues (Chayen et al, 1974).   In rat adipose tissue,
which has a high biosynthetic requirement, 97% of the NADPH
generated from the dehyrogenation of glucose-6-phosphate was of
Type II.   In contrast in the adrenal cortex 81% of the NADPH was of
Type I; the adrenal requires reducing equivalents for steroid
hydroxylation, reactions which occur through the mixed function
oxidase system.

Type I, Type II and NADPH reductase activities in lung cancer

Tissue section techniques have been used to study the reducing
equivalents generated from the oxidation of glucose-6-phosphate in
bronchogenic carcinomas.   Table III summarises the results from 14
lung carcinomas and from control epithelium taken from the same
lungs.   Whilst in control tissue the capacity to transfer electrons
from NADPH to O_2 is much greater than the demand, in cancer cells
this is not so.   Yet the total production of NADPH is vastly
increased in the carcinomas resulting in a much greater utilisation
of reducing equivalents in biosynthetic reactions.

TABLE III

Type I, Type II and NADPH cytochrome reductase activities
in control bronchial epithelium and in carcinoma of the bronchus

	$nmol H_2 . cm^{-2} . 10 min^{-1}$	
	CONTROL (14)	CARCINOMA (14)
NADPH reductase	13.4 + 2.4	11.0 + 3.4
Type I	6.0 + 1.4	10.9 + 3.2
Type II	22.5 + 16.5	204.2 + 87.9

Histological examination of serial sections of the control
tissues indicated that some epithelia showed no abnormalities
("normal" controls) whilst others displayed hyperplastic and/or

metaplastic changes ("abnormal" controls). A separation of the
results into these two groups is shown in Fig. 2. The greatest
difference was seen in the type II activity; the amount of NADPH
used for biosynthetic functions was considerably increased in the
actively growing hyper- and metaplastic epithelia.

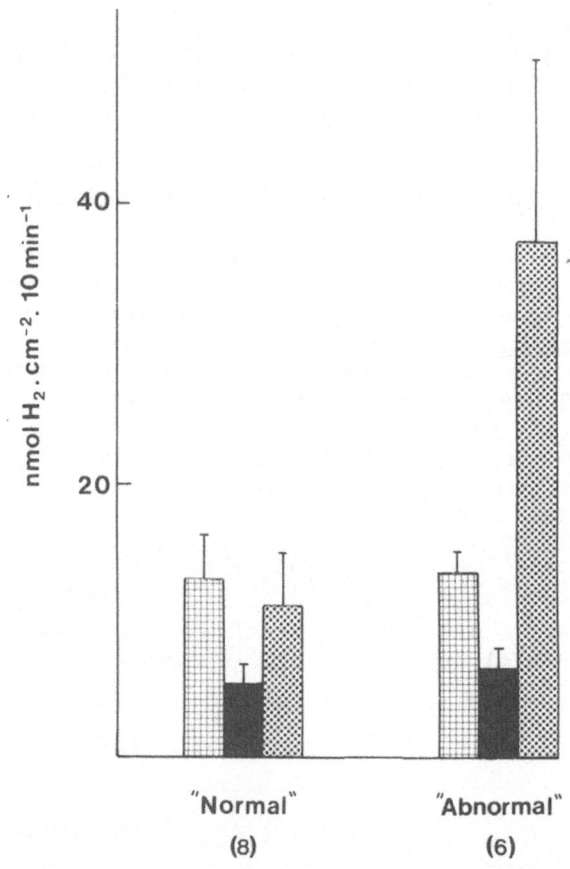

Fig. 2.   Activities in "normal" and "abnormal" control bronchial
          epithelia: ▣ NADPH cytochrome reductase;   ■ Type I;
          ▨ Type II.

        Decreased levels of NADPH cytochrome reductase activity have
been shown in several studies of rat hepatomas (e.g. Kato et al,
1968; Miyake et al, 1974). A more detailed analysis of the
individual results indicated that this was true also in 5 of the 14
lung cancers studied here (Fig. 3). More striking was the finding
that in nearly all the bronchogenic carcinomas, the rate of
transfer, along the microsomal transport chain, of NADPH generated

from the dehydrogenation of glucose-6-phosphate, was equal to the maximum capacity of that pathway as measured by the addition of exogenous NADPH. This would lead, as postulated by Apffel (1978), to a greater influx of reducing equivalents into protein and nucleic acid synthesis.

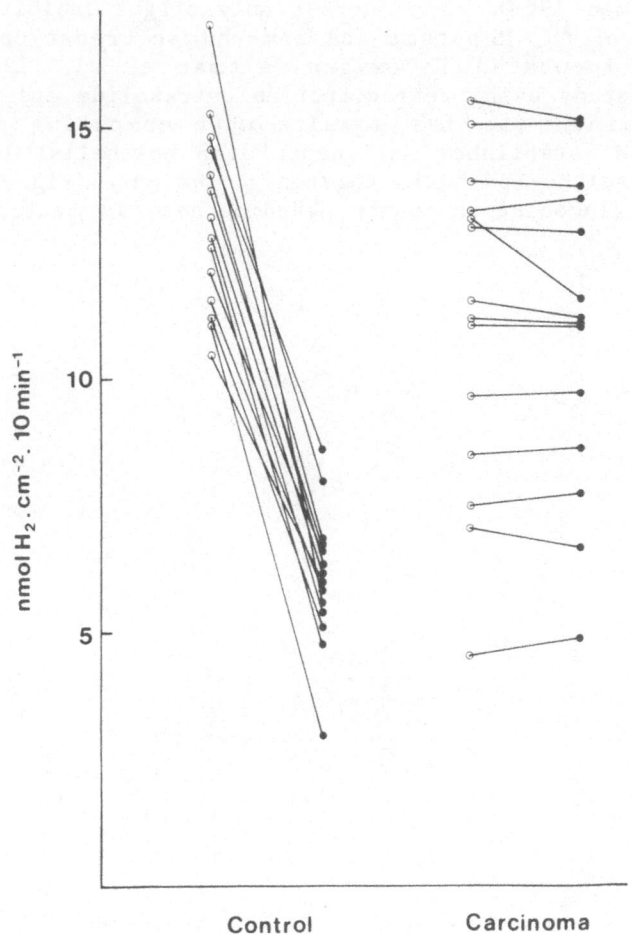

Fig. 3. The individual activities of NADPH cytochrome reductase (O) and the utilisation of NADPH through the microsomal transport pathway (Type I, ●) in control epithelium and in bronchogenic carcinomas.

The Effect of Oxygen

The total production of NADPH (Type I and Type II) is
visualised in tissue sections by the addition of both
neotetrazolium (NT) and phenazine methosulphate (PMS) to the
incubation medium (Table 2; reaction 5).   Incubation media
containing these compounds are usually saturated with nitrogen
since the production of formazan in inhibited by oxygen in normal
tissues (Altman 1969).   In contrast only slight inhibition occured
in sections of rat hepatoma and some human breast cancers when
these were incubated in oxygen (Altman et al, 1970).   In a
qualitative study using tetranitroblue tetrazolium and PMS, Heyden
(1974) obtained similar results when comparing potentially
malignant and established malignant human epithelial lesions with
clinically healthy epithelia and benign lesions.   Fig. 4 shows the
activity of glucose-6-phosphate dehydrogenase in sections from 67

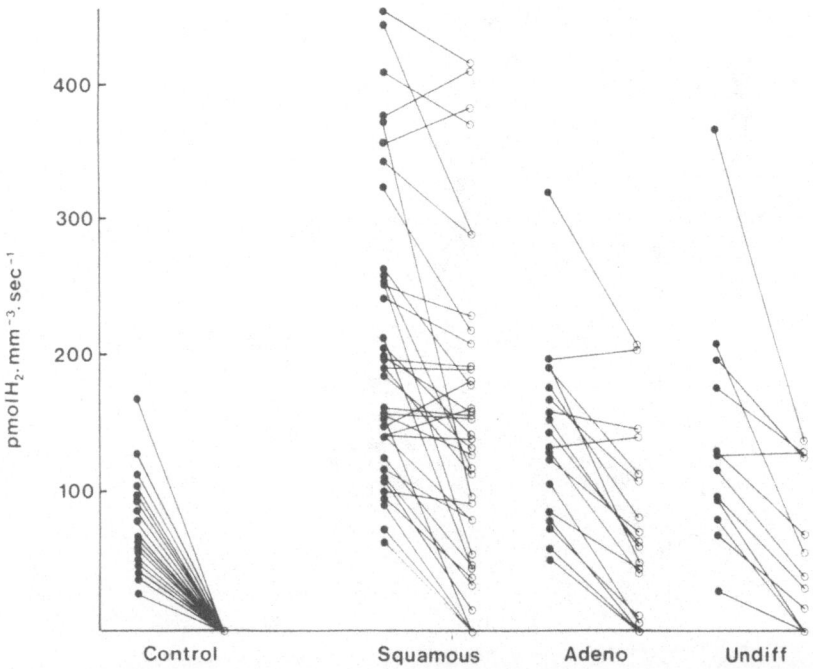

Fig. 4.   Glucose-6-phosphate dehydrogenase activity in control
           epithelia and in bronchogenic carcinomas:- (●) in ni-
           trogen; (○) in oxygen. 5 min incubation. (From Butcher,
           1979: reproduced with permission.)

human bronchogenic carcinomas and in control bronchial epithelium (Butcher, 1979). All the control epithelia, irrespective of their anaerobic activity, proved sensitive to oxygen. Only ten of the carcinomas were similarly sensitive; in the other 57 specimens, cancer cells were active when treated in a medium saturated with oxygen (Fig. 4). Analysis of the mechanism of inhibition in normal tissues (Butcher 1978) has indicated that this is due to direct competition between oxygen and the tetrazolium salt for the reducing equivalents from NADPH. The exact nature of this inhibition is unknown but may involve the formation of free radicals (Nishikimi et al, 1972; Sato and Iwaizumi, 1969) and be dependent on the presence of cytochrome P-450. Certainly many observations indicate a decrease in or total deletion of microsomal cytochromes in cancer cells (see review by Apffel, 1978).

REFERENCES

Altman, F.P. (1969).   The use of eight different tetrazolium salt
    for a quantitative study of pentose shunt dehydrogenation
    Histochemie 19, 363-379.
Altman, F.P. (1980).   "Tissue stabilizer methods in histochemistry
    In: "Trends in Enzyme Histochemistry and Cytochemistry."   Cib
    Foundation Symposium 73. Excepta. Medica. Amsterdam 81-101.
Altman, F.P. and Chayen, J. (1970).   Evidence for two types o
    hydrogen atom in reduced micotinamide-adenine dinucleotid
    phosphate arising from glucose-6-phosphate oxidation, based o
    the inhibitory action of certain steroids. Biochem. J. 118, 6P
Altman, F.P., Bitensky, L., Butcher, R.G. and Chayen, J. (1970)
    Integrated cellular chemistry applied to malignant cells.   I
    "Cytology Automation" (ed. D.M.D. Evans), Livingstone
    Edinburgh.   82-97.
Apffel, C.A. (1978).   The endoplasmic reticulum membrane system an
    malignant neoplasia. Prog. exp. Tumor Res. 22, 317-362.
Butcher, R.G. (1972).   Precise cytochemical measurement o
    neotetrazolium formazan by scanning and integratin
    microdensitometry.   Histochemie 32, 171-190.
Butcher, R.G. (1978).   Oxygen and the production of formazan fro
    neotetrazolium chloride.   Histochemistry 56, 329-340.
Butcher, R.G. (1979).   "The oxygen insensitivity phenomenon as
    diagnostic aid in carcinoma of the bronchus".   In "Quantitativ
    cytochemistry and its applications".   (eds. J.R. Pattison
    L. Bitensky and J. Chayen) Academic Press.   London, 241-251.
Butcher, R.G. and Altman, F.P. (1973).   Studies on the reduction o
    tetrazolium salts. II.   The measurement of the half reduced an
    fully reduced formazans of neotetrazolium chloride in tissu
    sections.   Histochemie 37, 351-363.
Chayen, J., Altman, F.P. and Butcher, R.G. (1974).   "The effect o
    certain drugs on the production and possible utilization o
    reducing equivalents outside the mitochondria".   In
    "Fundamentals of cell Pharmacology" (ed. S. Dickstein) Thomas
    Springfield.   196-230.
Cohen, G.M. (1981).   "The metabolism of chemicals by the lung'
    This volume.
Estabrook, R.W., Werringloer, J., Capdevila, J. and Prough, R.A
    (1978).   "The role of cytochrome P-450 and the microsoma
    electron transport system: The oxidative metabolism of Benzo (
    pyrene.   In "Polycyclin Hydrocetoes and Cancer." Vol. I. (eds
    H. V. Gelboin and P.O.P. Ts'o).   Academic Press. N.Y. 285-319.
Heyden, G. (1974).   Histochemical investigation of malignant cells
    Histochemistry 39, 327-334.
Kato, R., Takanaka, A. and Takahashi, A. (1968).   Drug metabolis
    in tumor-bearing rats. I. Activities of NADPH-linked electro
    transport and drug metabolizing enzyme systems in live
    microsomes of tumor-bearing rats. Jap. J. Pharmacol. 18, 224
    244.

Livni, N. and Laufer, A. (1975). Histochemical studies of human breast tumours. Activity of alkaline phosphatase, acid phosphatase and glucose-6-phosphate dehydrogenase. Pathol. Microbiol. (Basel) 42, 159-170.

Miyake, Y., Gaylor, J.L. and Morris, H.P. (1974). Abnormal microsomal cytochromes and electron transport in Morris hepatomas. J. biol. Chem. 249 1980-1987.

Nishikimi, M., Appaji Roco, N. and Yagi, K. (1972). The occurence of superoxide anion in the reaction of reduced phenazine methosulphate and molecular oxygen. Biochem. Biophys. Res. Comm. 46, 849-854.

Pette, D., Wasmund, H. and Winmer, M. (1979). Principle and method of microphotometric enzyme activity determination. Histochemistry 64, 1-10.

Remmer, H. (1975). "Pulmonary drug-metabolizing enzymes". In "Lung Metabolism" (eds. A.E. Junod and R. de Haller) Academic Press. London. 133-158.

Sato, S. and Iwaizumi, M. (1969). Free radical mechanism by which triphenyl-tetrazolium chloride stimulates aerobic oxidation of NADPH by microsomes. Biochim. Biophys. Acta. 172, 30-36.

Weber, G. (1977a). Enzymology of cancer cells. Part I. New England J. Med. 296, 486-493.

Weber, G. (1977b). Enzymology of cancer cells. Part II. New England J. Med. 296, 541-551.

DISCUSSION

LECTURER: Butcher   CHAIRMAN: Cumming

CUMMING: This paper is now open for discussion.

BAUM: Firstly congratulations on some very very elegant
                techniques, which were most impressive, only a
                biochemist who has struggled to measure these
                things other ways can fully appreciate I think how
                elegant it is. A specific technical question:
                what happens to your controls if instead of using
                nitrogen you do it in oxygen, but in the presence
                of a respiratory inhibitor such as an antimycin?

BUTCHER: The results are extremely confusing, I haven't
                tried antimycin, I have tried cyanide and it
                depends to a certain extent on what electron
                acceptor you were using.

BAUM: Because if that is the case then it does imply that
                your reduced PMS is being oxidised not directly
                non-catalytic by oxygen, but through the
                respiratory chain.

BUTCHER: Yes, I am myself completely convinced that the
                mechanisms of a reaction between PMS in a test
                tune is very different to what happens in the
                tissue. One of the problems that we have is to get
                a direct comparison on two sections between that
                reaction and the amount of static P 450 that we
                have in the tissue because I think that is what we
                are picking up, we are picking up a measure of how
                much P 450 is available and of course there is a
                lot of experimental evidence which shows that in
                carcinoma the levels of P 450 are very much reduced
                and so if we are dependent on this for our reaction
                and there isn't any in the carcinoma, then what we
                would expect is to get an inhibition. We can't do
                that in the reticulum system. We can measure a
                similar thing by comparing dehydrogenase and
                cytochrome oxidase in the mithochondria in the
                presence and absence of oxygen and there is a
                remarkable correlation between the amount of
                inhibition that one gets and the level of an
                oxidase activity, so I think it's probably a true
                phenomenon and it is probably measuring the amount

of cytochrome P 450 but I would love to have a technique that would allow us to do that directly in the tissue.

BAUM: Alternatively, and I think this was almost implied by what you said, it might be that what one is measuring is the amount of oxidase. In other words you might be measuring how many functional mitochondria you have. The more functional mitochondria you would have, the less sensitive and the more nearly you are to Barbar's view.

BUTCHER: May I ask how easy would it be in that system to get NADPH generated through mithochondria and transferring via cytochrome oxidase.

BAUM: I have looked at hydrogenation activities and it's very likely.

BUTCHER: But what would be required for that to happen?

BAUM: There are plenty of enzymes which could do that.

KAY: I wanted to ask you if you could clarify a point about accompanying histological features. Some of the tumours that you have been studying have varying degrees of acute and chronic inflammatory cells as well as degrees of necrosis and differentiation. That being so, you have other sources of electrons for reducing your P 450. Do you agree that this in fact could be a problem? Do you find any difference between the degree of inflammation and your biochemical findings?

BUTCHER: I haven't seen any effect, but of course one of the advantages of this technique is that one can measure a specific area so one can exclude such cells from the measurement and you can be sure that you are measuring cancer cells and not some other group of cells. This is one of the beauties of the technique. Having said that I agree with you that insofar as histological classification is concerned it is extremely difficult because within any one section you do not have a biochemical moderator which gives a single answer. There is a tremendous range of cells, some of which may be more necrotic than others, some of which may be mixed or as you say, inflammatory cells and so on. At the moment all we can do is get a mean measure, the mean

measure of the activity of that particular specimen
by doing random measurements throughout the
section. Of course ideally it would be nice to do
two things: one, have a direct relationship
between that measurement and its histological
findings and ultimately I would hope to measure
that particular type of cell in normal bronchial
epithelium under a certain activity. At the moment
we are at the level of a group of cells within the
epithelium or a group of cells within the
carcinoma. I would love to refine the techniques
to be able to talk about activity per cell. It's
unfortunate that the tumour cells required to
produce discrete localisation which would allow us
to do that are not the tumour cells which are
easily quantified and are not the sort of cells
which would allow us to get this differentiation
between for example type 1 and type 2 hydrogen, but
I think in time we will be able to do that.

LAURENT: Could I also congratulate you on the methodology
and tell you that despite some of your problems I
think it is well worth pursuing this method; it's
so nice to be able not only to make the
measurements but see where these changes are
occurring. Two comments: firstly a technical one,
the fit that you showed with the very large trend
in deviation meant to me that you must be getting a
lot of measurements of 0 or near 0 because your
number was in fact fairly small and about the same
size as your deviation.

BUTCHER: Sorry, may I stop you? This is from hydrogenase
activity in the control group?

LAURENT: It suggests to me that you have some tissues in
which nothing is occurring. Do you have problems
with proteolytic enzymes interfering with your
enzyme systems and normally in vitro one tries to
get around this problem using inhibitors, you used
inhibitors in your system. Secondly, would you
like to comment on the significance if you are
brave enough to have a shot at it, at this oxygen
inhibition that we were seeing in the carcinoma?

BUTCHER: Let me take the second point first. As I said in
answer to Harold Baum's question, I am convinced
that it is almost certainly due to one of the
cytochromes or part of the cytochrome system. It

would be nice to think that in fact it was
cytochrome P 450 because that would fit with the
liaterature and as Harold says it could well be
cytochrome oxidase because there are a number of
things which reduce mitochondrial oxidase in
carcinoma. I wouldn't like to speculate on that.
To go back to the first point, that large deviation
was caused not so much by having tissues with zero
activity but by just one specimen which happened to
be the one that displayed squamous metaplasia, and
that had a considerably higher activity than the
rest. That specimen had, I think, a mean of around
50 and that one had something around 150/170 in
total of about 10 or 12 results. Certainly it
would be nice to do control experiments in terms of
adding specific inhibitors but we haven't done
that. The reactions are controlled in the sense
that one eliminates non-specific reactions due to
just adding a NDA or glucose phosphate.

JEFFERY: Just two questions. One of them, have you had any
              information about any defects in metastases?
              That's the first question, the second thing is: is
              there any variations within the cells of the
              malignant tissues that you examined between the
              concentration and evidence of enzyme activity.   In
              other words, do the cells actually vary?

BUTCHER: One of the problems of this study is that of the
              availability of material and one of my jobs here is
              to look around and talk to all of you and see where
              I can get material. One of the things I would love
              to look at is metastases and I haven't looked at
              that. Of more interest, I would like to get hold
              of all sorts of earlier conditions and perhaps look
              and see just how early these changes do occur. I
              think I have partially answered your second
              question, which was similar to that put by Barry
              Kay. Yes, we do have variation, I don't know quite
              how to interpret this. One of the problems is in
              choosing the unit of reference by which one
              expresses activity, I have expressed activity per
              unit area and unit volume of tissue. One of the
              problems is that the size and the shape of the cell
              differs between a control and a carcinoma and of
              course if one expresses results in terms of
              activity per cell, one sometimes gets extremely
              different answers to express an activity per unit
              of tissue. All biochemists have this problem, so

some of the differences that I see may in fact be
due to that sort of phenomenon.   if one
extrapolates results per cell, one might in fact
get a wider variation.   I don't get to that stage.

CUMMING: Thanks Roger, I think we will call the discussion
to a halt at that point and go on to Martin Hetzel
who is going to talk to us.

INBORN SUSCEPTIBILITY/RESISTANCE TO LUNG CANCER

M.R. Hetzel*, M. Law*, E.E. Keal*,
T.P. Sloan[+], J.R. Idle[+] and R.L. Smith[+]

Brompton Hospital, London, England*
Department of Pharmacology, St. Mary's Hospital
Medical School, London, England[+]

INTRODUCTION

The association between bronchial carcinoma and cigarette smoking is well known (Doll and Hill, 1952; 1964) but, while smokers carry a much higher risk of lung cancer than non-smokers, the majority of them escape this disease. A substantial variation in individual susceptibility to the carcinogenic effects of smoking must therefore be present in the population.

The process of chemical carcinogenesis is still incompletely understood, but recent work supports the view that the suspected agents in cigarette smoke such as polycyclic aromatic hydrocarbons and nitrosamines require metabolic activation by oxidation to realise their carcinogenic potential (Miller, 1974). If this is indeed the case the host's ability to perform such oxidative reactions may be a major determinant in individual susceptibility to smoking-induced bronchial carcinoma.

It is now well established that a number of oxidative reactions show marked inter-individual variation and are controlled by a single gene locus of major effect.

Following the introduction of the adrenergic neurone blocking drug, debrisoquine in 1966, marked variations were noted in individual dose requirements and a minority of patients were unusually sensitive to its hypotensive effect. Angelo et al. (1975) demonstrated wide variation in urinary recovery of unchanged drug which correlated with clinical response. Debrisoquine is predominantly metabolised by hydroxylation of the alicyclic and aromatic rings as shown in figure 1 (Allen et al. 1976; Idle et al. 1979a). The major urinary metabolite is the 4-hydroxy derivative.

457

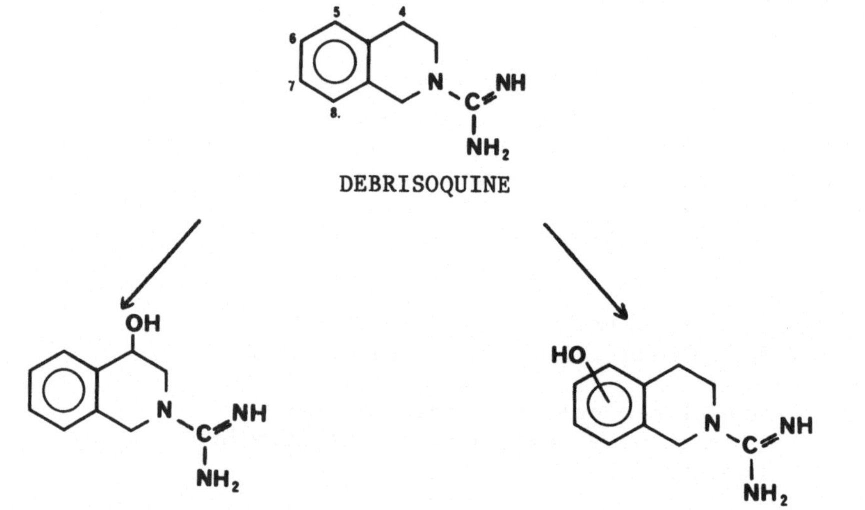

DEBRISOQUINE

4-HYDROXYDEBRISOQUINE   5-, 6-, 7-, 8-HYDROXYDEBRISOQUINE

Fig. 1.   The metabolism of debrisoquine in man.

A reliable measure of metabolic capacity for debrisoquine is the metabolic ratio (MR) i.e. :

% dose excreted as unchanged debrisoquine
-- in urine 0-8 h.
% dose excreted as 4-hydroxydebrisoquine

Polymorphism for debrisoquine metabolism is demonstrable in the normal population (Mahgoub et al., 1977; Idle and Smith, 1979; Evans et al., 1980). Extensive metabolisers (EM) and poor metabolisers (PM) constitute 91% and 9% of the population respectively. These two phenotypes are identified by their metabolic ratios as seen in figure 2. This shows the frequency distribution of MR in 258 normal British white Caucasian subjects on a logarithmic scale (Evans et al., 1980). The distribution is bimodal with an antimode at log MR 1.1. Poor metabolisers are defined as subjects with log MR >1.1 (or metabolic ratio $>$ 12.58). Family studies of 23 poor metabolisers show that this is an autosomal recessive trait (Evans et al., 1980). Thus at least two alleles, termed D^H (extensive metabolism of debrisoquine) and D^L (low oxidative metabolism) occur in the population. Individuals homozygous for the D^L allele are unable to metabolise more than 1-2% of a 10 mg oral dose, and thus constitute the PM phenotype. Individuals homozygous or heterozygous for the D^H allele readily metabolise 40-80% of the same dose. The frequency of the D^L allele in the British white Caucasian population has been estimated at

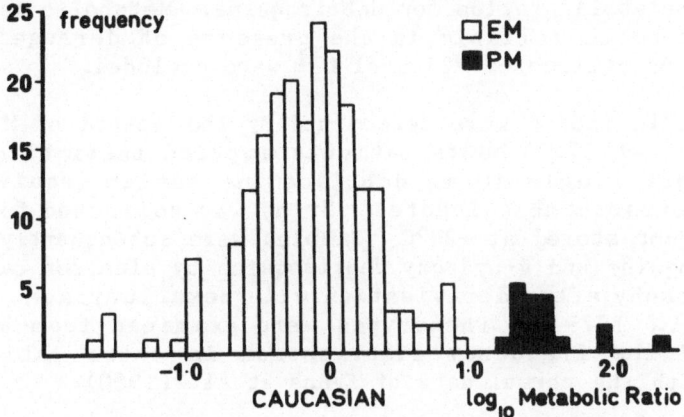

Fig. 2.   The frequency distribution of the $\log_{10}$ metabolic ratio
in 258 unrelated British white Caucasian subjects.
EM; extensive metaboliser, PM; poor metaboliser.

0.299 ± 0.030 (s.e.), giving rise to expected genotype frequencies
for $D^H D^H, D^H D^L$ and $D^L D^L$ of 49.2%, 41.9% and 8.9% of the population
respectively.   Studies of 21 $D^H D^L$ subjects identified as relatives
of propositi, indicated that the degree of dominance of D^H over D^L
was approximately 30%.   Thus whilst the $D^H D^H$ genotype cannot be
detected by the determination of metabolic ratio, very low ratios
should be more common in the homozygous genotype.

    Our study set out to compare the pheontype frequencies for
debrisoquine metabolism in smokers with bronchial carcinoma with
the normals described by Evans et al., (1980).   A significant
discrepency between the phenotype frequencies of the 2 groups would
then lend some support to the hypothesis that genetically
determined capacity for oxidative metabolism of debrisoquine also
influences ability to produce the proximate carcinogens from
cigarette tar, and hence cancer risk.

    British white Caucasian patients with bronchial carcinoma and
a smoking history were recruited.   The diagnosis was based on
clinical and radiological findings supported by positive histology
or cytology.   Smoking histories were documented as accurately as
possible and recorded in pack years.   Drugs known to be metabolised
in the same way as debrisoquine were avoided.   Inevitably, however,
the majority of hospitalised patients with bronchial carcinoma were
taking medication, particularly analgesics, which could not
ethically be discontinued.   Patients taking drugs not known to
compete with debrisoquine for metabolic oxidation or to induce
oxidising enzyme activity were accepted for study.   Drug histories
were carefully documented to facilitate detection of any consistent

effect on metabolic ratios for debrisoquine. Metabolic ratios have been shown to be reliable in the presence of deranged hepatic metastases or bilirubin >25 mmol L^{-1} were excluded.

Metabolic ratios were determined by the method of Mahgoub et al. (1977). At 07.00 hours patients emptied their bladders and were given a single 10 mg debrisoquine tablet (equivalent to 12.8 mg debrisoquine sulphate). Urine was collected for 8 hours and an aliquot stored at $-20^{\circ}C$. Samples were subsequently analysed for debrisoquine and 4-hydroxy debrisoquine by electron capture gas chromatography after derivisation with hexafluoroacetylacetone (Idle et al., 1979a). Phenotypes were examined from metabolic ratios and the frequency distribution in these patients was compared with the normal data of Evans et al. (1980).

106 patients (76 males and 30 females) gave informed consent to the study. Their mean age was 63.1 years (S.D. 8.7). 40 patients were defined as current smokers in that they were still smoking at the time of study or had given up within 3 months prior to study. Mean cigarette consumption for the whole group was 53.7 pack years (S.D. 34.9 years). Diagnosis was confirmed by histology at bronchoscopy, thoracotomy or mediastinal or peripheral node biopsy in 92 cases and by cytology alone in the remaining 14 cases. Cell types were reported as squamous (39 cases), large cell undifferentiatied (22), small cell (30), adenocarcinoma (11), undifferentiated (3), and bronchio-alveolar cell (1).

The distribution of log MR is shown for these 106 patients in comparison with the normal data of Evans et al. (1980) in figure 3. The observed incidence of PM cancer patients of 1.9% (2 in 106 patients) was significantly lower than the 8.9% (23 in 258 normals) PM frequency in normal controls ($2p < 0.02$). Moreover the 95% confidence limits for PM frequency in the lung cancer population (0.2-5.3%) were significantly lower ($2p= 0.0069$) than for the random population (5.7-12.7%).

Within the EM range for log MR the distribution of metabolic ratios in figure 3 in cancer patients is also different from the normal data; showing higher frequencies for both EM patients with very low MR and for EM patients with relatively high MR near the antimode. This different pattern of frequency distribution probably at least partly relates to the effect of concurrent treatment with other drugs. Several different drugs were taken during the study and numbers were often too small to analyse for any potential interference with the result for metabolic ratio. A significant effect was indicated in 23 patients taking distalgesic whose median MR (by rank) was 2.3 (log MR 0.36) which was significantly higher than a median MR of 1.1 (log MR 0.04) for all patients ($2p < 0.001$). Four patients also received cytotoxic drugs shortly before the study and had relatively high metabolic ratios.

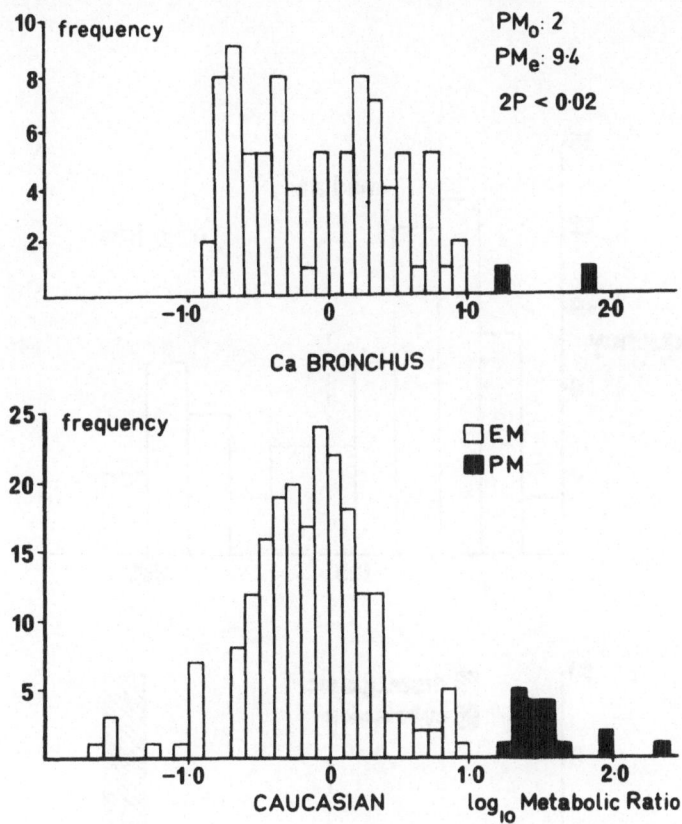

Fig. 3.   Comparison of the frequency distribution of the $\log_{10}$
          metabolic ratio in 106 British white Caucasian smokers
          with bronchial carcinoma with the normal population shown
          in Fig. 2.

Figure 4 shows metabolic ratios in rank order (for all
patients and controls). In the bottom half of the figure are shown
these 27 patients taking distalgesic or cytotoxic drugs. If these
patients are excluded, we can identify a 'drug free' group shown in
the top half of the figure who either took no treatment or were
taking other drugs which did not appear to influence metabolic
ratios. Their results are skewed towards very low metabolic
ratios. Their median MR (by rank) (0.7) is significantly lower
($2p= 0.027$) than the median for all patients and controls (1.1).

The results of these preliminary studies support the
hypothesis that genetically determined oxidation capacity may be a
contributory factor in the aetiology of bronchial carcinoma in
cigarette smokers. The frequency of poor metabolisers in the
cancer group (1.9%) was significantly lower than in the normal
control group (8.9%). If these results are borne out by wider

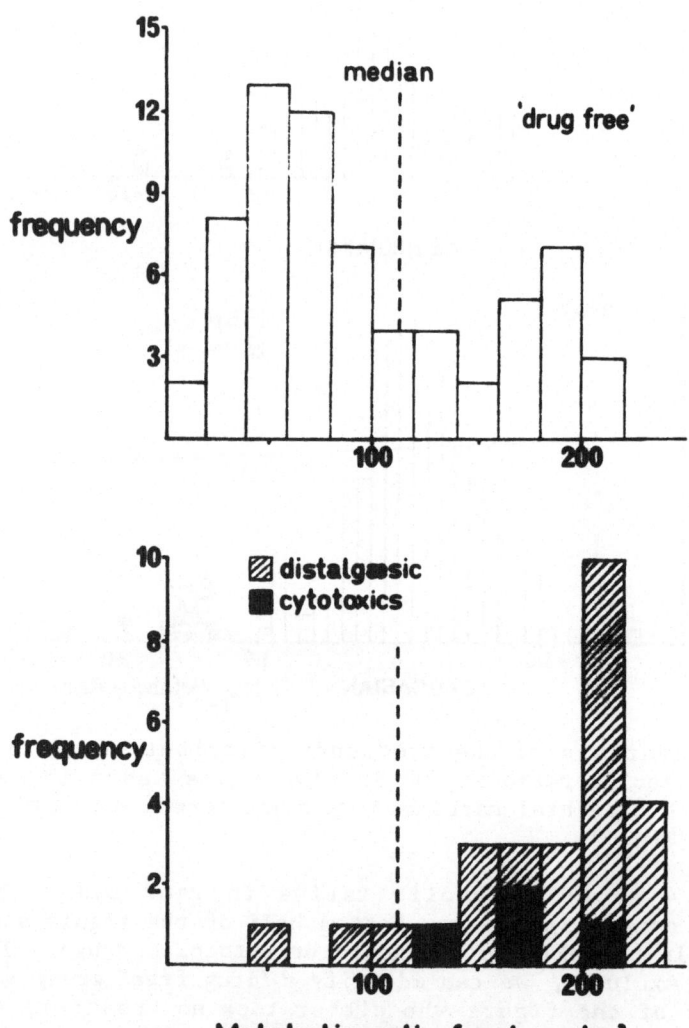

Fig. 4.   Analysis for effect of concurrent treatment on metabolic
          ratio;
          Rank-order frequency distributions for 27 patients taking
          distalgesic® or cytotoxic drugs compared with the re-
          maining 79 patients ('drug free').

studies currently in progress, the PM phenotype apparently enjoys a 4-5 fold lower incidence of bronchial carcinoma, a finding which may be relevant to an estimated 2 million smokers in the U.K. who are phenotypically poor metabolisers.

Of much greater importance to the clinician, however, is the possible corollary that EM smokers with low metabolic ratios are a 'high-risk' group; should this be the case the debrisoquine phenotyping procedure is potentially a valuable screening test in the early detection of bronchial carcinoma, which is the single most important factor in successful treatment at the present time.

The validity of such an approach depends entirely on the degree to which debrisoquine oxidation predicts an individual's capacity to activate cigarette smoke carcinogens. The concept that measurement of the rate of oxidation of a model drug may be predictive in this context is not new. Kellerman et al. (1973 a and b) proposed that antipyrine half-life and arylhydrocarbon hydroxylase (AHH) inducibility in cultured mitogen-stimulated lymphocytes were predictive of susceptibility to bronchial carcinoma. However, the evidence to this effect remains inconclusive, partly owing to methodological difficulties in the AHH determination, and also the finding that antipyrine half-life is determined by overall elimination rates rather than the activity of specific oxidation pathways (Danhof, 1980).

The evidence that the debrisoquine oxidation phenotyping procedure measures the activity of a specific pathway is, to date, encouraging. The D^H/D^L locus has been shown to influence the metabolism of a wide variety of substrates, including phenacetin (Sloan et al., 1978), phenytoin (Idle et al., 1979 b), and phenformin (Shah et al., 1980), whilst the metabolism of others, such as acetanilide (Wakile et al., 1979) and tolbutamide (Idle et al., 1979 b) is apparently unaffected. Thus whilst the exact biochemical basis of the debrisoquine oxidation polymorphism is unknown, the drug apparently interacts with the product of a single gene locus, a factor lending the phenotyping procedure a greater specificity and predictive value than the use of compounds eliminated by multiple pathways.

The role of genetic factors in determining the responsiveness of certain inbred mouse strains to environmental contaminants has been emphasized by Nebert (1979). The vast majority of foreign compounds, including carcinogens, are oxidatively metabolised by cytochrome P450 mono-oxygenase enzyme systems present in most tissues, particularly the liver, kidney and lung. On exposure to many drugs and carcinogens these enzyme systems undergo an adaptive enhancement of activity termed 'enzyme induction'. It has been demonstrated that the ability of some inbred mouse strains to undergo this deficiency is genetically determined (Nebert, 1979).

A more definitive interpretation of the results of the present study is hampered by the fact that the control population used was not age and smoking matched with the bronchial carcinoma group. Whilst age or smoking cannot, by definition, alter genotype, it is conceivable that factors such as smoking could cause sufficient enzyme induction to alter the pattern of metabolic ratios seen in relation to largely non-smoking controls. The inevitable effects of smoking in increasing the incidence of pulmonary, cardiac and peripheral vascular disease make it difficult to recruit subjects in their seventh decade who are fit and taking no medication; nor is it possible to be sure that a 'smoking control' has not developed a bronchial carcinoma which is not yet clinically detectable. This notwithstanding, preliminary results from 40 subjects suggest that the phenotype distribution in older, healthy smokers is no different from that seen in younger non-smokers.

We conclude that oxidation phenotype may be an important factor in the susceptibility of the individual cigarette smoker to bronchial carcinoma, and that the role of host factors in the aetiology of this disease is worthy of more intensive study. This field of research clearly has important clinical implications if it can lead to improved treatment through earlier detection of bronchial carcinoma.

ACKNOWLEDGEMENT

We thank the physicians of Brompton Hospital for permission to study patients under their care. We are also grateful to the Cancer Research Campaign and Wellcome Trust for financial support.

REFERENCES

Allen, J.G., East, P.B., Francis, R.J. and Haigh, J.L., 1975, Metabolism of debrisoquine: identification of some urinary metabolites in rat and man. Drug Metab. Disp., 3: 332.

Angelo, M., Dring, L.G., Lancaster, R., Latham, A. and Smith, R.L., 1975, The metabolism of debrisoquine in rat and man. Br. J. Pharmacol., 55: 264P.

Danhof, M., 1980, Antipyrine metabolite profile as a tool in the assessment of the activity of different drug oxidizing enzymes in man. Ph.D. Thesis, Univ. of Leiden.

Doll, R. and Hill, A.B., 1952, A study of the aetiology of carcinoma of the lung, Br. Med. J., 2: 1271.

Ibid., 1964, Mortality in relation to smoking, Br. Med. J., 1: 1399.

Evans, D.A.P., Mahgoub, A., Sloan, T.P., Idle, J.R. and Smith, R.L., 1980, A family and population study of the genetic polymorphism of debrmsoquine metabolism in a British White population, J. Med. Genet. 17: 102.

Idle, J.R., Mahgoub, A., Angelo, M.M., Dring, L.G., Lancaster, R. and Smith, R.L., 1979a, The metabolism of ^{14}C debrisoquine in man, Br. J. Clin. Pharmacol, 7: 257.

Idle, J.R., Sloan, T.P., Smith, R.L. and Wakile, L.A., 1979b, The application of the phenotyped panel approach to the detection of polymorphism of drug oxidation in man. Br. J. Pharmacol., 66: 430P.

Idle, J.R. and Smith, R.L., 1979, Polymorphism of oxidation at carbon centres of drugs and their clinical significance, Drug Metab. Rev., 9: 301.

Kellerman, G., Kellerman, L.M. and Shaw, C.R., 1973a, Genetic variation of aryl hydrocarbon hydroxylase in human lymphocytes. Am. J. Hum. Genet. 25: 327.

Kellerman, G., Shaw, C.R. and Kellerman, L.M., 1973b, Aryl hydrocarbon hydroxylase inducibility and bronchogenic carcinoma. N. Engl. J. Med. 289: 934.

Mahgoub, A., Idle, J.R., Dring, L.G., Lancaster, R. and Smith, R.L. 1977, Polymorphic hydroxylation of debrisoquine in man, Lancet, ii: 584.

Miller, J.A., 1974, in: "The induction of Drug metabolism, Symposia Medica Hoechst 14", R.W. Estabrook, E. Lindenlaub, eds., Shattauer Verlag, Stuttgart-New York.

Shah, R.R., Oates, N.S., Idle, J.R. and Smith, R.L., 1980, Genetic impairment of phenformin metabolism. Lancet i: 1147.

Sloan, T.P., Mahgoub, A., Lancaster, R., Idle, J.R. and Smith, R.L. 1978, Polymorphism of carbon oxidation of drugs and clinical implications, Br. Med. J., 2: 655.

Wakile, L.A., Sloan, T.P., Idle, J.R. and Smith, R.L., 1979, Genetic evidence for the involvement of different oxidative mechanisms in drug oxidation. J. Pharm. Pharmacol. 31:350.

DISCUSSION

LECTURER: Hetzel   CHAIRMAN: Cumming

CUMMING: Thank you Martin. This paper is open for
                    discussion.

MITCHELL-HEGGS: Thank you very much for that very nice paper,
                    just a few questions. Firstly have you had any
                    opportunities to do any family studies, have
                    there been any patients who have had bronchial
                    carcinoma in their family? The second question,
                    have you done any altering of those figures to
                    try and work out which patients had a squamous
                    carcinoma and which had oat cell and to try and
                    position them?   And thirdly, is there any
                    relationship between the defect of metabolism
                    and the age of onset or the age of diagnosis of
                    the cancer in your subjects?

HETZEL: The answer to that is that while we thought of
                    all three things, we found that we simply hadn't
                    enough subjects to make any reliable statistics
                    or statistical comparison. The big problem is
                    that the only thing we can see with any clarity
                    at the moment is the phenotype and because we
                    were looking at a poor metabolised frequency
                    which is only about 10%, then 100 subjects are
                    about the lowest number in which one can be
                    reasonably confident that you have really picked
                    up the difference. We are hoping to look at
                    these relationships as well as whether there is
                    any correlation between the carcinogen loading
                    dose and the total number of cigarettes smoked.

COHEN: I enjoyed your paper and I admire what you try
                    to do because I think it is extremely important
                    to try and identify those people at risk. I
                    know you are well aware, but I don't think the
                    audience is, of the attempts to do something
                    similar with lymphocytes, but most people have
                    not been able to reproduce that work. Now, if I
                    can go on to your particular paper there are two
                    questions, one I think is a fairly simple one
                    and the second one is more difficult. There are
                    many types of P 450, I don't think many people
                    would agree with Nibbot who says that there are

thousands of P 450's. Most people think they have only good evidence for 3 at the moment, possibly 6. But specifically in terms of benzpyrene to which you are trying to relate it, you must have a cytochrome P 480 or cytochrome P 450 present. Is the debresoquine methabolized by that particular cytochrome because if it is not then is it really giving you the measure that you want?

HETZEL: We don't know for certain that this is the way which debresoquine is metabolised, I don't think anybody has fully identified how this metabolism takes place but this is a popular belief at the moment. Your point is well taken, that's another factor in the whole circumstantial story that we are looking at.

COHEN: The second is a very important conceptual point. You are metabolising predominantly in the liver, because most debresoquine metabolism takes place in the liver. If you are generating reactive metabolites and they are inducing cancer in the lung, the greatest possibility is that those metabolites would be generated in situ in the bronchial epithelium rather than be metabolised in the lung and then transported to the lung because most of these metabolites are too reactive and it's much more likely that they would be generated in the target tissue. If you accept that, then we are looking for metabolic variation within the lung, and does your study reflect metabolic variation in the lung? A study of respiratory tract tissue metabolism in patients with cancer and patients without cancer showed no correlation whatsoever, so I would agree that based on these data and other data in the literature including our own, that there is not a good foundation for the type of study that yu are doing.

HETZEL: We fully accept all these arguments, this is just an attempt to try and look at this in a way applicable to man.

COHEN: I appreciate that and I think it is correct to do some studies but I think the unfortunate thing from your point of view is that you are trying to see differences in the poor metabolisers but statistical significance doesn't always imply biological significance.

KAY:

I think you have got some jolly interesting data and that takes you a long way and the conclusions are very fascinating. I'd just like to ask two things, not being a geneticistst it is quite new to me that you could make a genetic assumption without doing a family study. I didn't appreciate this and maybe you would like to enlarge on that. The second question is: do you have any animal model for this? Because if you have, you could get around a lot of questions which have been posed to you already.

HETZEL:

In answer to your first question, yes, there were family studies done on the original group of normal subjects. The 23 metabolisers had studies of their families done and studies of other subjects in this group were also done and there were enough studies done to show this was poor metabolisers' recessive inheritance behaving as a single gene locus. The second question about metabolism in animals. At the moment we don't appear to have an animal model that we can use in the same way.

KAY:

I understand of course that metabolism is under genetic control, you made that quite clear, but it was the conclusion that the susceptibility to cancer amongst smokers was generally controlled. That's where you surely have to do the family study.

HETZEL:

You are implying that we study families with lung cancer. I think that's going to be very difficult to get sufficient numbers.

CUMMING:

I think the point Barry is trying to make and with which I agree is that the demonstration of genetic association with metabolic processes is accepted, but there is also a genetic association with the susceptibility to carcinoma, it may be that a tracer of one function but it is not necessarily a tracer of the second function. That's the point Barry is trying to make, you have to establish that this jump is genetically viable.

ROSSI: You mentioned some studies on mice, in which the metabolic function was related to the H2 locus or something like that and also the susceptibility to tumours.

HETZEL: No, I was just trying to comment on this question of P 450 mediator mono-oxygenase systems, the work that Nibbon has done in different strains of mice. He has studied the ability of hydrocarbons to induce products of P 450 mediated systems, hydrocarbon hydroxylase. You can measure the amount of enzyme that's induced in the individual animal on exposure and he has been able to show that the degree of enzyme induction produced in some strains of mice is very different from others, so he has some strains of mice which should appear to be behaving in the way that we are suggesting. The other point that we feel is particularly encouraging to us is that he has shown that the response he gets in an individual type of animal is very much the same response he gets in the lung and in other tissues as well.

ROSSI: That correlates with H2 locus or not in mice?

HETZEL: No, I think its' the AH locus.

KAY: To go further on this point, you see, if Nibbon has that model in which C57 is induceable and DBA2 is not induceable, then if it is metabolised by the same system you have an animal model.

HETZEL: I believe that this doesn't work. I think that's something that has already been tried.

KAY: If I can take it a step further in Nibbon animal model, if you put the hydrocarbon onto the mouse skin of the highly inducable animal and then you promote with TPA you do not get tumours; it is resistent. In fact the opposite is what we are proposing.

HETZEL: Tumours in the skin and not in the mice's lung.

KAY: Yes, but we are arguing just now that the response in the lung, the liver, the skin, all these tissues is the same.

HETZEL: But if all that is wrong we still have to explain it.

BAUM: This is really a small question and follows from what Gerry was asking before. Is there any evidence at all that hydroxylase in man is induceable by any other drug?

HETZEL: I don't know, all I know is that one can show the same response to these other drugs. This is the response, the first time you take it you don't see any significant induction nor if you repeat the study, so I think the answer is no.

BAUM: Nobody looks, to take the most naive case, to see whether phenobarbitone causes any increased hydroxylation.

HETZEL: I think we have avoided this drug as an obvious enzyme inducer, but I don't know how much evidence there is that it works in that way. I think we just tried to remove an extra variable by avoiding people who were taking it.

CUMMING: Thank you. I think that note of scientific uncertainty I should draw the proceedings for this afternoon to a close and that brings the proceedings for the week to a close. I would like to say that the speakers have shown great equanimity in the face of the many difficulties that I have produced for them and I would like to thank them all for their collaboration and cheerful acceptance of difficult situations. Also our thanks is due to Giuliana and Philip, our two translators, who, I think you agree, have done a magnificent job throughout the whole week. To the people who have provided the milieu in which we have debated, Mr. Gaspari and all his assistants for providing the sound, the light and the vision, whilst we have provided the heat. In the background the staff of the School of Ettore Majorana led so capably by Alberto Gabrielli and all his assistants who have made our week so pleasant and I hope also so informative. Lastly, and most importantly, yourselves. Thank you and good-bye.